FUZZY SETS AND THEIR APPLICATIONS TO COGNITIVE AND DECISION PROCESSES

Academic Press Rapid Manuscript Reproduction

Proceedings of the U.S.—Japan Seminar on
Fuzzy Sets and Their Applications, held at
The University of California, Berkeley, California
July 1-4, 1974

FUZZY SETS AND THEIR APPLICATIONS TO COGNITIVE AND DECISION PROCESSES

EDITED BY

Lotfi A. Zadeh
University of California, Berkeley
Berkeley, California

King-Sun Fu
Purdue University
West Lafayette, Indiana

Kokichi Tanaka
Osaka University
Toyonaka, Osaka, Japan

Masamichi Shimura
Osaka University
Toyonaka, Osaka, Japan

Academic Press, Inc. *New York San Francisco London 1975*
A Subsidiary of Harcourt Brace Jovanovich, Publishers

ACADEMIC PRESS, INC.
111 Fifth Avenue, New York, New York 10003

United Kingdom Edition published by
ACADEMIC PRESS, INC. (LONDON) LTD.
24/28 Oval Road, London NW1

Library of Congress Cataloging in Publication Data

U. S.--Japan Seminar on Fuzzy Sets and Their Applications, University of California, Berkeley, 1974.
Fuzzy sets and their applications to cognitive and decision processes.

Bibliography: p.
Includes index.
1. Set theory--Congresses. 2. Pattern perception--Congresses. 3. Decision-making--Congresses. 4. Problem solving--Congresses. I. Zadeh, Lotfi Asker. II. Title.
QA248.U54 1974 511'.3 75-15772
ISBN 0-12-775260-9

PRINTED IN THE UNITED STATES OF AMERICA

CONTENTS

CONTRIBUTORS

K. Asai, Department of Industrial Engineering, University of Osaka Prefecture, Sakai, Osaka 591, Japan

S. Y. Bang, Department of Computer Science, University of Texas, Austin, Texas

C. L. Chang, IBM Research Laboratory, San Jose, California 95193

S. S. L. Chang, Department of Electrical Sciences, State University of New York, Stony Brook, New York 11794

G. F. DePalma, Departments of Computer Sciences and Electrical Engineering, Northwestern University, Evanston, Illinois 60201

K. S. Fu, School of Electrical Engineering, Purdue University, West Lafayette, Indiana 47907

T. Fukumura, Department of Information Science, Faculty of Engineering, Nagoya University, Nagoya 464, Japan

L. W. Fung, School of Electrical Engineering, Purdue University, West Lafayette, Indiana 47907

J. A. Goguen, Computer Science Department, University of California, Los Angeles, California 90024

N. Honda, Research Institute of Electrical Communication, Tohoku University, Sendai 980, Japan

Y. Inagaki, Department of Information Science, Faculty of Engineering, Nagoya University, Nagoya 464, Japan

T. Kitagawa, Kyushu University, Fukuoka 812, Japan

M. Kochen, Mental Health Research Institute, University of Michigan, Ann Arbor, Michigan 48014

M. Kokawa, Automatic Control Laboratory, Nagoya University, Nagoya 464, Japan

M. Mizumoto, Department of Information and Computer Sciences, Faculty of Engineering Science, Osaka University, Toyonaka, Osaka 560, Japan

K. Nakamura, Automatic Control Laboratory, Nagoya University, Nagoya 464, Japan

M. Nasu, Research Institute of Electrical Communication, Tohoku University Sendai 980, Japan

M. Oda, Automatic Control Laboratory, Nagoya University, Nagoya 464, Japan

T. Okuda, Department of Industrial Engineering, University of Osaka Prefecture, Sakai, Osaka 591, Japan

A. Rosenfeld, Computer Science Center, University of Maryland, College Park, Maryland 20742

M. Shimura, Faculty of Engineering Science, Osaka University, Toyonaka, Osaka 560, Japan

M. Sugeno, Tokyo Institute of Technology, Tokyo 152, Japan

H. Tanaka, Department of Industrial Engineering, University of Osaka Prefecture, Sakai, Osaka 591, Japan

K. Tanaka, Department of Information and Computer Sciences, Faculty of Engineering Science, Osaka University, Toyonka, Osaka 560, Japan

T. Terano, Tokyo Institute of Technology, Tokyo 152, Japan

L. Uhr, Computer Sciences Department, The University of Wisconsin, Madison, Wisconsin 53706

C. K. Wong, Computer Sciences Department, IBM Thomas J. Watson Research Center, Yorktown Heights, New York 10598

S. S. Yau, Departments of Computer Sciences and Electrical Engineering, Northwestern University, Evanston, Illinois 60201

R. T. Yeh, Department of Computer Science, University of Texas, Austin, Texas

L. A. Zadeh, Department of Electrical Engineering and Computer Sciences, University of California, Berkeley, California 94720

PREFACE

The papers presented in this volume were contributed by participants in the U.S.–Japan Seminar on Fuzzy Sets and Their Applications, held at the University of California, Berkeley, in July 1974. These papers cover a broad spectrum of topics related to the theory of fuzzy sets, ranging from its mathematical aspects to applications in human cognition, communication, decision-making, and engineering systems analysis.

Basically, a fuzzy set is a class in which there may be a continuum of grades of membership as, say, in the class of *long* objects. Such sets underlie much of our ability to summarize, communicate, and make decisions under uncertainty or partial information. Indeed, fuzzy sets appear to play an essential role in human cognition, especially in relation to concept formation, pattern classification, and logical reasoning.

Since its inception about a decade ago, the theory of fuzzy sets has evolved in many directions, and is finding applications in a wide variety of fields in which the phenomena under study are too complex or too ill defined to be analyzed by conventional techniques. Thus, by providing a basis for a systematic approach to approximate reasoning, the theory of fuzzy sets may well have a substantial impact on scientific methodology in the years ahead, particularly in the realms of psychology, economics, law, medicine, decision analysis, information retrieval, and artificial intelligence.

The U.S.–Japan Seminar on Fuzzy Sets was sponsored by the U.S.–Japan Cooperative Science Program, with the joint support of the National Science Foundation and the Japan Society for the Promotion of Science. In organizing the seminar, the co-chairmen received considerable help from J.E. O'Connell and L. Trent of the National Science Foundation; the staff of the Japan Society for the Promotion of Science; and D. J. Angelakos and his staff at the University of California, Berkeley. As co-editors of this volume, we wish also to express our heartfelt appreciation to Terry Brown for her invaluable assistance in the preparation of the manuscript, and to Academic Press for undertaking its publication.

For the convenience of the reader, a brief introduction to the theory of fuzzy sets is provided in the Appendix of the first paper in this volume. An up-to-date bibliography on fuzzy sets and their applications is included at the end of the volume.

CALCULUS OF FUZZY RESTRICTIONS

L. A. Zadeh*
Department of Electrical Engineering
and Computer Sciences
University of California
Berkeley, California 94720

ABSTRACT

A fuzzy restriction may be visualized as an elastic constraint on the values that may be assigned to a variable. In terms of such restrictions, the meaning of a proposition of the form "x is P," where x is the name of an object and P is a fuzzy set, may be expressed as a relational assignment equation of the form R(A(x)) = P, where A(x) is an implied attribute of x, R is a fuzzy restriction on x, and P is the unary fuzzy relation which is assigned to R. For example, "Stella is young," where young is a fuzzy subset of the real line, translates into R(Age(Stella))= young.

The calculus of fuzzy restrictions is concerned, in the main, with (a) translation of propositions of various types into relational assignment equations, and (b) the study of transformations of fuzzy restrictions which are induced by linguistic modifiers, truth-functional modifiers, compositions, projections and other operations. An important application of the calculus of fuzzy restrictions relates to what might be called approximate reasoning, that is, a type of reasoning which is neither very exact nor very inexact. The

---

*This work was supported in part by the Naval Electronics Systems Command, Contract N00039-75-0034, The Army Research Office, Grant DAHC04-75-G-0056, and the National Science Foundation, Grant GK-43024X.

main ideas behind this application are outlined and illustrated by examples.

I. INTRODUCTION

During the past decade, the theory of fuzzy sets has developed in a variety of directions, finding applications in such diverse fields as taxonomy, topology, linguistics, automata theory, logic, control theory, game theory, information theory, psychology, pattern recognition, medicine, law, decision analysis, system theory and information retrieval.

A common thread that runs through most of the applications of the theory of fuzzy sets relates to the concept of a fuzzy restriction - that is, a fuzzy relation which acts as an elastic constraint on the values that may be assigned to a variable. Such restrictions appear to play an important role in human cognition, especially in situations involving concept formation, pattern recognition, and decision-making in fuzzy or uncertain environments.

As its name implies, the calculus of fuzzy restrictions is essentially a body of concepts and techniques for dealing with fuzzy restrictions in a systematic fashion. As such, it may be viewed as a branch of the theory of fuzzy relations, in which it plays a role somewhat analogous to that of the calculus of probabilities in probability theory. However, a more specific aim of the calculus of fuzzy restrictions is to furnish a conceptual basis for fuzzy logic and what might be called approximate reasoning [1], that is, a type of reasoning which is neither very exact nor very inexact. Such reasoning plays a basic role in human decision-making because it provides a way of dealing with problems which are too complex for precise solution. However, approximate reasoning is more than a method of last recourse for coping with insurmountable complexities. It is, also, a way of simplifying the performance of tasks in which a high degree of precision is neither needed nor required. Such tasks pervade much of what we do on both conscious and subconscious levels.

What is a fuzzy restriction? To illustrate its meaning in an informal fashion, consider the following proposition (in which italicized words represent fuzzy concepts):

Tosi is young (1.1)

Ted has gray hair (1.2)

Sakti and Kapali are approximately equal in height. (1.3)

Starting with (1.1), let Age (Tosi) denote a numerically-valued variable which ranges over the interval [0,100]. With this interval regarded as our universe of discourse U, young may be interpreted as the label of a fuzzy subset[1] of U which is characterized by a compatibility function, μ_{young}, of the form shown in Fig. 1.1. Thus, the degree to which a numerical age, say u = 28, is compatible with the concept of young is 0.7, while the compatibilities of 30 and 35 with young are 0.5 and 0.2, respectively. (The age at which the compatibility takes the value 0.5 is the crossover point of young.) Equivalently, the function μ_{young} may be viewed as the membership function of the fuzzy set young, with the value of μ_{young} at u representing the grade of membership of u in young.

Since young is a fuzzy set with no sharply defined boundaries, the conventional interpretation of the proposition "Tosi is young," namely, "Tosi is a member of the class of young men," is not meaningful if membership in a set is

[1]A summary of the basic properties of fuzzy sets is presented in the Appendix.

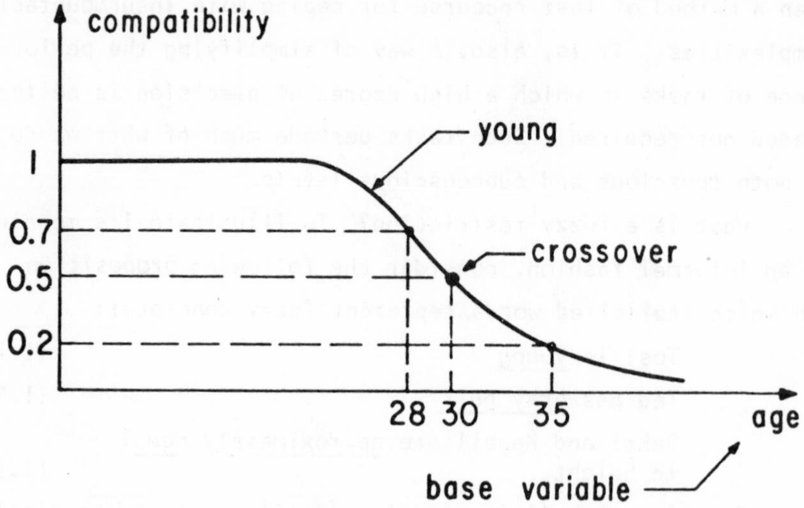

Figure 1.1 Compatibility Function of young.

interpreted in its usual mathematical sense. To circumvent this difficulty, we shall view (1.1) as an assertion of a restriction on the possible values of Tosi's age rather than as an assertion concerning the membership of Tosi in a class of individuals. Thus, on denoting the restriction on the age of Tosi by R(Age(Tosi)), (1.1) may be expressed as an assignment equation

R(Age(Tosi)) = young (1.4)

in which the fuzzy set young (or, equivalently, the unary fuzzy relation young) is assigned to the restriction on the variable Age(Tosi). In this instance, the restriction R(Age(Tosi)) is a fuzzy restriction by virtue of the fuzziness of the set young.

Using the same point of view, (1.2) may be expressed as

R(Color(Hair(Ted))) = gray (1.5)

Thus, in this case, the fuzzy set gray is assigned as a value to the fuzzy restriction on the variable Color(Hair(Ted)).

In the case of (1.1) and (1.2), the fuzzy restriction

has the form of a fuzzy set or, equivalently, a unary fuzzy relation. In the case of (1.3), we have two variables to consider, namely, Height(Sakti) and Height(Kapali). Thus, in this instance, the assignment equation takes the form

R(Height(Sakti), Height(Kapali)) = approximately equal (1.6)

in which approximately equal is a binary fuzzy relation characterized by a compatibility matrix $\mu_{\text{approximately equal}}(u,v)$ such as shown in Table 1.2.

Table 1.2. Compatibility matrix of the fuzzy Relation approximately equal.

u \ v	5'6	5'8	5'10	6	6'2	6'4
5'6	1	0.8	0.6	0.2	0	0
5'8	0.8	1	0.9	0.7	0.3	0
5'10	0.6	0.9	1	0.9	0.7	0
6	0.2	0.7	0.9	1	0.9	0.8
6'2	0	0.3	0.7	0.9	1	0.9
6'4	0	0	0	0.8	0.9	1

Thus, if Sakti's height is 5'8 and Kapali's is 5'10, then the degree to which they are approximately equal is 0.9.

The restrictions involved in (1.1), (1.2) and (1.3) are unrelated in the sense that the restriction on the age of Tosi has no bearing on the color of Ted's hair or the height of Sakti and Kapali. More generally, however, the restrictions may be interrelated, as in the following example.

u is small (1.7)

u and v are approximately equal (1.8)

In terms of the fuzzy restrictions on u and v, (1.7) and (1.8) translate into the assignment equations

R(u) = small (1.9)

R(u,v) = approximately equal (1.10)

where R(u) and R(u,v) denote the restrictions on u and (u,v), respectively.

As will be shown in Section 2, from the knowledge of a fuzzy restriction on u and a fuzzy restriction on (u,v) we can deduce a fuzzy restriction on v. Thus, in the case of (1.9) and (1.10), we can assert that

$$R(v) = R(u) \circ R(u,v) = \underline{\text{small}} \circ \underline{\text{approximately equal}} \quad (1.11)$$

where $\circ$ denotes the composition[2] of fuzzy relations.

The rule by which (1.11) is inferred from (1.9) and (1.10) is called the compositional rule of inference. As will be seen in the sequel, this rule is a special case of a more general method for deducing a fuzzy restriction on a variable from the knowledge of fuzzy restrictions on related variables.

In what follows, we shall outline some of the main ideas which form the basis for the calculus of fuzzy restrictions and sketch its application to approximate reasoning. For convenient reference, a summary of those aspects of the theory of fuzzy sets which are relevant to the calculus of fuzzy restrictions is presented in the Appendix.

2. CALCULUS OF FUZZY RESTRICTIONS

The point of departure for our discussion of the calculus of fuzzy restrictions is the paradigmatic proposition[1]

$$p \triangleq x \text{ is } P \quad (2.1)$$

which is exemplified by

[2] If A is a unary fuzzy relation in U and B is a binary fuzzy relation in U x V, the membership function of the composition of A and B is expressed by $\mu_{A \circ B}(v) = \vee_u(\mu_A(u) \wedge \mu_B(u,v))$, where $\vee_u$ denotes the supremum over $u \in U$. A more detailed discussion of the composition of fuzzy relations may be found in [2] and [3].

[1] The symbol $\triangleq$ stands for "denotes" or "is defined to be."

x is a positive integer (2.2)

Soup is hot (2.3)

Elvira is blond (2.4)

If P is a label of a nonfuzzy set, e.g., $P \triangleq$ set of positive integers, then "x is P," may be interpreted as "x belongs to P," or, equivalently, as "x is a member of P." In (2.3) and (2.4), however, P is a label of a fuzzy set, i.e., $P \triangleq$ hot and $P \triangleq$ blond. In such cases, the interpretation of "x is P," will be assumed to be characterized by what will be referred to as a relational assignment equation. More specifically, we have

Definition 2.5 The meaning of the proposition

$$p \triangleq x \text{ is } P \qquad (2.6)$$

where x is a name of an object (or a construct) and P is a label of a fuzzy subset of a universe of discourse U, is expressed by the relational assignment equation

$$R(A(x)) = P \qquad (2.7)$$

where A is an implied attribute of x, i.e., an attribute which is implied by x and P; and R denotes a fuzzy restriction on A(x) to which the value P is assigned by (2.7). In other words, (2.7) implies that the attribute A(x) takes values in U and that R(A(x)) is a fuzzy restriction on the values that A(x) may take, with R(A(x)) equated to P by the relational assignment equation.

As an illustration, consider the proposition "Soup is hot." In this case, the implied attribute is Temperature and (2.3) becomes

$$R(\text{Temperature(Soup)}) = \underline{\text{hot}} \qquad (2.8)$$

with hot being a subset of the interval [0,212] defined by, say, a compatibility function of the form (see Appendix)

$$\mu_{\underline{\text{hot}}}(u) = S(u; 32,100,200) \qquad (2.9)$$

Thus, if the temperature of the soup is u = 100°, then the

degree to which it is compatible with the fuzzy restriction hot is 0.5, whereas the compatibility of 200° with hot is unity. It is in this sense that R(Temperature(Soup)) plays the role of a fuzzy restriction on the soup temperature which is assigned the value hot, with the compatibility function of hot serving to define the compatibilities of the numerical values of soup temperature with the fuzzy restriction hot.

In the case of (2.4), the implied attribute is Color(Hair), and the relational assignment equation takes the form

$$R(\text{Color}(\text{Hair}(\text{Elvira}))) = \underline{\text{blond}} \tag{2.10}$$

There are two important points that are brough out by this example. First, the implied attribute of x may have a nested structure, i.e., may be of the general form

$$A_k(A_{k-1}(\ldots A_2(A_1(x)) \ldots)); \tag{2.11}$$

and second, the fuzzy set which is assigned to the fuzzy restriction (i.e., blond) may not have a numerically-valued base variable, that is, the variable ranging over the universe of discourse U. In such cases, we shall assume that P is defined by exemplification, that is, by pointing to specific instances of x and indicating the degree (either numerical or linguistic) to which that instance is compatible with P. For example, we may have $\mu_{\underline{\text{blond}}}(\text{June}) = 0.2$, $\mu_{\underline{\text{blond}}}(\text{Jurata})$ = very high, etc. In this way, the fuzzy set blond is defined in an approximate fashion as a fuzzy subset of a universe of discourse comprised of a collection of individuals U = {x}, with the restriction R(x) playing the role of a fuzzy restriction on the values of x rather than on the values of an implied attribute A(x).[2] (In the sequel,

[2] A more detailed discussion of this and related issues may be found in [3], [4] and [5].

we shall write R(x) and speak of the restriction on x rather than on A(x) not only in those cases in which P is defined by exemplification but also when the implied attribute is not identified in an explicit fashion.)

So far, we have confined our attention to fuzzy restrictions which are defined by a single proposition of the form "x is P." In a more general setting, we may have n constituent propositions of the form

$$x_i \text{ is } P_i,\ i = 1,\ldots,n \tag{2.12}$$

in which P_i is a fuzzy subset of U_i, $i = 1,\ldots,n$. In this case, the propositions "x_i is P_i," $i = 1,\ldots,n$, collectively define a fuzzy restriction on the n-ary object $(x_1,\ldots,x_n)$. The way in which this restriction depends on the P_i is discussed in the following.

The Rules of Implied Conjunction and Maximal Restriction

For simplicity we shall assume that n = 2, with the constituent propositions having the form

$$x \text{ is } P \tag{2.13}$$

$$y \text{ is } Q \tag{2.14}$$

where P and Q are fuzzy subsets of U and V, respectively. For example,

Georgia is very warm (2.15)

George is highly intelligent (2.16)

or, if x = y,

Georgia is very warm (2.17)

Georgia is highly intelligent (2.18)

The rule of implied conjunction asserts that, in the absence of additional information concerning the constituent propositions, (2.13) and (2.14) taken together imply the composite proposition "x is P and y is Q;" that is,

$$\{x \text{ is } P,\ y \text{ is } Q\} \Longrightarrow x \text{ is } P \text{ and } y \text{ is } Q \tag{2.19}$$

Under the same assumption, the rule of maximal restriction asserts that

$$x \text{ is } P \text{ and } y \text{ is } Q \Longrightarrow (x,y) \text{ is } P \times Q \tag{2.20}$$

and, if $x = y$,

$$x \text{ is } P \text{ and } x \text{ is } Q \Longrightarrow x \text{ is } P \cap Q \tag{2.21}$$

where $P \times Q$ and $P \cap Q$ denote, respectively, the cartesian product and the intersection of P and Q.[3]

The rule of maximal restriction is an instance of a more general principle which is based on the following properties of n-ary fuzzy restrictions.

Let R be a n-ary fuzzy relation in $U_1 \times \ldots \times U_n$ which is characterized by its membership (compatibility) function $\mu_R(u_1,\ldots,u_n)$. Let $q = (i_1,\ldots,i_k)$ be a subsequence of the index sequence $(1,\ldots,n)$ and let q' denote the complementary subsequence $(j_1,\ldots,j_\ell)$. (E.g., if $n = 5$ and $q = (2,4,5)$, then $q' = (1,3)$.). Then, the projection of R on $U_{(q)} \triangleq U_{i_1} \times \ldots \times U_{i_k}$ is a fuzzy relation, R_q, in $U_{(q)}$ whose membership function is related to that of R by the expression

$$\mu_{R_q}(u_{i_1},\ldots u_{i_k}) = \vee_{u_{(q')}} \mu_R(u_1,\ldots,u_n) \tag{2.22}$$

where the right-hand member represents the supremum of $\mu_R(u_1,\ldots,u_n)$ over the u's which are in $u_{(q')}$.

If R is interpreted as a fuzzy restriction on $(u_1,\ldots,u_n)$ in $U_1 \times \ldots \times U_n$, then its projection on $U_{i_1} \times \ldots \times U_{i_k}$, R_q, constitutes a marginal restriction which is induced by R in $U_{(q)}$. Conversely, given a fuzzy restriction R_q in $U_{(q)}$, there exist fuzzy restrictions in $U_1 \times \ldots \times U_n$ whose projection on $U_{(q)}$

[3] The cartesian product of P and Q is a fuzzy subset of $U \times V$ whose membership function is expressed by $\mu_{P \times Q}(u,v) = \mu_P(u) \wedge \mu_Q(v)$. The membership function of $P \cap Q$ is given by $\mu_{P \cap Q}(u) = \mu_P(u) \wedge \mu_Q(u)$. The symbol $\wedge$ stands for min. (See the Appendix for more details.)

is R_q. From (2.22), it follows that the largest[4] of these restrictions is the <u>cylindrical</u> <u>extension</u> of R_q, denoted by $\bar{R}_q$, whose membership function is given by

$$\mu_{\bar{R}_q}(u_1,\ldots,u_n) = \mu_{R_q}(u_{i_1},\ldots,u_{i_k}) \qquad (2.23)$$

and whose <u>base</u> is R_q. ($\bar{R}_q$ is referred to as the cylindrical extension of R_q because the value of $\mu_{\bar{R}_q}$ at any point $(u_1',\ldots,u_n')$ is the same as at the point $(u_1,\ldots,u_n)$ so long as $u_{i_1}' = u_{i_1},\ldots,u_{i_k}' = u_{i_k}$.)

Since $\bar{R}_q$ is the largest restriction in $U_1 \times\ldots\times U_n$ whose base is R_q, it follows that

$$R \subset \bar{R}_q \qquad (2.24)$$

for all q, and hence that R satisfies the <u>containment relation</u>

$$R \subset \bar{R}_{q_1} \cap \bar{R}_{q_2} \cap\ldots\cap \bar{R}_{q_r} \qquad (2.25)$$

which holds for arbitrary index subsequences $q_1,\ldots,q_r$. Thus, if we are given the marginal restrictions $R_{q_1},\ldots,R_{q_r}$, then the restriction

$$R_{MAX}(R_{q_1},\ldots,R_{q_r}) \triangleq \bar{R}_{q_1} \cap\ldots\cap \bar{R}_{q_r} \qquad (2.26)$$

is the <u>maximal</u> (i.e., least restrictive) restriction which is consistent with the restrictions $R_{q_1},\ldots,R_{q_r}$. It is this choice of R_{MAX} given $R_{q_1},\ldots,R_{q_r}$ that constitutes a general selection principle of which the rule of maximal restriction is a special case.[5]

By applying the same approach to the disjunction of two propositions, we are led to the rule

---

[4]A fuzzy relation R in U is <u>larger</u> than S (in U) iff $\mu_R(u) \geq \mu_S(u)$ for all u in U.

[5]A somewhat analogous role in the case of probability distributions is played by the minimum entropy principle of E. Jaynes and M. Tribus [6], [7].

$$x \text{ is } P \text{ or } y \text{ is } Q \Rightarrow (x,y) \text{ is } \overline{P} + \overline{Q} \quad (2.27)$$

or, equivalently,

$$x \text{ is } P \text{ or } y \text{ is } Q \Rightarrow (x,y) \text{ is } (P' \times Q')' \quad (2.28)$$

where P' and Q' are the complements of P and Q, respectively, and + denotes the union.[6]

As a simple illustration of (2.27), assume that

$$U = 1 + 2 + 3 + 4$$

and that

$$P \stackrel{\Delta}{=} \underline{\text{small}} \stackrel{\Delta}{=} 1/1 + 0.6/2 + 0.2/3 \quad (2.29)$$

$$\underline{\text{large}} \stackrel{\Delta}{=} 0.2/2 + 0.6/3 + 1/4 \quad (2.30)$$

$$Q \stackrel{\Delta}{=} \underline{\text{very large}} = 0.04/2 + 0.36/3 + 1/4 \quad (2.31)$$

Then

$$P' = 0.4/2 + 0.8/3 + 1/4 \quad (2.32)$$

$$Q' = 1/1 + 0.96/2 + 0.64/3$$

and

$$\overline{P} + \overline{Q} = (P' \times Q')' = 1/((1,1) + (1,2) + (1,3) + (1,4) + (2,4) + (3,4) + (4,4)) + 0.6/((2,1) + (2,2) + (2,3)) + 0.3/((3,1) + (3,2)) + 0.36/((3,3) + (4,3)) + 0.04/(4,2) \quad (2.33)$$

Conditional Propositions

In the case of conjunctions and disjunctions, our intuition provides a reasonably reliable guide for defining the form of the dependence of R(x,y) on R(x) and R(y). This is less true, however, of conditional propositions of the form

$$p \stackrel{\Delta}{=} \text{If } x \text{ is } P \text{ then } y \text{ is } Q \text{ else } y \text{ is } S \quad (2.34)$$

---

[6] The membership function of P' is related to that of P by $\mu_{P'}(u) = 1 - \mu_P(u)$. The membership function of the union of P and Q is expressed by $\mu_{P+Q}(u) = \mu_P(u) \vee \mu_Q(u)$, where $\vee$ denotes max.

and

$$q \triangleq \text{If } x \text{ is } P \text{ then } y \text{ is } Q \tag{2.35}$$

where P is a fuzzy subset of U, while Q and S are fuzzy subsets of V.

With this qualification, two somewhat different definitions for the restrictions induced by p and q suggest themselves. The first, to which we shall refer as the <u>maximin rule of conditional propositions</u>, is expressed by

$$\text{If } x \text{ is } P \text{ then } y \text{ is } Q \text{ else } y \text{ is } S \Longrightarrow (x,y) \text{ is } P \times Q + P' \times S, \tag{2.36}$$

which implies that the meaning of P is expressed by the relational assignment equation

$$R(x,y) = P \times Q + P' \times S \tag{2.37}$$

The conditional proposition (2.35) may be interpreted as a special case of (2.34) corresponding to S = V. Under this assumption, we have

$$\text{If } x \text{ is } P \text{ then } y \text{ is } Q \Longrightarrow (x,y) \text{ is } P \times Q + P' \times V \tag{2.38}$$

As an illustration, consider the conditional proposition

$$p \triangleq \text{If Maya is } \underline{\text{tall}} \text{ then Turkan is } \underline{\text{very tall}} \tag{2.39}$$

Using (2.38), the fuzzy restriction induced by p is defined by the relational assignment equation

$$R(\text{Height(Maya)}, \text{Height(Turkan)}) = \underline{\text{tall}} \times \underline{\text{very tall}} + + \underline{\text{not tall}} \times V$$

where V might be taken to be the interval [150,200] (in centimeters), and <u>tall</u> and <u>very tall</u> are fuzzy subsets of V defined by their respective compatibility functions (see Appendix)

$$\mu_{\underline{\text{tall}}} = S(160, 170, 180) \tag{2.40}$$

and

$$\mu_{\underline{\text{very tall}}} = S^2(160, 170, 180) \tag{2.41}$$

in which the argument u is suppressed for simplicity.

An alternative definition, to which we shall refer as the arithmetic rule of conditional propositions, is expressed by

$$\text{If } x \text{ is } P \text{ then } y \text{ is } Q \text{ else } y \text{ is } S \Longrightarrow (x,y) \text{ is } ((P \times V \ominus U \times Q) + (P' \times V \ominus U \times S))' \qquad (2.42)$$

or, equivalently and more simply,

$$\text{If } x \text{ is } P \text{ then } y \text{ is } Q \text{ else } y \text{ is } S \Longrightarrow (x,y) \text{ is } (\overline{P}' \oplus \overline{Q}) \cap (\overline{P} \oplus \overline{S}) \qquad (2.43)$$

where $\oplus$ and $\ominus$ denote the bounded-sum and bounded-difference operations,[7] respectively; $\overline{P}$ and $\overline{Q}$ are the cylindrical extensions of P and Q; and + is the union. This definition may be viewed as an adaptation to fuzzy sets of Lukasiewicz's definition of material implication in L_{aleph_1} logic, namely [8]

$$v(r \to s) \triangleq \min(1, 1 - v(r) + v(s)) \qquad (2.44)$$

where v(r) and v(s) denote the truth-values of r and s, respectively, with $0 \leq v(r) \leq 1$, $0 \leq v(s) \leq 1$.

In particular, if S is equated to V, then (2.43) reduces to

$$\text{If } x \text{ is } P \text{ then } y \text{ is } Q \Longrightarrow (x,y) \text{ is } (\overline{P}' \oplus \overline{Q}) \qquad (2.45)$$

Note that in (2.42), P x V and U x Q are the cylindrical extensions, $\overline{P}$ and $\overline{Q}$, of P and Q, respectively.

Of the two definitions stated above, the first is somewhat easier to manipulate but the second seems to be in closer accord with our intuition. Both yield the same result when P, Q and S are nonfuzzy sets.

As an illustration, in the special case where x = y and P = Q, (2.45) yields

---

[7]The membership functions of the bounded-sum and-difference of P and Q are defined by $\mu_{P \oplus Q}(u) = \min(1, \mu_P(u) + \mu_Q(u))$ and $\mu_{P \ominus Q}(u) = \max(0, \mu_P(u) - \mu_Q(u)$, $u \in U$, where + denotes the arithmetic sum.

$$\text{If } x \text{ is } P \text{ then } x \text{ is } P \Longrightarrow \begin{array}{l} x \text{ is } (P' \oplus P) \\ x \text{ is } V \end{array} \tag{2.46}$$

which implies, as should be expected, that the proposition in question induces no restriction on x. The same holds true, more generally, when $P \subset Q$.

Modification of Fuzzy Restrictions

Basically, there are three distinct ways in which a fuzzy restriction which is induced by a proposition of the form

$$p \triangleq x \text{ is } P$$

may be modified.

First, by a combination with other restrictions, as in

$$r \triangleq x \text{ is } P \text{ and } x \text{ is } Q \tag{2.47}$$

which transforms P into $P \cap Q$.

Second, by the application of a modifier m to P, as in

Hans is very kind (2.48)

Maribel is highly temperamental (2.49)

Lydia is more or less happy (2.50)

in which the operators very, highly and more or less modify the fuzzy restrictions represented by the fuzzy sets kind, temperamental and happy, respectively.

And third, by the use of truth-values, as in

(Sema is young) is very true (2.51)

in which very true is a fuzzy restriction on the truth-value of the proposition "Sema is young."

The effect of modifiers such as very, highly, extremely, more or less, etc., is discussed in greater detail in [9], [10] and [11]. For the purposes of the present discussion, it will suffice to observe that the effect of very and more or less may be approximated very roughly by the operations CON (standing for CONCENTRATION) and DIL (standing for DILATION) which are defined respectively by

$$\mathrm{CON}(A) = \int_U (\mu_A(u))^2/u \tag{2.52}$$

and

$$\mathrm{DIL}(A) = \int_U (\mu_A(u))^{0.5}/u \tag{2.53}$$

where A is a fuzzy set in U with membership function μ_A, and

$$A = \int_U \mu_A(u)/u \tag{2.54}$$

is the integral representation of A. (See the Appendix.) Thus, as an approximation, we assume that

$$\underline{\text{very}}\ A = \mathrm{CON}(A) \tag{2.55}$$

and

$$\underline{\text{more or less}}\ A = \mathrm{DIL}(A) \tag{2.56}$$

For example, if

$$\underline{\text{young}} = \int_0^{100} \left(1 + \left(\frac{u}{30}\right)^2\right)^{-1}/u \tag{2.57}$$

then

$$\underline{\text{very young}} = \int_0^{100} \left(1 + \left(\frac{u}{30}\right)^2\right)^{-2}/u \tag{2.58}$$

and

$$\underline{\text{more or less young}} = \int_0^{100} \left(1 + \left(\frac{u}{30}\right)^2\right)^{-0.5}/u \tag{2.59}$$

The process by which a fuzzy restriction is modified by a fuzzy truth-value is significantly different from the point-transformations expressed by (2.55) and (2.56). More specifically, the rule of truth-functional modification, which defines the transformation in question, may be stated in symbols as

$$(x \text{ is } Q) \text{ is } \tau \Longrightarrow x \text{ is } \mu_Q^{-1} \circ \tau \tag{2.60}$$

where τ is a linguistic truth-value (e.g., true, very true, false, not very true, more or less true, etc.); μ_Q^{-1} is a relation inverse to the compatibility function of A, and $\mu_Q^{-1} \circ \tau$ is the composition of the nonfuzzy relation μ_Q^{-1} with the unary fuzzy relation τ. (See footnote 2 in Section 1 for the definition of composition.)

As an illustration, the application of this rule to the proposition

(Sema is young) is very true (2.61)

yields

$$\text{Sema is } \mu_{\text{young}}^{-1} \circ \text{very true} \quad (2.62)$$

Thus, if the compatibility functions of young and very true have the form of the curves labeled μ_{young_1} and $\mu_{\text{very true}}$ in Fig. 2.1, then the compatibility function of $\mu_{\text{young}} \circ$ very true is represented by the curve μ_{young_2}. The ordinates of μ_{young_2} can readily be determined by the graphical procedure illustrated in Fig. 2.1.

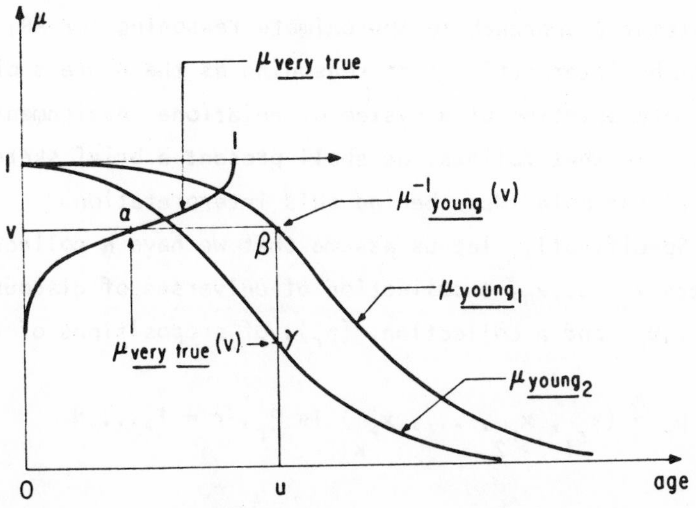

Figure 2.1 Illustration of Truth-Functional Modification.

The important point brought out by the foregoing discussion is that the association of a truth-value with a proposition does not result in a proposition of a new type; rather, it merely modifies the fuzzy restriction induced by that proposition in accordance with the rule expressed by (2.60). The same applies, more generally, to nested propositions of the form

$$(\ldots(((x \text{ is } P_1) \text{ is } \tau_1) \text{ is } \tau_2)\ldots \text{ is } \tau_n) \tag{2.63}$$

in which $\tau_1,\ldots,\tau_n$ are linguistic or numerical truth-values. It can be shown[8] that the restriction on x which is induced by a proposition of this form may be expressed as

$$x \text{ is } P_{n+1}$$

where

$$P_{k+1} = \mu_{P_k}^{-1} \circ \tau_k\,, \quad k = 1, 2, \ldots, n \tag{2.64}$$

3. APPROXIMATE REASONING (AR)

The calculus of fuzzy restrictions provides a basis for a systematic approach to approximate reasoning (or AR, for short) by interpreting such reasoning as the process of approximate solution of a system of relational assignment equations. In what follows, we shall present a brief sketch of some of the main ideas behind this interpretation.

Specifically, let us assume that we have a collection of objects $x_1,\ldots,x_n$, a collection of universes of discourse $U_1,\ldots,U_n$, and a collection, $\{p_r\}$, of propositions of the form

$$p_r \triangleq (x_{r_1}, x_{r_2}, \ldots, x_{r_k}) \text{ is } P_r,\ r = 1,\ldots,N \tag{3.1}$$

[8] A more detailed discussion of this and related issues may be found in [4].

in which P_r is a fuzzy relation in $U_{r_1} \times \ldots \times U_{r_k}$.[1] E.g.,

$$p_1 \triangleq x_1 \text{ is } \underline{\text{small}} \tag{3.2}$$

$$p_2 \triangleq x_1 \text{ and } x_2 \text{ are } \underline{\text{approximately equal}} \tag{3.3}$$

in which $U_1 = U_2 \triangleq (-\infty,\infty)$; small is a fuzzy subset of the real line $(-\infty,\infty)$; and approximately equal is a fuzzy binary relation in $(-\infty,\infty) \times (-\infty,\infty)$.

As stated in Section 2, each p_r in $\{p_r\}$ may be translated into a relational assignment equation of the form

$$R(A_{r_1}(x_{r_1}),\ldots,A_{r_k}(x_{r_k})) = P_r, \quad r = 1,\ldots,N \tag{3.4}$$

where A_{r_i} is an implied attribute of x_{r_i}, $i = 1,\ldots,k$, (with k dependent on r). Thus, the collection of propositions $\{p_r\}$ may be represented as a system of relational assignment equations (3.4).

Let $\overline{P}_r$ be the cylindrical extension of P_r, that is,

$$\overline{P}_r = P_r \times U_{s_1} \times \ldots \times U_{s_\ell} \tag{3.5}$$

where the index sequence $(s_1,\ldots,s_\ell)$ is the complement of the index sequence $(r_1,\ldots,r_k)$ (i.e., if $n = 5$, for example, and $(r_1, r_2, r_3) = (2,4,5)$, then $(s_1,s_2) = (1,3)$).

By the rule of the implied conjunction, the collection of propositions $\{p_r\}$ induces a relational assignment equation of the form

$$R(A_1(x_1),\ldots,A_n(x_n)) = \overline{P}_1 \cap \ldots \cap \overline{P}_N \tag{3.6}$$

which subsumes the system of assignment equations (3.4). It is this equation that forms the basis for approximate inferences from the given propositions $p_1,\ldots,p_N$.

Specifically, by an inference about $(x_{r_1},\ldots,x_{r_k})$ from

---

[1] In some cases, the proposition "$(x_{r_1},\ldots,x_{r_k})$ is P_r," may be expressed more naturally in English as "x_{r_1} and ... x_{r_k} are P_r."

$\{p_r\}$, we mean the fuzzy restriction resulting from the projection of $P \triangleq \overline{P}_1 \cap \ldots \cap \overline{P}_N$ on $U_{r_1} \times \ldots \times U_{r_k}$. Such an inference will, in general, be approximate in nature because of (a) approximations in the computation of the projection of P; and/or (b) linguistic approximation to the projection of P by variables whose values are linguistic rather than numerical.[2]

As a simple illustration of (3.6), consider the propositions

$$x_1 \text{ is } P_1 \tag{3.7}$$

$$x_1 \text{ and } x_2 \text{ are } P_2 \tag{3.8}$$

In this case, (3.6) becomes

$$R(A(x_1), A(x_2)) = \overline{P}_1 \cap P_2 \tag{3.9}$$

and the projection of $\overline{P}_1 \cap P_2$ on U_2 reduces to the composition of P_1 and P_2. In this way, we are led to the <u>compositional rule of inference</u> which may be expressed in symbols as

$$\begin{array}{l} x_1 \text{ is } P_1 \\ x_1 \text{ and } x_2 \text{ are } P_2 \\ \hline x_2 \text{ is } P_1 \circ P_2 \end{array} \tag{3.10}$$

or, more generally,

$$\begin{array}{l} x_1 \text{ and } x_2 \text{ are } P_1 \\ x_2 \text{ and } x_3 \text{ are } P_2 \\ \hline x_1 \text{ and } x_3 \text{ are } P_1 \circ P_2 \end{array} \tag{3.11}$$

in which the respective inferences are shown below the horizontal line.

---

[2] A linguistic variable is a variable whose values are words or sentences in a natural or artificial language. For example, Age is a linguistic variable if its values are assumed to be <u>young</u>, <u>not young</u>, <u>very young</u>, <u>more or less young</u>, etc. A more detailed discussion of linguistic variables may be found in [3], [4] and [11]. (See also Appendix.)

As a more concrete example, consider the propositions

x_1 is small (3.12)

x_1 and x_2 are approximately equal (3.13)

where

$$U_1 = U_2 \triangleq 1 + 2 + 3 + 4 \tag{3.14}$$

$$\underline{\text{small}} \triangleq 1/1 + 0.6/2 + 0.2/3 \tag{3.15}$$

and

$$\underline{\text{approximately equal}} = 1/((1,1) + (2,2) + (3,3) + (4,4)) + 0.5/((1,2) + (2,1) + (2,3) + (3,2) + (3,4) + (4,3)) \tag{3.16}$$

In this case, the composition small ∘ approximately equal may be expressed as the max-min product of the relation matrices of small and approximately equal. Thus

$$\underline{\text{small}} \circ \underline{\text{approximately equal}} = [1 \quad 0.6 \quad 0.2 \quad 0] \circ \begin{bmatrix} 1 & 0.5 & 0 & 0 \\ 0.5 & 1 & 0.5 & 0 \\ 0 & 0.5 & 1 & 0.5 \\ 0 & 0 & 0.5 & 1 \end{bmatrix} = [1 \quad 0.6 \quad 0.5 \quad 0.2] \tag{3.17}$$

and hence the fuzzy restriction on x_2 is given by

$$R(x_2) = 1/1 + 0.6/2 + 0.5/3 + 0.2/4 \tag{3.18}$$

Using the definition of more or less (see (2.56)), a rough linguistic approximation to (3.18) may be expressed as

$$LA(1/1 + 0.6/2 + 0.5/3 + 0.2/4) = \underline{\text{more or less small}} \tag{3.19}$$

where LA stands for the operation of linguistic approximation. In this way, from (3.12) and (3.13) we can deduce the approximate conclusion

x_2 is more or less small (3.20)

which may be regarded as an approximate solution of the relational assignment equations

$$R(x_1) = \underline{\text{small}} \quad (3.21)$$

and

$$R(x_1, x_2) = \underline{\text{approximately equal}} \quad (3.22)$$

Proceeding in a similar fashion in various special cases, one can readily derive one or more approximate conclusions from a given set of propositions, with the understanding that the degree of approximation in each case depends on the definition of the fuzzy restrictions which are induced by the propositions in question. Among the relatively simple examples of such approximate inferences are the following.

x_1 is <u>close</u> to x_2 (3.23)
x_2 is <u>close</u> to x_3

---

x_1 is <u>more or less close</u> to x_3

<u>Most</u> Swedes are <u>tall</u> (3.24)
Nils is a Swede

It is <u>very likely</u> that Nils is <u>tall</u>

<u>Most</u> Swedes are <u>tall</u> (3.25)
<u>Most</u> <u>tall</u> Swedes are <u>blond</u>
Karl is a Swede

It is <u>very likely</u> that Karl is <u>tall</u> and it is <u>more or less</u> (<u>very likely</u>) that Karl is <u>blond</u>.

It should be noted that the last two examples involve a fuzzy quantifier, <u>most</u>, and fuzzy linguistic probabilities <u>very likely</u> and <u>more or less</u> (<u>very likely</u>). By defining <u>most</u> as a fuzzy subset of the unit interval, and <u>tall</u> as a fuzzy subset of the interval [150,200], the proposition $p \triangleq$ <u>Most</u>

Swedes are tall induces a fuzzy restriction on the distribution of heights of Swedes, from which the conclusion "It is very likely that Nils is tall," follows as a linguistic approximation. The same applies to the last example, except that the probability very likely is dilated in the consequent proposition because of the double occurrence of the quantifier most among the antecedent propositions. The goodness of the linguistic approximation in these examples depends essentially on the degree to which very likely approximates to most.

A more general rule of inference which follows at once from (2.45) and (3.10) may be viewed as a generalization of the classical rule of modus ponens. This rule, which will be referred to as the compositional modus ponens, is expressed by

$$\begin{array}{l} x \text{ is } P \\ \text{If } x \text{ is } Q \text{ then } y \text{ is } S \\ \hline y \text{ is } P \circ (\overline{Q'} \oplus \overline{S}) \end{array} \qquad (3.26)$$

where $\oplus$ is the bounded-sum operation, $\overline{Q'}$ is the cylindrical extension of the complement of Q, and $\overline{S}$ is the cylindrical extension of S. Alternatively, using the maximin rule for conditional propositions (see (2.36)), we obtain

$$\begin{array}{l} x \text{ is } P \\ \text{If } x \text{ is } Q \text{ then } y \text{ is } S \\ \hline y \text{ is } P \circ (Q \times S + \overline{Q'}) \end{array} \qquad (3.27)$$

where + is the union and $\overline{Q'} \triangleq Q' \times V$.

Note 3.28. If P = Q and P and S are nonfuzzy, both (3.26) and) (3.27) reduce to the classical modus ponens

$$\begin{array}{l} x \text{ is } P \\ \text{If } x \text{ is } P \text{ then } y \text{ is } S \\ \hline y \text{ is } S \end{array} \tag{3.29}$$

However, if P = Q and P is fuzzy, we do not obtain (3.29) because of the <u>interference effect</u> of the implied part of the conditional proposition "If x is P then y is S," namely "If x is P' then y is V." As a simple illustration of this effect, let U = 1 + 2 + 3 + 4 and assume that

$$P = 0.6/2 + 1/3 + 0.5/4 \tag{3.30}$$

and

$$S = 1/2 + 0.6/3 + 0.2/4 \tag{3.31}$$

In this case,

$$\overline{P'} \oplus \overline{S} = \begin{bmatrix} 1 & 1 & 1 & 1 \\ 0.4 & 1 & 1 & 0.6 \\ 0 & 1 & 0.6 & 0.6 \\ 0.5 & 1 & 1 & 0.7 \end{bmatrix} \tag{3.32}$$

$$P \times S + \overline{P'} = \begin{bmatrix} 1 & 1 & 1 & 1 \\ 0.4 & 0.6 & 0.6 & 0.6 \\ 0 & 1 & 0.6 & 0.2 \\ 0.5 & 0.5 & 0.5 & 0.5 \end{bmatrix} \tag{3.33}$$

and both (3.26) and (3.27) yield

$$y = 0.5/1 + 1/2 + 0.6/3 + 0.6/4 \tag{3.34}$$

which differs from S at those points at which $\mu_S(v)$ is below 0.5.

The compositional form of the <u>modus ponens</u> is of use in the formulation of fuzzy algorithms and the execution of fuzzy instructions [11]. The paper by S. K. Chang [12] and the recent theses by Fellinger [13] and LeFaivre [14] present a number of interesting concepts relating to such instructions and contain many illustrative examples.

4. CONCLUDING REMARKS

In the foregoing discussion, we have attempted to convey some of the main ideas behind the calculus of fuzzy restrictions and its application to approximate reasoning. Although our understanding of the processes of approximate reasoning is quite fragmentary at this juncture, it is very likely that, in time, approximate reasoning will become an important area of study and research in artificial intelligence, psychology and related fields.

REFERENCES

1. L. A. Zadeh, "Fuzzy Logic and Approximate Reasoning," ERL Memorandum M-479, November 1974. To appear in Synthese.

2. L. A. Zadeh, "Similarity Relations and Fuzzy Orderings," Inf. Sci., Vol. 3, pp. 177-200, 1971.

3. L. A. Zadeh, "The Concept of a Linguistic Variable and Its Application to Approximate Reasoning," ERL Memorandum M-411, October 1973. To appear in Information Sciences.

4. L. A. Zadeh, "A Fuzzy-Algorithmic Approach to the Definition of Complex or Imprecise Concepts," ERL Memorandum M-474, October 1974.

5. R. E. Bellman, R. Kalaba and L. A. Zadeh, "Abstraction and Pattern Classification," Jour. Math. Analysis and Appl., Vol. 13, pp. 1-7, 1966.

6. E. T. Jaynes, "Maximum Entropy for Hypothesis Formulation," Annals. of Math. Stat., Vol. 34, pp. 911-930, 1963.

7. M. Tribus, Rational Descriptions, Decisions and Designs, Pergamon Press, New York and Oxford, 1969.

8. N. Rescher, Many-Valued Logic, McGraw-Hill, New York, 1969.

9. L. A. Zadeh, "A Fuzzy-Set-Theoretic Interpretation of Linguistic Hedges," Jour. of Cybernetics, Vol. 2, pp. 4-34, 1972.

10. G. Lakoff, "Hedges: A Study of Meaning Criteria and the Logic of Fuzzy Concepts," Jour. of Philosophical Logic, Vol. 2, pp. 458-508, 1973.

11. L. A. Zadeh, "Outline of a New Approach to the Analysis of Complex Systems and Decision Processes," IEEE Trans. on Systems, Man and Cybernetics, Vol. SMC-3, pp. 28-44, January 1973.

12. S. K. Chang, "On the Execution of Fuzzy Programs Using Finite State Machines," IEEE Trans. Elec. Comp., Vol. C-21, pp. 241-253, 1972.

13. W. L. Fellinger, "Specifications for a Fuzzy System Modeling Language," Ph.D. Thesis, Oregon State University, 1974.

14. R. LeFaivre, "Fuzzy Problem Solving," Tech. Report 37, Madison Academic Computing Center, University of Wisconsin, August 1974.

15. J. Goguen, "L-Fuzzy Sets," Jour. Math. Analysis and Appl., Vol. 18, pp. 145-174, 1967.

16. J. G. Brown, "A Note on Fuzzy Sets," Inf. Control, Vol. 18, pp. 32-39, 1971.

17. A De Luca and S. Termini, "Algebraic Properties of Fuzzy Sets," Jour. Math. Analysis and Appl., Vol. 40, pp. 373-386, 1972.

APPENDIX

Fuzzy Sets - Notation, Terminology and Basic Properties

The symbols U, V, W,..., with or without subscripts, are generally used to denote specific universes of discourse, which may be arbitrary collections of objects, concepts or mathematical constructs. For example, U may denote the set of all real numbers; the set of all residents in a city; the set of all sentences in a book; the set of all colors that can be perceived by the human eye, etc.

Conventionally, if A is a fuzzy subset of U whose elements are $u_1,\ldots,u_n$, then A is expressed as

$$A = \{u_1,\ldots,u_n\} \tag{A1}$$

For our purposes, however, it is more convenient to express A as

$$A = u_1 + \ldots + u_n \tag{A2}$$

or

$$A = \sum_{i=1}^{n} u_i \tag{A3}$$

with the understanding that, for all i,j,

$$u_i + u_j = u_j + u_i \tag{A4}$$

and

$$u_i + u_i = u_i \tag{A5}$$

As an extension of this notation, a finite <u>fuzzy</u> subset of U is expressed as

$$F = \mu_1 u_1 +\ldots+ \mu_n u_n \tag{A6}$$

or, equivalently, as

$$F = \mu_1/u_1 +\ldots+ \mu_n/u_n \tag{A7}$$

where the μ_i, $i=1,\ldots,n$, represent the grades of membership of the u_i in F. Unless stated to the contrary, the μ_i are assumed to lie in the interval [0,1], with 0 and 1 denoting no membership and full membership, respectively.

Consistent with the representation of a finite fuzzy set as a linear form in the u_i, an arbitrary fuzzy subset of U may be expressed in the form of an integral

$$F = \int_U \mu_F(u)/u \tag{A8}$$

in which $\mu_F : U \to [0,1]$ is the membership or, equivalently, the compatibility function of F; and the integral $\int_U$ denotes the union (defined by (A28)) of fuzzy singletons $\mu_F(u)/u$ over the universe of discourse U.

The points in U at which $\mu_F(u) > 0$ constitute the support of F. The points at which $\mu_F(u) = 0.5$ are the crossover points of F.

Example A9. Assume

$$U = a + b + c + d \tag{A10}$$

Then, we may have

$$A = a + b + d \tag{A11}$$

and

$$F = 0.3a + 0.9b + d \tag{A12}$$

as nonfuzzy and fuzzy subsets of U, respectively.

If

$$U = 0 + 0.1 + 0.2 + \ldots + 1 \tag{A13}$$

then a fuzzy subset of U would be expressed as, say,

$$F = 0.3/0.5 + 0.6/0.7 + 0.8/0.9 + 1/1 \tag{A14}$$

If $U = [0,1]$, then F might be expressed as

$$F = \int_0^1 \frac{1}{1+u^2} / u \qquad \text{(A15)}$$

which means that F is a fuzzy subset of the unit interval [0,1] whose membership function is defined by

$$\mu_F(u) = \frac{1}{1+u^2} \qquad \text{(A16)}$$

In many cases, it is convenient to express the membership function of a fuzzy subset of the real line in terms of a standard function whose parameters may be adjusted to fit a specified membership function in an approximate fashion. Two such functions, of the form shown in Fig. A1, are defined below.

$$\begin{aligned} S(u;\ \alpha,\beta,\gamma) &= 0 && \text{for } u \leq \alpha && \text{(A17)} \\ &= 2\left(\frac{u-\alpha}{\gamma-\alpha}\right)^2 && \text{for } \alpha \leq u \leq \beta \\ &= 1 - 2\left(\frac{u-\gamma}{\gamma-\alpha}\right)^2 && \text{for } \beta \leq u \leq \gamma \\ &= 1 && \text{for } u \geq \gamma \end{aligned}$$

$$\begin{aligned} \pi(u;\ \beta,\gamma) &= S(u;\ \gamma-\beta,\ \gamma-\tfrac{\beta}{2},\ \gamma) \text{ for } u \leq \gamma && \text{(A18)} \\ &= 1 - S(u;\ \gamma,\ \gamma+\tfrac{\beta}{2},\ \gamma+\beta) \text{ for } u \geq \gamma \end{aligned}$$

In $S(u;\ \alpha,\ \beta,\ \gamma)$, the parameter β, $\beta = \frac{\alpha+\gamma}{2}$, is the crossover point. In $\pi(u;\ \beta,\gamma)$, β is the bandwidth, that is, the separation between the crossover points of π, while γ is the point at which π is unity.

In some cases, the assumption that μ_F is a mapping from U to [0,1] may be too restrictive, and it may be desirable to allow μ_F to take values in a lattice or, more particularly,

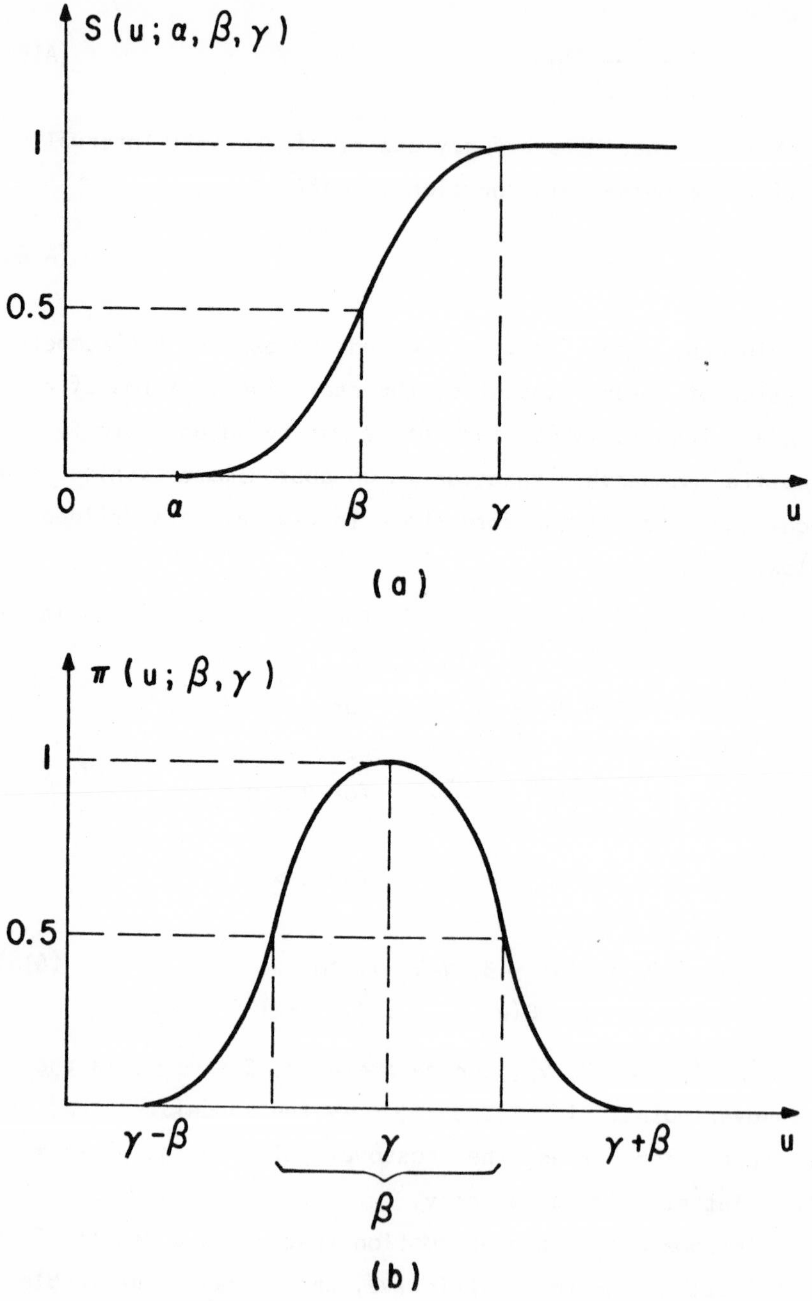

Figure A1 Plots of S and π Functions.

in a Boolean algebra [15], [16], [17]. For most purposes, however, it is sufficient to deal with the first two of the following hierarchy of fuzzy sets.

Definition A19. A fuzzy subset, F, of U is of type 1 if its membership function, μ_F, is a mapping from U to [0,1]; and F is of type n, n = 2,3,..., if μ_F is a mapping from U to the set of fuzzy subsets of type n-1. For simplicity, it will always be understood that F is of type 1 if it is not specified to be of a higher type.

Example A20. Suppose that U is the set of all nonnegative integers and F is a fuzzy subset of U labeled small integers. Then F is of type 1 if the grade of membership of a generic element u in F is a number in the interval [0,1], e.g.,

$$\mu_{\underline{\text{small integers}}}(u) = (1 + (\frac{u}{5})^2)^{-1} \qquad u = 0,1,2,... \quad \text{(A21)}$$

On the other hand, F is of type 2 if for each u in U, $\mu_F(u)$ is a fuzzy subset of [0,1] of type 1, e.g., for u = 10,

$$\mu_{\underline{\text{small integers}}}(10) = \underline{\text{low}} \quad \text{(A22)}$$

where low is a fuzzy subset of [0,1] whose membership function is defined by, say,

$$\mu_{\underline{\text{low}}}(v) = 1 - S(v;\ 0,\ 0.25,\ 0.5), \quad v \in [0,1] \quad \text{(A23)}$$

which implies that

$$\underline{\text{low}} = \int_0^1 (1 - S(v;\ 0,\ 0.25,\ 0.5))/v \quad \text{(A24)}$$

If F is a fuzzy subset of U, then its α-level-set, F_α, is a nonfuzzy subset of U defined by ([18])

$$F_\alpha = \{u | \mu_F(u) \geq \alpha\} \quad \text{(A25)}$$

for $0 < \alpha \leq 1$.

If U is a linear vector space, the F is convex iff for all $\lambda \varepsilon [0,1]$ and all u_1, u_2 in U,

$$\mu_F(\lambda u_1 + (1-\lambda)u_2) \geq \min(\mu_F(u_1),\mu_F(u_2)) \qquad \text{(A26)}$$

In terms of the level-sets of F, F is convex iff the F_α are convex for all $\alpha \varepsilon (0,1]$.[1]

The realtion of containment for fuzzy subsets F and G of U is defined by

$$F \subset G \Longleftrightarrow \mu_F(u) \leq \mu_G(u), \qquad u \varepsilon U \qquad \text{(A27)}$$

Thus, F is a fuzzy subset of G if (A27) holds for all u in U.

Operations on Fuzzy Sets

If F and G are fuzzy subsets of U, their union, $F \cup G$, intersection, $F \cap G$, bounded-sum, $F \oplus G$, and bounded-difference, $F \ominus G$, are fuzzy subsets of U defined by

$$F \cup G \triangleq \int_U \mu_F(u) \vee \mu_G(u)/u \qquad \text{(A28)}$$

$$F \cap G = \int_U \mu_F(u) \wedge \mu_G(u)/u \qquad \text{(A29)}$$

$$F \oplus G = \int_U 1 \wedge (\mu_F(u) + \mu_G(u))/u \qquad \text{(A30)}$$

$$F \ominus G = \int_U 0 \vee (\mu_F(u) - \mu_G(u))/u \qquad \text{(A31)}$$

where $\vee$ and $\wedge$ denote max and min, respectively. The complement of F is defined by

$$F' = \int_U (1 - \mu_F(u))/u \qquad \text{(A32)}$$

or, equivalently,

$$F' = U \ominus F \qquad \text{(A33)}$$

[1]This definition of convexity can readily be extended to fuzzy sets of type 2 by applying the extension principle (see (A75)) to (A26).

It can readily be shown that F and G satisfy the identities

$$(F \cap G)' = F' \cup G' \tag{A34}$$

$$(F \cup G)' = F' \cap G' \tag{A35}$$

$$(F \oplus G)' = F' \ominus G \tag{A36}$$

$$(F \ominus G)' = F' \oplus G \tag{A37}$$

and that F satisfies the resolution identity [2]

$$F = \int_0^1 \alpha F_\alpha \tag{A38}$$

where F_α is the α-level-set of F; αF_α is a set whose membership function is $\mu_{\alpha F_\alpha} = \alpha\mu_{F_\alpha}$, and $\int_0^1$ denotes the union of the αF, with $\alpha \in (0,1]$.

Although it is traditional to use the symbol ∪ to denote the union of nonfuzzy sets, in the case of fuzzy sets it is advantageous to use the symbol + in place of ∪ where no confusion with the arithmetic sum can result. This convention is employed in the following example, which is intended to illustrate (A28), (A29), (A30), (A31) and (A32).

Example A39. For U defined by (A10) and F and G expressed by

$$F = 0.4a + 0.9b + d \tag{A40}$$

$$G = 0.6a + 0.5b \tag{A41}$$

we have

$$F + G = 0.6a + 0.9b + d \tag{A42}$$

$$F \cap G = 0.4a + 0.5b \tag{A43}$$

$$F \oplus G = a + b + d \tag{A44}$$

$$F \ominus G = 0.4b + d \tag{A45}$$

$$F' = 0.6a + 0.1b + c \tag{A46}$$

The linguistic connectives *and* (conjunction) and *or* (disjunction) are identified with $\cap$ and +, respectively. Thus,

$$F \text{ \underline{and} } G \triangleq F \cap G \tag{A47}$$

and

$$F \text{ \underline{or} } G \triangleq F + G \tag{A48}$$

As defined by (A47) and (A48), *and* and *or* are implied to be *noninteractive* in the sense that there is no "trade-off" between their operands. When this is not the case, *and* and *or* are denoted by <*and*> and <*or*>, respectively, and are defined in a way that reflects the nature of the trade-off. For example, we may have

$$F \langle\underline{and}\rangle G \triangleq \int_U \mu_F(u)\ \mu_G(u)/u \tag{A49}$$

$$F \langle\underline{or}\rangle G \triangleq \int_U (\mu_F(u) + \mu_G(u) - \mu_F(u)\ \mu_G(u))/u \tag{A50}$$

whose + denotes the arithmetic sum. In general, the interactive versions of *and* and *or* do not possess the simplifying properties of the connectives defined by (A47) and (A48), e.g., associativity, distributivity, etc. (See [4].)

If α is a real number, then F^α is defined by

$$F^\alpha \triangleq \int_V (\mu_F(n))^\alpha/u \tag{A51}$$

For example, for the fuzzy set defined by (A40), we have

$$F^2 = 0.16a + 0.81b + d \tag{A52}$$

and

$$F^{1/2} = 0.63a + 0.95b + d \tag{A53}$$

These operations may be used to approximate, very roughly,

to the effect of the linguistic modifiers very and more or less. Thus,

$$\text{very } F \triangleq F^2 \tag{A54}$$

and

$$\text{more or less } F \triangleq F^{1/2} \tag{A55}$$

If $F_1,\ldots,F_n$ are fuzzy subsets of $U_1,\ldots,U_n$, then the cartesian product of $F_1,\ldots,F_n$ is a fuzzy subset of $U_1 \times \ldots \times U_n$ defined by

$$F_1 \times \ldots \times F_n = \int_{U_1\times\ldots\times U_n} (\mu_{F_1}(u_1) \wedge \ldots \wedge \mu_{F_n}(u_n))/(u_1,\ldots,u_n) \tag{A56}$$

As an illustration, for the fuzzy sets defined by (A40) and (A41), we have

$$\begin{aligned} F\times G &= (0.4a + 0.9b + d) \times (0.6a + 0.5b) \\ &= 0.4/(a,a) + 0.4/(a,b) + 0.6/(b,a) \\ &\quad + 0.5/(b,b) + 0.6/(d,a) + 0.5/(d,b) \end{aligned} \tag{A57}$$

which is a fuzzy subset of $(a + b + c + d)\times(a + b + c + d)$.

Fuzzy Relations

An n-ary fuzzy relation R in $U_1 \times \ldots \times U_n$ is a fuzzy subset of $U_1 \times \ldots \times U_n$. The projection of R on $U_{i_1} \times \ldots \times U_{i_k}$, where $(i_1,\ldots,i_k)$ is a subsequence of $(1,\ldots,n)$, is a relation in $U_{i_1} \times \ldots \times U_{i_k}$ defined by

$$\text{Proj } R \text{ on } U_{i_1} \times \ldots \times U_{i_k} \triangleq \int_{U_{i_1} \times \ldots \times U_{i_k}} \bigvee_{u_{j_1},\ldots,u_{j_\ell}} \mu_R(u_1,\ldots,u_n)/(u_1,\ldots,u_n) \tag{A58}$$

where $(j_1,\ldots,j_\ell)$ is the sequence complementary to $(i_1,\ldots,i_k)$

(e.g., if n=6 then (1,3,6) is complementary to (2,4,5)), and $\vee_{u_{j_1},\ldots,u_{j_\ell}}$ denotes the supremum over $U_{j_1} \times \ldots \times U_{j_\ell}$.

If R is a fuzzy subset of $U_{i_1},\ldots,U_{i_k}$, then its cylindrical extension in $U_1 \times \ldots \times U_n$ is a fuzzy subset of $U_1 \times \ldots \times U_n$ defined by

$$\bar{R} = \int_{U_1 \times \ldots \times U_n} \mu_R(u_{i_1},\ldots,u_{i_k})/(u_1,\ldots,u_n) \tag{A59}$$

In terms of their cylindrical extensions, the composition of two binary relations R and S (in $U_1 \times U_2$ and $U_2 \times U_3$, respectively) is expressed by

$$R \circ S = \text{Proj } \bar{R} \cap \bar{S} \text{ on } U_1 \times U_3 \tag{A60}$$

where $\bar{R}$ and $\bar{S}$ are the cylindrical extensions of R and S in $U_1 \times U_2 \times U_3$. Similarly, if R is a binary relation in $U_1 \times U_2$ and S is a unary relation in U_2, their composition is given by

$$R \circ S = \text{Proj } R \cap \bar{S} \text{ on } U_1 \tag{A61}$$

Example A62. Let R be defined by the right-hand member of (A57) and

$$S = 0.4a + b + 0.8d \tag{A63}$$

Then

$$\text{Proj } R \text{ on } U_1 (\underline{\underline{\Delta}}\ a + b + c + d) = 0.4a + 0.6b + 0.6d \tag{A64}$$

and

$$R \circ S = 0.4a + 0.5b + 0.5d \tag{A65}$$

Linguistic Variables

Informally, a linguistic variable, $\mathcal{X}$, is a variable whose values are words or sentences in a natural or artificial language. For example, if age is interpreted as a linguistic variable, then its term-set, $T(\mathcal{X})$, that is, the set of its linguistic values, might be

$$T(\underline{age}) = \underline{young} + \underline{old} + \underline{very\ young} + \underline{not\ young} + \underline{very\ old} + \underline{very\ very\ young} + \underline{rather\ young} + \underline{more\ or\ less\ young} + \ldots \quad \text{(A66)}$$

where each of the terms in T(age) is a label of a fuzzy subset of a universe of discourse, say U = [0,100].

A lingiustic variable is associated with two rules: (a) a syntactic rule, which defines the well-formed sentences in $T(\mathcal{X})$; and (b) a semantic rule, by which the meaning of the terms in $T(\mathcal{X})$ may be determined. If X is a term in $T(\mathcal{X})$, then its meaning (in a denotational sense) is a subset of U. A primary term in $T(\mathcal{X})$ is a term whose meaning is a primary fuzzy set, that is, a term whose meaning must be defined a priori, and which serves as a basis for the computation of the meaning of the nonprimary terms in $T(\mathcal{X})$. For example, the primary terms in (A66) are young and old, whose meaning might be defined by their respective compatibility functions $\mu_{\underline{young}}$ and $\mu_{\underline{old}}$. From these, then, the meaning - or, equivalently, the compatibility functions - of the non-primary terms in (A66) may be computed by the application of a semantic rule. For example, employing (A54) and (A55), we have

$$\mu_{\underline{very\ young}} = (\mu_{\underline{young}})^2 \quad \text{(A67)}$$

$$\mu_{\underline{more\ or\ less\ old}} = (\mu_{\underline{old}})^{1/2} \quad \text{(A68)}$$

$$\mu_{\underline{not\ very\ young}} = 1 - (\mu_{\underline{young}})^2 \quad \text{(A69)}$$

For illustration, plots of the compatibility functions of these terms are shown in Fig. A2.

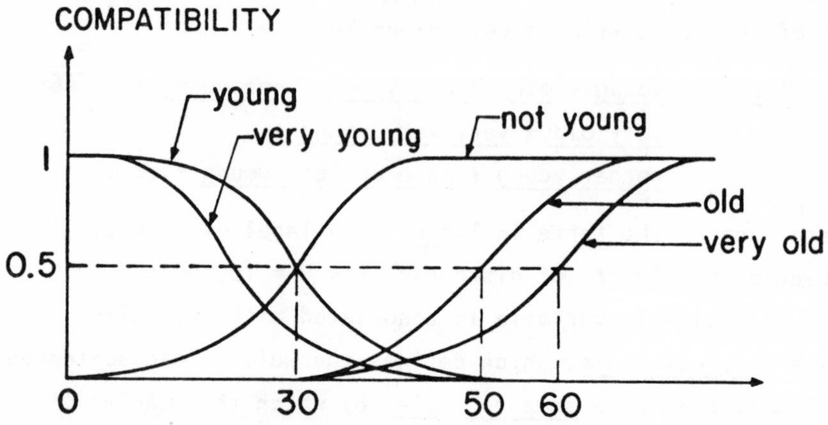

Figure A2 Compatibility Function of young and its Modifications.

<u>The Extension Principle</u>

Let f be a mapping from U to V. Thus,

$$v = f(u) \tag{A70}$$

where u and v are generic elements of U and V, respectively.

Let F be a fuzzy subset of U expressed as

$$F = \mu_1 u_1 + \ldots + \mu_n u_n \tag{A71}$$

or, more generally,

$$F = \int_U \mu_F(u)/u \tag{A72}$$

By the extension principle [3], the image of F under f is given by

$$f(F) = \mu_1 \; f(u_1) + \ldots + \mu_n \; f(u_n) \tag{A73}$$

or, more generally,

$$f(F) = \int_U \mu_F(u)/f(u) \tag{A74}$$

Similarly, if f is a mapping from U x V to W, and F and G are fuzzy subsets of U and V, respectively, then

$$f(F,G) = \int_W (\mu_F(u) \wedge \mu_G(v))/f(u,v) \tag{A75}$$

Example A76. Assume that f is the operation of squaring. Then, for the set defined by (A14), we have

$$f(0.3/0.5 + 0.6/0.7 + 0.8/0.9 + 1/1) = 0.3/0.25 + 0.6/0.49 + 0.8/0.81 + 1/1 \tag{A77}$$

Similarly, for the binary operation $\vee$ ($\triangleq$ max), we have

$$(0.9/0.1 + 0.2/0.5 + 1/1) \vee (0.3/0.2 + 0.8/0.6) = 0.3/0.2 + 0.2/0.5 + 0.8/1 + 0.8/0.6 + 0.2/0.6 \tag{A78}$$

It should be noted that the operation of squaring in (A77) is different from that of (A51) and (A52).

FUZZY PROGRAMS AND THEIR EXECUTION

K. Tanaka and M. Mizumoto
Department of Information
and Computer Sciences
Faculty of Engineering Science
Osaka University
Toyonaka, Osaka 560, Japan

1. INTRODUCTION

In our daily life, we often encounter situations where we shall not always need the exact and detailed information to execute the intended behavior. For instance, let us suppose the case where a person asks the way in a strange place. For example, he will receive such an instruction as: "go straight on this way and turn right at the signal, then you could find the spot after about a few minutes walk." Then he could get to the spot without trouble, if the instruction is true. However, if we want to make a machine execute such an instruction as mentioned above, just then we shall find it difficult to do.

In the real world, as a matter of fact, many ill-defined and inexact instructions, that is, the so-called fuzzy instructions exist which we want to translate and execute by a machine. Therefore, the execution of fuzzy instructions using a machine is of much interest and very useful in a wide variety of problems relating to pattern recognition, control, artificial intelligence, linguistics, information retrieval and decision processes involved in psychological, economical and social fields.

In this paper, a generalized automaton is proposed as an abstract model for a fuzzy machine which can translate and execute fuzzy programs and several methods which translate a given sequence of fuzzy instructions into another sequence of precise instructions called a machine program are also discussed.

In addition, the practical application is presented in a few interesting examples to demonstrate the usefulness of the foregoing proposal.

2. GENERALIZED FUZZY MACHINES

A finite-state automaton has been taken up as a fuzzy machine model which executes a fuzzy program by S. K. Chang [1].

Here formulated is an extended fuzzy machine based on a generalized automaton and a few procedures for execution of fuzzy programs are also presented.

Definition 1. A generalized machine M is a system given by

$$M = (k, X, \psi, x_0, T, V) \tag{1}$$

where (i) K is a finite set of machine instructions.

(ii) X is a finite set of internal states.

(iii) ψ is a function such that

$$\psi: \quad X \times K \times X \rightarrow V \tag{2}$$

and is called a state transition function. The value of ψ, $\psi(x, \mu, x') \in V$, designates a weight value controlling the transition from a state x to a new state x' for a given machine instruction μ.

(iv) x_0 is an initial state in X.

(v) T is a finite set of final states and is a subset of X.

(vi) V is a space of weight (or grade) controlling the state transition.

In the present paper an "L-fuzzy automaton" with the weight space defined in the lattice ordered semigroup is considered as a general machine. Several machines are also derived from L-fuzzy automata as their specific examples [2], [6].

Let us now define $V = (L,v,*,0,I)$ in a lattice ordered semigroup L. where v and * denote a least upper bound in L and an operation of semigroup, respectively; and 0 and I denote zero (least element) and identity (greatest element), respectively. Then the state transition of L-fuzzy automata can be formulated as follows. For a given string of machine instructions $\bar{\mu} = \mu_1\mu_2\cdots\mu_n$ in K* where K* denotes a set of all finite strings over K, the state transition function at each step of the machine instruction will be $\psi(x_0, \mu_1, x_1)$, $\psi(x_1, \mu_2, x_2),\ldots,\psi(x_{n-1}, \mu_n, x_n)$. Then the state of the L-fuzzy automaton is said to transit from x_0 through x_n one by one by the string of machine instruction $\bar{\mu}$ and the weight (or grade) corresponding to this state transition is simply given by

$$\psi(x_0, \psi_1, x_1)*\psi(x_1, \mu_2, x_2)*\cdots*\psi(x_{n-1}, \mu_n, x_n) \quad (3)$$

Thus the domain X x K x X of the state transition function ψ will be extended to X x K* x X and the weight (or grade) of the state transition for any input string $\bar{\mu} = \mu_1\mu_2\cdots\mu_n \in K^*$ can be given recursively as

$$\psi(x, e, x') = \begin{cases} I & \text{for} \quad x = x' \\ 0 & \text{for} \quad x \neq x' \end{cases} \quad (4)$$

$$\psi(x, \bar{\mu}, x') = \bigvee_{x_1,x_2,\ldots,x_{n-1}} [\psi(x, \mu_1, x_1)*\psi(x_1, \mu_2, x_2)* \cdots *\psi(x_{n-1}, \mu_n, x')] \quad (5)$$

where e denotes a null string.

Note: For the following algebraic structure of V, various types of automata can be derived as specific cases of L-fuzzy automata.

(i) For $V = ([0,1], \max, \min, 0, 1)$, fuzzy automata can be obtained.

(ii) For $V = (\{0,1\}, \max, \min, 0, 1)$, nondeterministic automata can be obtained.

(iii) For $V = (\{0,1\}, \max, \min, 0, 1)$, deterministic automata can be obtained under the constraint as follows: there exists x' uniquely such that $\psi(x, \mu, x') = 1$ for each pair of x and μ.

(iv) For $V = ([0,1], +, \times, 0, 1)$, probabilistic automata can be obtained under the constraint such that $\sum_{x' \varepsilon X} (x, \mu, x') = 1$.

Definition 2. A generalized fuzzy machine is a system

$$\hat{M} = (\Sigma, M, W) \tag{6}$$

where (i) Σ is a finite set of fuzzy instructions and each fuzzy instruction σ_i is a function such that

$$\sigma_i: \ X \times K \rightarrow W \tag{7}$$

(ii) M is a generalized automaton defined by Definition 1.

(iii) W is a space of weight (or grade) with respect to the selection of a machine instruction μ_i. The value of $\sigma_i(x_i, \mu_i) \ \varepsilon \ W$ designates the weight (or grade) of selecting the machine instruction μ_i when a generalized fuzzy machine $\hat{M}$ associated with a generalized automaton M in the state of x_i has received a fuzzy instruction σ_i.

2.1. Fuzzy Machines Derived From Deterministic Automata

This is the case where a deterministic automaton is chosen as an example of generalized machine M.

(a) For W = [0,1], a fuzzy-deterministic machine similar to that of S. K. Chang can be derived, where

$$\sigma_i(x_i, \mu_i) = \min[f(x_i, \sigma_i, \mu_i), \lambda(x_i, \sigma_i, \mu_i)]$$

shows the grade of selecting the machine instruction μ_i, when the machine M is in the state of x_i and receives the fuzzy instruction σ_i. Here note that $f(\cdot)$ and $\lambda(\cdot)$ in the above equation represent the feasibility function and the performance function, respectively [1].

(b) For W = [0,1], a probabilistic-deterministic machine can be derived under the condition that

$$\sum_{\mu_i} \sigma_i(x_i, \mu_i) = 1 \text{ for every } \sigma_i \in \Sigma \text{ and } x_i \in X.$$

This condition shows that a machine instruction μ_i is selected in a probabilistic way when the machine is in the state of x_i and receives a fuzzy instruction of σ_i.

(c) For W = {0,1} can be obtained a nondeterministic-deterministic machine, where $\sigma_i(x_i, \mu_i) = 1$ or $\sigma_i(x_i, \mu_i) = 0$.

The equation of $\sigma_i(x_i, \mu_i) = 1$ shows that a machine instruction μ_i is selected in a nondeterministic way when the machine is in the state x_i and receives a fuzzy instruction σ_i.

2.2. Another Type of Fuzzy Machines Derived From Various Classes of Automata

A variety of generalized fuzzy machines will be derived from various classes of automata.

For W = [0,1] and V = [0,1] can be exemplified the generalized fuzzy machines enumerated as below.

Fuzzy-fuzzy machine; Fuzzy-probabilistic machine; Fuzzy-nondeterministic machine; Fuzzy-deterministic machine; Probabilistic-fuzzy machine; Probabilistic-probabilistic machine; Probabilistic-nondeterministic machine; Probabilistic-deterministic machine; Nondeterministic-fuzzy machine; Nondeterministic-probabilistic machine; Nondeterministic-nondeterministic machine; Nondeterministic-deterministic machine

Note: The fuzzy machine defined by S. K. Chang is equivalent to a fuzzy-deterministic machine and that of Jakubowski [3] is equivalent to a fuzzy-nondeterministic machine.

Definition 3. A sequence of fuzzy instructions of $\sigma_1, \sigma_2, \ldots, \sigma_n \in \Sigma$ is called an elementary fuzzy program $\bar{\sigma} = \sigma_1\sigma_2 \ldots \sigma_n \in \Sigma^*$ and a fuzzy program is a regular expression over Σ. If every machine instruction μ_i is feasible with respect to the every situation of the machine M and the state of the machine will transit successively from x_0 through x_n, that is, if

$$\begin{array}{ll} \sigma_1(x_0, \mu_1) = w_1 & \psi(x_0, \mu_1, x_1) = v_1 \\ \sigma_2(x_1, \mu_2) = w_2 & \psi(x_1, \mu_2, x_2) = v_2 \\ \cdots\cdots & \cdots\cdots \\ \sigma_n(x_{n-1}, \mu_n) = w_n & \psi(x_{n-1}, \mu_n, x_n) = v_n, \end{array} \tag{8}$$

then the fuzzy program $\bar{\sigma}$ is said to be executable with respect to the fuzzy machine $\hat{M}$ and the machine program $\bar{\mu}$ is said to be an execution of the fuzzy program $\bar{\sigma}$. This statement will be represented in the form as

$$x_0 \frac{(\sigma_1, \mu_1)}{w_1} \frac{x_1}{v_1} \frac{(\sigma_2, \mu_2)}{w_2} \frac{x_2}{v_2} \cdots x_{n-1} \frac{(\sigma_n, \mu_n)}{w_n} \frac{x_n}{v_n}. \tag{9}$$

<u>Definition 4</u>. Let the weight (w,v) be defined by

$$(w,v) = (w_1 \cdot w_2 \cdot \ldots \cdot w_n,\ v_1 * v_2 * \ldots * v_n) \quad (10)$$

where the marks • and * denote the operation on the weight space W and V, respectively. Then the fuzzy program $\bar{\sigma}$ is said to be executable with the weight (w,v) if (w,v) > (0,0).

Example 1: Let the operation • on W = [0,1] be min (Λ) and the operation * on V = {0,1} be min or x(product). Then the generalized fuzzy machine will be a fuzzy-nondeterministic machine described previously and the corresponding weight (w,v) is given by

$$(w,v) = (w_1 \wedge w_2 \wedge \ldots \wedge w_n,\ 1),$$

when the fuzzy program $\bar{\sigma}$ is feasible. The 1 on the right side of the above expression means that $\psi(x_0, \mu_1, x_1) = \psi(x_1, \mu_2, x_2) = \ldots = \psi(x_{n-1}, \mu_n, x_n) = 1$.

Example 2: Let the operation • on W = [0,1] be min and the operation * on V = [0,1] be x(product). Then the generalized fuzzy machine will be a fuzzy-probabilistic machine and the corresponding weight (w,v) is given by

$$(w,v) = (w_1 \wedge w_2 \wedge \ldots \wedge w_n,\ v_1 \times v_2 \times \ldots \times v_n).$$

3. EXECUTION PROCEDURE OF FUZZY PROGRAMS

As one particular way of executing fuzzy programs using a finite state machine, Chang has given the way called simple execution procedure which selects the machine instruction μ_i with the highest grade $w_i = \sigma(x_{i-1}, \mu_i)$ with respect to the fuzzy instruction σ_i at each step i.

In this paper discussed is a more general way of executing fuzzy programs by making use of the generalized fuzzy machine described previously. The reason why the machine is to be given with such a generality as mentioned above with regard to not only the selection of the machine instruction, but also the mode of state transition will be easily approved in the following example. Suppose that a person has received such a fuzzy instruction as, for instance, "come to the school at about 9:30 a.m.". Then he will make up his mind to come to the school just on 9:20 a.m. This corresponds to the selection of a machine instruction. In fact, however, he will not be able to come to the school just on 9:20 a.m. as usual, but his arrival will shift slightly from 9:20 a.m. This may be interpreted as corresponding to the state transition. Thus, the generality of the state transition given to the machine will enlarge the executability of the fuzzy programs.

Furthermore, in such a case where the state aimed at (for instance, arrival to the destination in the example of simulation of human drivers) can not be attained, there will be needed to alter the way of selecting either the machine instruction or the state transition. That is to say, if a fuzzy instruction is received by the machine in any state, the machine instruction will be selected in a certain way and the state of the machine will change according to a certain manner. Then, if the state after the transition is not equivalent to the state aimed at, the successive transition will occur step by step according to the same manner as above until the attainment of the state aimed at. In this case, if the successive state will not be available, the machine instruction is updated and the same procedure is repeated for the same fuzzy instruction. If there is no machine instruction available for the given fuzzy instruction, a back-tracking procedure will be introduced. That is to say, the available machine instruction must be selected to the fuzzy instruction of one step prior to the last one and the desired state must be searched according to the same procedure described above.

The following Figure 1 shows the flow chart of the above mentioned procedure for execution of an elementary fuzzy program $\bar{\sigma} = \sigma_1\sigma_2 \ldots \sigma_n$ by making use of a generalized fuzzy machine $\hat{M}$ given by Equation (6). The way of selecting the machine instruction and that of transition of the state, which is labeled as ① and ② in Figure 1, respectively, will differ depending on the class of a generalized machine chosen. Then let us now explain that in detail.

3.1 Selection of Machine Instruction

(a) In case of fuzzy selection, the machine $\hat{M}$ selects the machine instruction $\mu \in K(i, x(i-1))$ with the highest grade at each step of the fuzzy instruction σ_i such that

$$\sigma_i(x(i-1),\mu) \geqq \sigma_i(x(i-1), \mu') \tag{11}$$

for all other μ' in K.

(b) In case of probabilistic selection, the machine $\hat{M}$ selects the machine instruction $\mu \in K(i, x(i-1))$ with the probability p at each step of the fuzzy instruction σ_i in proportion to the fuzzy grade $\sigma_i(x(i-1),\mu)$ such that

$$p = \frac{\sigma_i(x(i-1),\mu)}{\sum_{\mu' \in K(i,\, x(i-1))} \sigma_i(x(i-1),\mu')} \tag{12}$$

(c) In case of nondeterministic selection, the machine instruction μ is chosen in $K(i, x(i-1))$ in a nondeterministic manner.

3.2 State Transition

(a) In case of fuzzy transition, the state of the machine transits from $x(i-1)$ to x such that

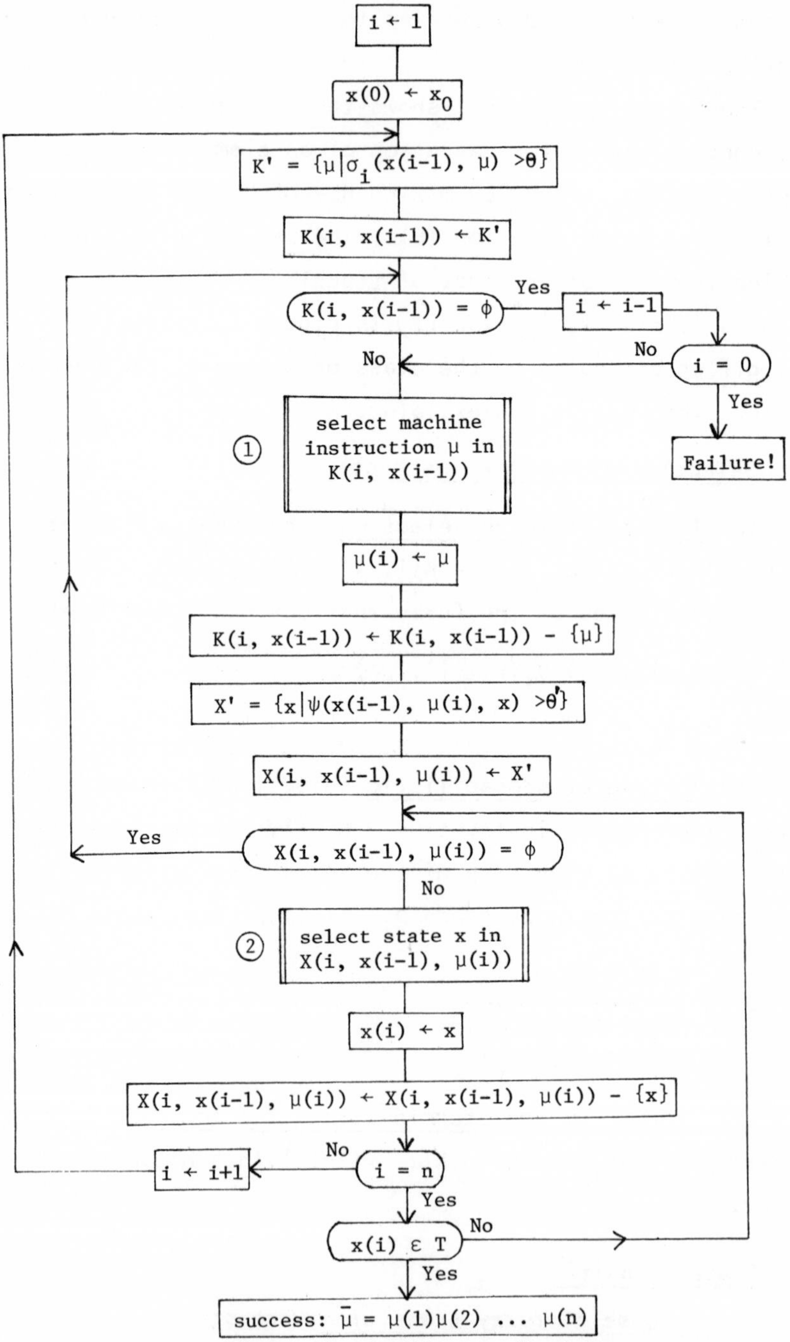

Figure 1 Caption on Next Page

$$\psi(x(i-1), \mu(i), x) \geqq \psi(x(i-1), \mu(i), x') \tag{13}$$

for all other $x' \in X(i, x(i-1), \mu(i))$.

(b) In case of probabilistic transition, the state of the machine transits to x from x(i-1) with the probability p, where

$$p = \frac{\psi(x(i-1), \mu(i), x)}{\sum\limits_{x' \in X(i,\, x(i-1),\, \mu(i))} \psi(x(i-1), \mu(i), x')}. \tag{14}$$

(c) In case of nondeterministic transition, the state of the machine transits from x(i-1) to x(i) in X(i, x(i-1), μ(i)) in a nondeterministic way.

(d) In case of deterministic transition, the state available for the machine is uniquely determined depending on the property of its state transition function ψ.

4. SIMULATION OF HUMAN DRIVER'S BEHAVIOR

A simulation was conducted so as to experiment the procedures for executing a fuzzy program by the use of either a fuzzy-deterministic machine or a probabilistic-deterministic machine.

Let the fuzzy instructions for a driver by the following five kinds:

(i) $\sigma_{go}(L)$: Go about L meters, (ii) σ_R: Turn right,
(iii) σ_L: Turn left, (iv) σ^*_{go}: Go straight, (15)
(v) $\sigma_{\{\sim\}}$: Until ∼,

Preceding to the execution of these fuzzy instructions, each of them has to be rewritten as a sequence of the following three kinds of quasi- fuzzy instructions, i.e.,

Figure 1 Flow Chart Illustrating Execution Procedure of Fuzzy Program

$$(i)\ \sigma_{go}:\ \text{Go ahead one step},\ (ii)\ \sigma_R:\ \text{Turn right},\ (iii)\ \sigma_L:\ \text{Turn left}, \tag{16}$$

by making use of the three methods of MAX-method, PROB-method and *-method for rewriting. The quasi-fuzzy instructions thus obtained are then interpreted by the MAX-method into the eight kinds of machine instructions which are composed of the elementary movements given by the eight directions.

4.1. Fuzzy Instructions, Quasi-fuzzy Instructions, and Quasi-internal States

A computer experiment was made to simulate the behavior of a driver who is directed the way by a sequence of fuzzy instructions. The initial position (the coordinates and the direction) is given and a typical set of fuzzy driving instructions and a set of quasi-fuzzy instructions are shown respectively by (15) and (16) mentioned previously.

Assuming the quasi-fuzzy instructions just as the machine instructions, the execution procedure discussed in the Section 3 can now be available to rewrite the fuzzy instructions into a sequence of the quasi-fuzzy instructions.

The internal state of a fuzzy machine is given as a pair of the coordinate and the direction with respect to each position of the driver on a digitized map shown in Figure 7. However, let us now introduce the notion of a quasi-internal state so as to reduce the number of the internal states. The roads on the map are classified according to the shape of the node and the branch, and the quasi-internal state of the fuzzy machine is designated as a pair of the shape of the node or the branch on the map and the direction of the driver as shown in Figure 4.

Then let us compose an evaluation table for selecting a machine instruction at each step of the quasi-fuzzy instruction and the quasi-internal state as follows. Let the machine instructions be assigned by the following eight instructions of μ_j's(j=1,2,...,8) where μ_j denotes the instruction with respect to the j-th direction in Figure 7. If the evaluation value $\Phi(\beta,s)$ defined below is given for selecting a machine instruction, the evaluation table can be made for a given pair (β,s) of a quasi-fuzzy instruction β and a quasi internal-state s as shown in Figure 4.

$$\Phi(\beta,s) = \begin{cases} 0, & \text{if there is no machine instruction available.} \\ i, & \text{if } \mu_i \text{ is available.} \\ i\times 10+j, & \text{if } \mu_i \text{ is more available than } \mu_j. \end{cases} \tag{17}$$

where the last equation means that the grade of selecting the machine instruction μ_i is higher than that of selecting μ_j when the fuzzy machine is in the quasi-internal state s, i.e., $\beta(s, \mu_i) > \beta(s, \mu_j)$.

4.2. Execution Procedures of Fuzzy Instructions

There are two cases of giving fuzzy instructions, that is, (a) the case where fuzzy instructions are given step by step, and (b) the case where a sequence of fuzzy instructions is given a priori. A practical example of (a) is supposed to be the case where a fellow passenger gives the fuzzy instruction step by step to the driver who has to memorize all the past fuzzy instructions given at each step as well as the past fuzzy instructions given at each step as well as the present one and has to judge and behave by himself. On the other hand, the case of (b) will be illustrated by such an example that the driver is given a note showing a route and he can know the state of the route beforehand.

Let us consider the execution procedure of, for instance, a fuzzy instruction as "Go about L meters" given in the expression (15). The idea of "about L meters" may be dealt with the concept of a fuzzy set and it will be characterized by a membership function as shown in Figure 2. The mode of selection of the distance for a fuzzy instruction named "Go about L meters" willbe specified by the following three types.

① Type 1 where a threshold α is set and the distance with the highest grade of membership among all the distances whose grades of membership are larger than α is selected.

② Type 2 where the distance is selected with the probability proportional to the grade of membership which is larger than a specified threshold α.

③ Type 3 where any distance whose grade of membership is larger than a specific threshold α is permissible.

Here let the fuzzy instruction be given in the same way as in the case of (a) mentioned previously, and let us discuss in more detail on the execution procedures for this case.

[a-1] MAX-Method

A set of fuzzy instructions named "Go about L meters" can be specified by the membership function. Here, setting a threshold $\alpha(1 \geq \alpha \geq 0)$, the membership function $w(\ell)$ is truncated at a point η where $w(\eta) = \alpha$ and at another point η' where $w(\eta') = \alpha$ as shown in Figure 2.

The driver goes ahead η meters without condition. Upon arriving at $\ell = \eta$ he selects the distance with the highest grade of membership in the interval $[\eta,\eta']$. If it is not available, the distance with the second highest grade is selected. Such a method is designated MAX-Method and η is named a "Lowest Bound."

[a-2] PROB-Method

The fuzzy instruction is chosen with a probability $p(\ell)$

which is proportional to the grade of membership $w(\ell)$ in the interval $[\eta,\eta']$. This method is called PROB-Method.

[a-3] *-Method

The drivers goes ahead step by step to the point where the next instruction will be executable. This is named *-Method and is a specific type of 3 mentioned previously.

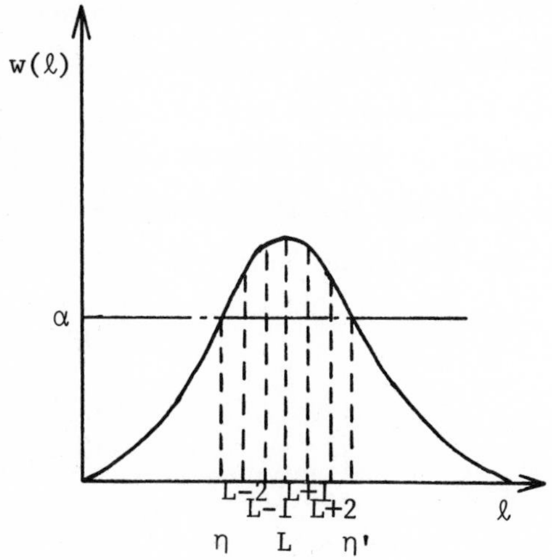

Figure 2 Membership Function w(ℓ) for a Set of Fuzzy Instructions named "Go about L Meters"

In the next section, we shall present an example of simulation of human drivers to whom the driving instructions are given step by step as in the case of (a).

4.3 Simulation of Human Driver's Behavior Directed by Fuzzy Instructions

In our example, the map is digitized as shown in Fig. 7, where the unit scale of the coordinates is equal to 10 meters, the direction allowable to the driver is quantized in the eight directions and the symbols on the map are illustrated in Fig. 3. The initial position is given by a triplet (x, y; d), where (x, y) is the coordinates of the initial position of the driver and d indicates the direction of the driver.

The procedure to rewrite a fuzzy instruction into a quasi-fuzzy instruction is as follows.

(i) "Go about L meters" $\sigma_{go}(L)$

The fuzzy instruction $\sigma_{go}(L)$ is divided into two instructions such as $\sigma_{go}^{\eta}(L)$ and $\sigma_{go}^{Y-\eta}(L)$, where $\sigma_{go}^{\eta}(L)$ is the instruction such that the driver goes on until the lowest bound η meters and $\sigma_{go}^{Y-\eta}(L)$ is the instruction such that the driver goes $Y - \eta$ meters.* Here note that $\sigma_{go}^{\eta}(L)$ is executed by the η times of quasi-fuzzy instruction σ_{go} named "Go ahead one step", $\sigma_{go}^{Y-\eta}(L)$ is executed by the $Y - \eta$ times of σ_{go}, and Y is determined by making use of one of the MAX-Method, PROB-Method and *-Method.

Hereupon the membership function for the set named "Go about L meters" is given by the following equation (18),

$$w(\ell) = \frac{1}{1 + \left(\frac{\ell - L}{a}\right)^2} \tag{18}$$

where $a = kL$ $(1 > k > 0)$, (19)

k may seem to denote a "Parameter Representing Distance-Sense" in the meaning that the driver is said to be sensitive to distance when k is small.

(ii) "Turn right" σ_R

(iii) "Turn left" σ_L

σ_R and σ_L are both found also in the quasi-fuzzy instructions defined already.

(iv) "Go straight" $\sigma_{go}*$

$\sigma_{go}*$ can be assumed to be the succession of the quasi-fuzzy instruction σ_{go} named "Go ahead one step" until the next State-testing Fuzzy Instruction such as σ_R, σ_L and $\sigma\{\sim\}$ in our case will be executable.

(v) "Until $\sim$" $\sigma\{\sim\}$

This fuzzy instruction is rather regarded as a state-

* In our experiment, one step is made equal to one meter.

testing instruction and does not need to be converted into a quasi-fuzzy instruction. The interpretation of this instruction is to examine whether the present location of the driver is coincident with the destination or not by comparing the present coordinate with that of the destination stored in the machine. Fig. 3 shows a table of the destination to be stored in the machine.

```
+-------- *** SYMBOLS ON THE MAP *** --------+
I                                           I
I ( H ) HALT POINT          ( + ) CROSSING  I
I ( S ) STARTING POINT      ( P ) PASSED POINT I
I ( * ) OBJECT POINT                        I
I   SG / TRAFFIC SIGNAL      RE / RESTAURANT I
I   BK / BANK                SC / SCHOOL    I
I   GS / GAS STATION         MP / PARKING LOT I
I                                           I
+-------------------------------------------+
```

Destination	Coordinates
Traffic Signal	(20,05),(40,12), (20,35),(50,35)
Restaurant	(50,07),(40,09), (40,15)
Gas Station	(31,12),(50,29)
Parking Lot	(20,20),(50,25)
Bank	(43,24),(35,40), (45,40)
School	(10,05),(50,55)

Figure 3 The Symbols and the Coordinates of Destinations on the Map

Thus the five kinds of fuzzy instructions are converted into a sequence of the three kinds of quasi-fuzzy instructions in (16).

Then the machine instruction μ_i $(i=1,2,\ldots,8)$ which is really executable in the machine is selected by reference to the evaluation value $\Phi(\beta, s)$ accompanied with the pair of the present quasi-fuzzy instruction β and the quasi-internal state s depending on the present position of the driver. The following Fig. 4 gives such an evaluation table where the integer of two figures shows that the machine instruction indicated by the second order figure is more preferable than that in the first order, and the integer of only one figure shows that the machine instruction indicated by that figure can be executable.

After all, summarizing the above argument, the execution procedure for a simulation of human driver's behavior directed by fuzzy instruction is illustrated by the flow chart shown in Fig. 5.

4.4 Computer Simulation Example for Human Driver's Behavior

Let such a sequence of fuzzy instructions as shown in Fig. 6 be given to a driver. This instruction means that (i) the driver starts from the point (50, 55), (ii) turns left at the branch point of y = 31, (iii) stops in the bank at (43, 24), (iv) turns right at the crossing of (40, 21), (v) drops in the restaurant at (40,15), (vi) and then goes on until the school at (10, 5) after turning left at the three-fork of (40, 5).

The computer simulation executed by the respective method of MAX-, PROB- and *-Method for the sequence of fuzzy instructions given above is exemplified in Figures 7, 8, and 9.

As can be seen from Figures 7, 8, and 9, the MAX-Method is most efficient to get to the destination, while the PROB-Method causes the driver to loiter around the same point and the *-Method lets him try such points as seem not to be concerned.

Such relative qualities of the three methods as mentioned above are supposed to be true from some results of simulation conducted with respect to some different kind of sequences of

Quasi-interval State s		Quasi-fuzzy Instruction β σ_{go}	σ_R	σ_L
1	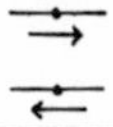	01	0	0
		05	0	0
2	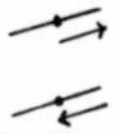	02	0	0
		06	0	0
3		03	0	0
		07	0	0
4		04	0	0
		08	0	0
5	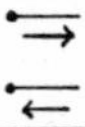	01	0	0
		0	0	0
6		03	0	0
		0	0	0
7	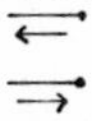	05	0	0
		0	0	0
8		0	0	0
		07	0	0
9		0	07	0
		0	0	05
		05	0	07
		07	05	0
10	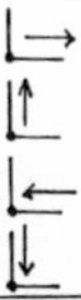	01	0	03
		03	01	0
		0	03	0
		0	0	01
11		02	07	0
		02	0	0
		07	0	0
		07	0	0
12	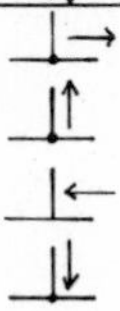	01	0	03
		03	01	05
		05	03	0
		0	05	01

Quasi-interval State s		Quasi-fuzzy Instruction β σ_{go}	σ_R	σ_L
13	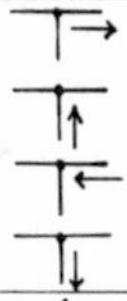	01	07	0
		0	01	05
		05	0	07
		07	05	01
14		0	07	03
		03	0	05
		05	03	07
		07	05	0
15		34	0	04
		43	03	07
		07	04	0
		07	07	03
16		18	08	0
		05	01	05
		05	0	08
		81	05	01
17	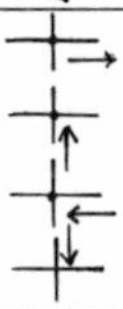	01	07	03
		03	01	05
		05	03	07
		07	05	01
18		34	08	04
		43	03	07
		78	04	08
		87	07	03
19		01	76	03
		13	71	03
		03	01	06
		06	03	76
		67	03	71
		76	06	01

Mark ↑ shows the direction of the driver.

Figure 4 Example of Evaluation Table in the Simulation

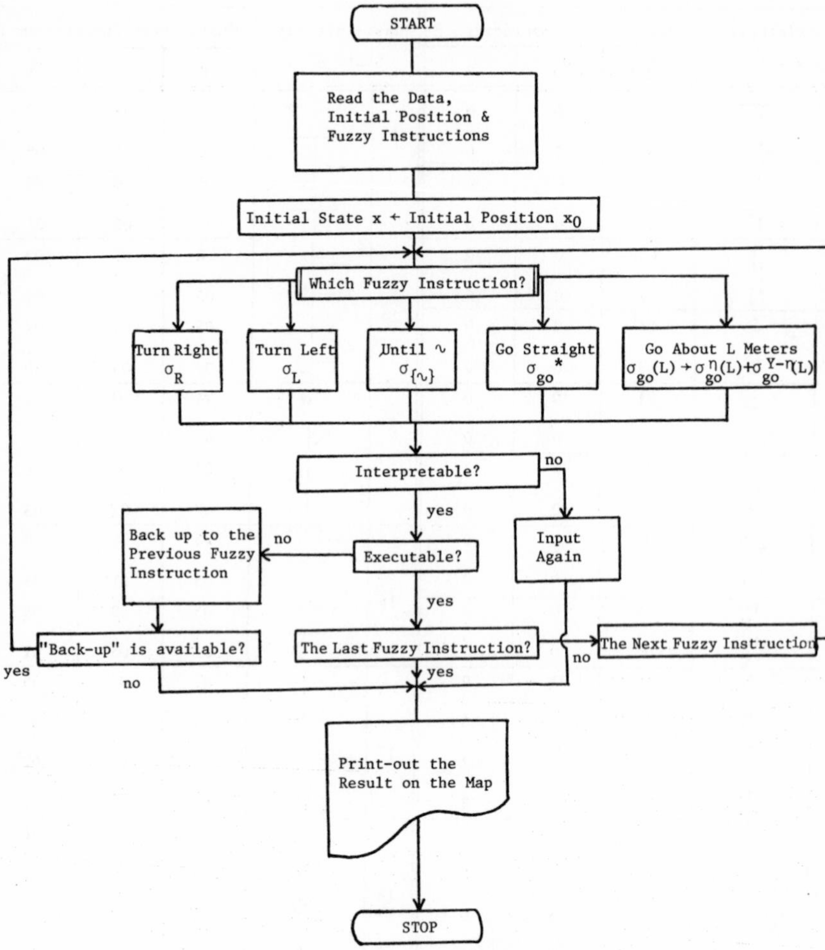

Figure 5 *Flow Chart for Simulation of Human Drivers's Behavior (Remark: This chart can be available to MAX-, PROB-, and *-Method, but here note that $\sigma_{go}(L) \rightarrow \sigma_{go}^{\eta}(L) + \sigma_{go}^{*}(L)$ is used for the *-Method.)*

```
*** INITIAL POSITION ***

     ( 50, 55, 3 )


** THE FUZZY INSTRUCTIONS **

    ( 1)   GO ABOUT 200 METERS
    ( 2)   TURN LEFT
    ( 3)   GO ABOUT 100 METERS
    ( 4)   UNTIL BANK
    ( 5)   GO ABOUT  50 METERS
    ( 6)   TURN RIGHT
    ( 7)   GO ABOUT  50 METERS
    ( 8)   UNTIL RESTAURANT
    ( 9)   GO ABOUT 100 METERS
    (10)   TURN LEFT
    (11)   GO STRAIGHT
    (12)   UNTIL SCHOOL
```

Figure 6 Initial Position and Fuzzy Instructions given in the Simulation

fuzzy instructions and various values of parameter representing distance-sense.

5. SIMULATION OF CHARACTER GENERATION

Let us now consider the process that a child starts to learn how to write characters and he will be good at writing. At the beginning, he is taught to write characters by his parents or his teachers. As is usual with this case, the teacher will teach him how to write a character in a rough way without measuring length, inclination and other features of strokes in the character or will set him a copy of a correct character. The above statement may seem to show that a child learns how to write a character based on a kind of rough rather than complete and correct informations about the character.

Thus a child writes a character following the instructions of his teacher, and then his teacher lets him correct the character by giving such ambiguous instructions as "make here a little shorter", "write round a little" and so on.

```
*************************************************
*  THE WANDERING PATH OF THE DRIVER ON THE MAP  *
*************************************************
```

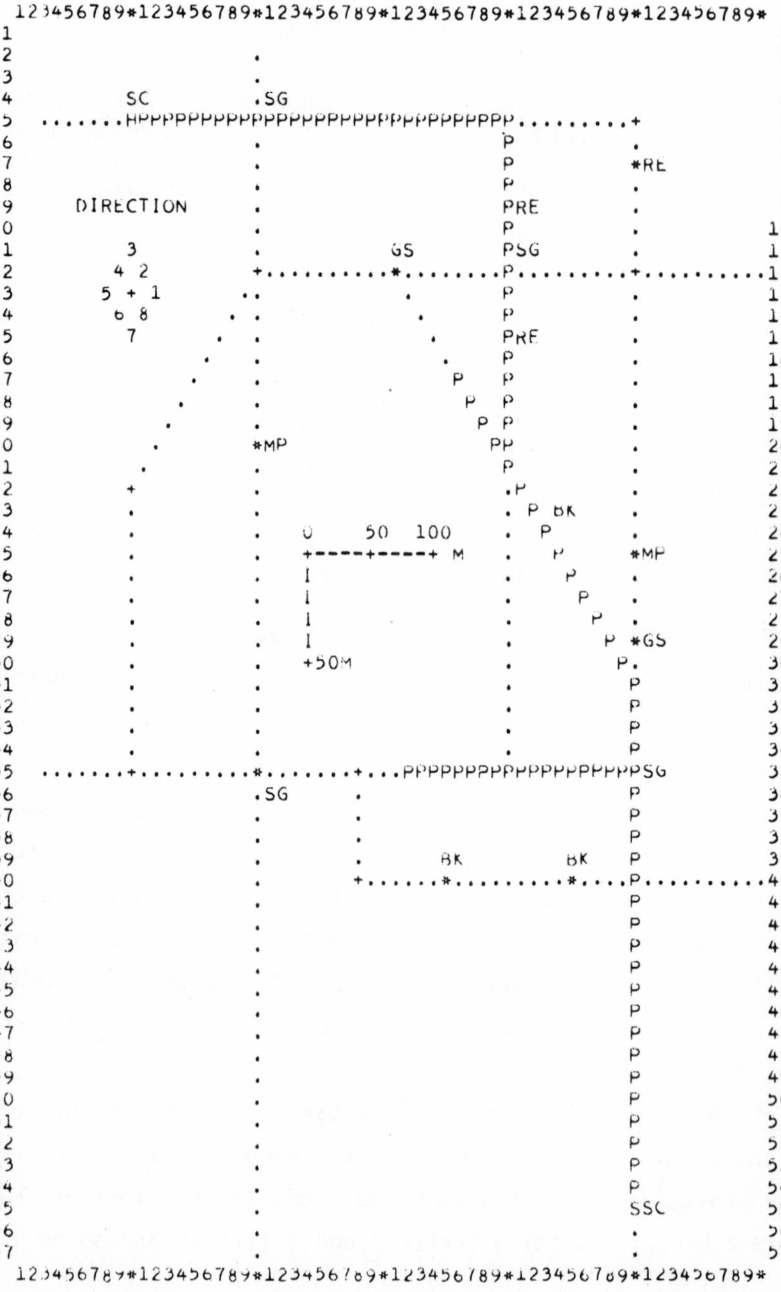

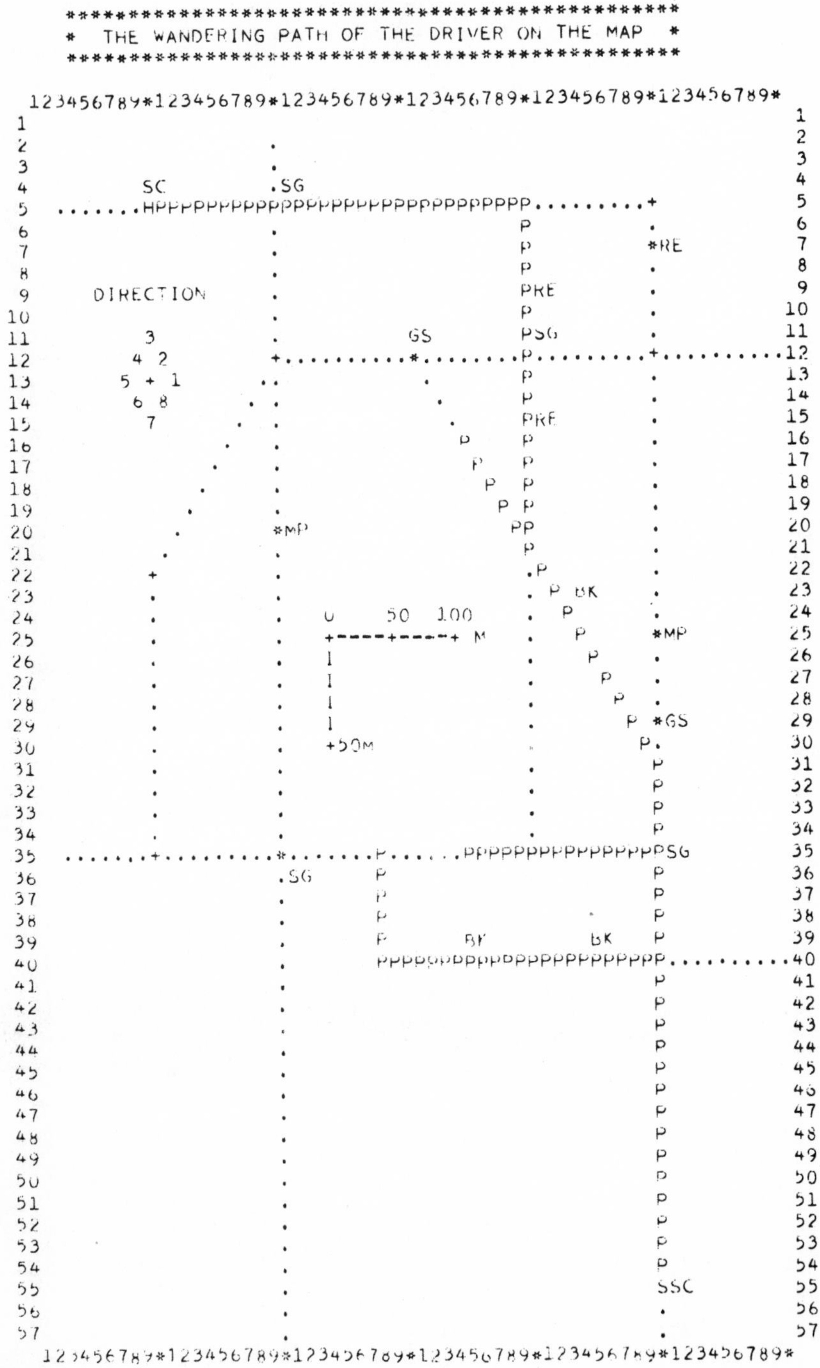

THE WANDERING PATH OF THE DRIVER ON THE MAP
123456789*123456789*123456789*123456789*123456789*123456789*
SC
SG
RE
DIRECTION
3
4 2
5 + 1
6 8
7
GS
PRE
PSG
MP
BK
0 50 100
M
+50M
GS
SG
BK
SSC

```
      ************************************************
      *  THE WANDERING PATH OF THE DRIVER ON THE MAP  *
      ************************************************

  123456789*123456789*123456789*123456789*123456789*123456789*
 1                                                             1
 2                   P                                         2
 3                   P                                         3
 4         SC        PSG                                       4
 5  .......HPPPPPPPPPPPPPPPPPPPPPPPPPPPPPPPPPPPPPPPP           5
 6                   P                   P         .           6
 7                   P                   P         *RE         7
 8                   P                   P         .           8
 9     DIRECTION     P                   PRE       .           9
10                   P                   P         .          10
11         3         P          GS       PSG       .          11
12        4 2        PPPPPPPPPPPPPPPPPPPPPPPPPPPPPPPPPPPPPPPPP12
13       5 + 1      P.           P       P         .          13
14        6 8      P .            P      P         .          14
15         7      P  .             P     PRE       .          15
16               P   .              P    P         .          16
17              P    .               P   P         .          17
18             P     .                P  P         .          18
19            P      .                 P P         .          19
20           P       *MP                PP         .          20
21          P        .                   P         .          21
22         P         .                   .P        .          22
23         P         .                   . P  BK   .          23
24         P         .   0    50  100    .  P      .          24
25         P         .   +----+----+ M   .   P     *MP        25
26         P         .   I               .    P    .          26
27         P         .   I               .     P   .          27
28         P         .   I               .      P  .          28
29         P         .   I               .       P *GS        29
30         P         .   +50M            .        P.          30
31         P         .                   .         P          31
32         P         .                   .         P          32
33         P         .                   .         P          33
34         P         .                   .         P          34
35  PPPPPPPPPPPPPPPPPPPPPPPPPPPPPPPPPPPPPPPPPPPPPPPPSG        35
36                   .SG     P                     P          36
37                   .       P                     P          37
38                   .       P                     P          38
39                   .       P      BK        BK   P          39
40                   .       PPPPPPPPPPPPPPPPPPPPPPP..........40
41                   .                             P          41
42                   .                             P          42
43                   .                             P          43
44                   .                             P          44
45                   .                             P          45
46                   .                             P          46
47                   .                             P          47
48                   .                             P          48
49                   .                             P          49
50                   .                             P          50
51                   .                             P          51
52                   .                             P          52
53                   .                             P          53
54                   .                             P          54
55                   .                             SSC        55
56                   .                             .          56
57                   .                             .          57
  123456789*123456789*123456789*123456789*123456789*123456789*
```

Through repeated practice in this manner, the child will gradually become to be able to write the characters in a correct way.

In this chapter, we shall conduct a computer simulation of the process of human learning stated above by making use of the concept of fuzzy program and learning algorithm. Let us provide the four kinds of fuzzy instruction as follows.

(i) Start (to write) from a point nearby (x, y).

(ii) Turn by about ρ degrees.

(iii) Draw about κ steps.

(iv) If the end point is not close to (x', y') then do "Back-up".

Then let us adopt the following three types of procedure for executing a fuzzy program composed of the fuzzy instructions given above.

(I) MAX-Method,

(II) PROB-Method with Simple Modification, (21)

(III) PROB-Method with Reinforced Modification.

5.1 Execution Procedures of Fuzzy Instructions

The execution procedure of a fuzzy program for generating a character will be similar in general to that in a simulation of human driver. However, there exists a slight discrepancy that fuzzy instructions in this case are converted directly into machine instructions while those in a simulation of human drivers are translated into machine instructions after converting once into quasi-fuzzy instructions.

As a matter of fact, in the latter case there exist some constraints that the driver has to go ahead step by step

---

Figure 7 Trail of Driver on Map Simulated by MAX-Method

Figure 8 Trail of Driver on Map Simulated by PROB-Method

*Figure 9 Trail of Driver on Map Simulated by *-Method*

looking up the destination on the map and that he can not start to walk from quite a different point from the present location, but he must go on the road consequtively without skip. On the other hand, in the former case there is no constraint excepting the space limit for writing a character and also there is no necessity for drawing a line step by step. Therefore, in a simulation of human driver, if there is a fuzzy instruction which is not executable, the driver is forced to turn back to that previous to the present fuzzy instruction, while in a generation of a character there is no such a constraint but the interpretation may proceed from any instruction which is not always previous to the present instruction. This matter may seem to correspond to the fact that if there is an incorrect portion in a character, we can erase that portion and rewrite it in a correct way.

The mode to select the machine instruction is specified as follows in the same way as in the simulation of human drivers, that is,

(a) MAX-Method

(b) PROB-Method

(c) Non-deterministic Method.

In case where a fuzzy instruction is not executable, the following Back-up procedures are provided.

(1) Turn back to the fuzzy instruction previous to the present one.

(2) Turn back to the fuzzy instruction corresponding to the machine instruction with the lowest grade of membership in a series of machine instructions selected consequtively up to now.

(3) The "Back-up" procedure is the same as (2). However, as for the selection of the machine instruction corresponding to the fuzzy instruction to which

"Back-up" was done, the machine instruction with the higher grade of membership than before is to be selected, though in (2) the selection is made without constraint as above.

Procedure (1) is the same as used in a simulation of human driver. Procedure (2) is available to the case where correction is made from the worst portion in a character written. And procedure (3) is used in the case where we try to rewrite better than before.

By combining the selection mode of machine instructions with the "Back-up" methods, there can be obtained a variety of execution procedures of a fuzzy program. For instance, combining the MAX-Method (a) with the Back-up Method (1), we can obtain the MAX-Method in Equation (22) and also, combining the PROB-Method (b) with the Back-up Method (1), we can obtain the PROB-Method as discussed in the previous chapter.

In this chapter, as is shown in Equation (21), let us use the MAX-Method, PROB-Method with Simple Modification which is a combination of (b) and (2), and PROB-Method with Reinforced Modification which is a combination of (b) and (3).

5.2 Simulation of Character Generation

A sypical set of fuzzy instructions for generating, for example, a character "A" is given in Fig. 10. The meaning of this sequence is illustrated in Fig. 11, where the mark of wavy underline "$\underset{\sim}{x}$" shows that $\underset{\sim}{x}$ is approximate to x and the degree of an angle measured anti-clockwise is positive and that measured clockwise is negative. And also the instruction named "TURN BY" indicates to turn by ρ degrees from the present direction.

```
NO.   FUZZY INSTRUCTIONS

 1    START FROM A POINT NEARBY (0,75).
 2    TURN BY ABOUT (150) DEGREES.
 3    DRAW ABOUT (200) STEPS.
 4    START FROM A POINT NEARBY (-5,75).
 5    TURN BY ABOUT ( 60) DEGREES.
 6    DRAW ABOUT (190) STEPS.
 7    START FROM A POINT NEARBY (-50,-10).
 8    TURN BY ABOUT ( 60) DEGREES.
 9    DRAW ABOUT ( 90) STEPS.
10    IF THE POINT IS NOT CLOSE TO (45,-10) THEN BACK UP.
11    END.
```

Figure 10 Fuzzy Instructions for Generation of "A"

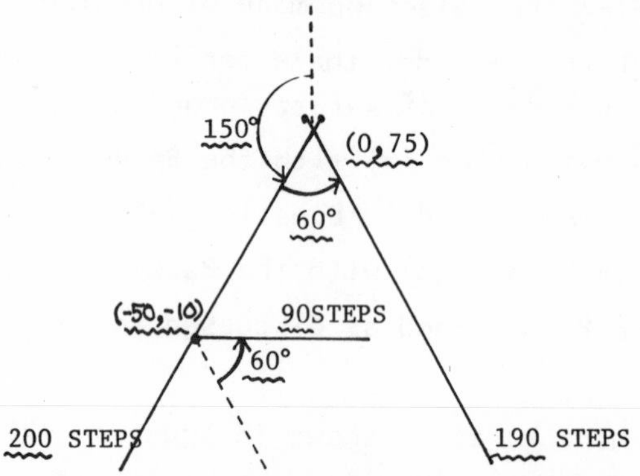

Figure 11 Example Illustrating the Meaning of Fuzzy Instructions

Plotted is in Figs. 12, 13, and 14, respectively, an example of simulation for generating a character, say, "A" by making use of the three types of execution procedures of fuzzy instructions shown in Equation (21). The only one conditional statement among the fuzzy instructions for generation of a character "A" is No. 10 in Fig. 10. If this condition is not satisfied, then "Back-up" is to be conducted. From Figs. 12, 13, and 14 we can see how "Back-up" is conducted in the respective method.

As is suggested from those figures, the execution of fuzzy programs employing the MAX-Method does not consume much

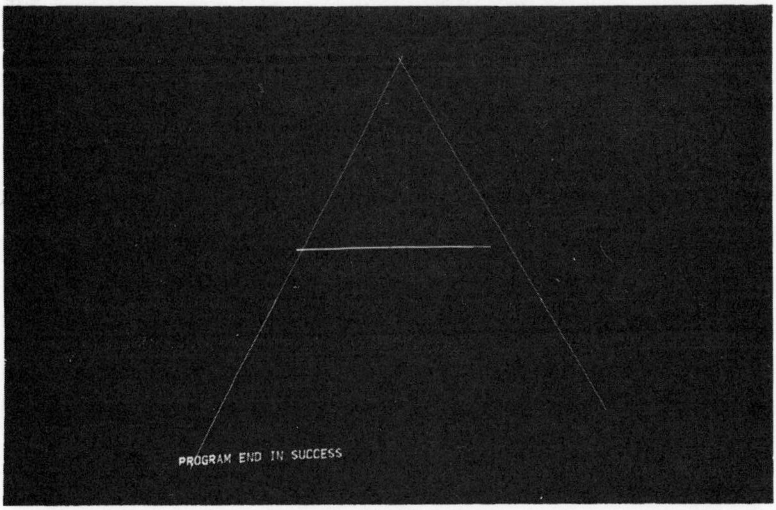

Figure 12 MAX-Method

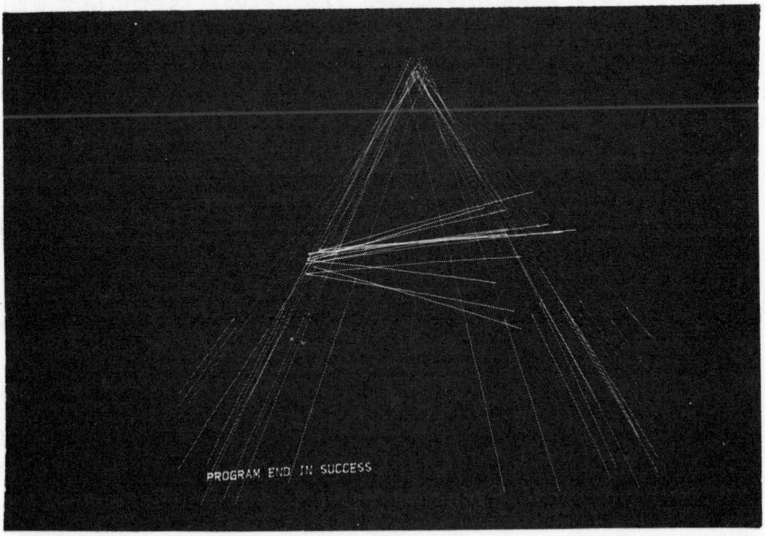

Figure 13 PROB-Method with Simple Modification

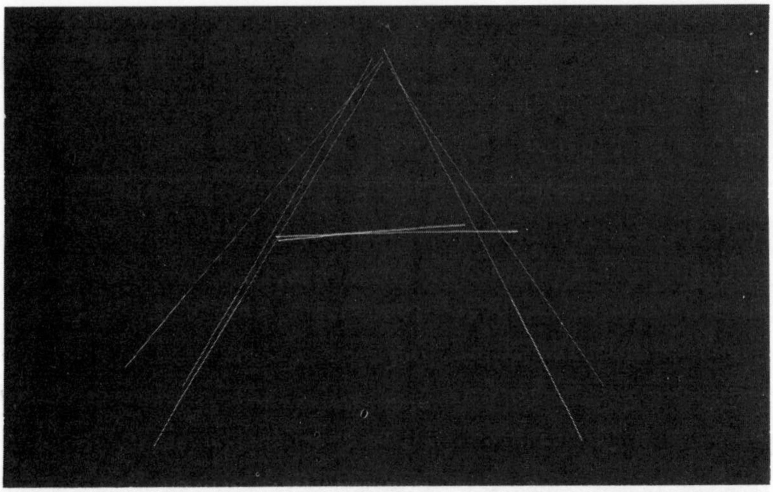

Figure 14 PROB-Method with Reinforced Modification

time in general if the parameters in the fuzzy instructions are correct. The reason is as follows. There is no contraint in the case of character generation unlike the case of simulation of driver's behavior where the driver is subject to some restriction depending on the road and others. Therefore the machine instructions selected in this case are equivalent to such parameters as coordinate, angle, number of steps and so on in the fuzzy instruction. On the other hand, the MAX-Method is not adequate to generation of characters in a free way. Form the fact that human does not always write the completely same character, this method may be said to be not so much suitable for character generation.

The PROB-Method with Simple Modification will be able to generate characters most freely. However, it consumes much time to execute fuzzy programs, because there is a possibility to select the machine instruction with a lower grade of membership than the previous one when the "Back-up" is conducted.

Finally, the PROB-Method with Reinforced Modification may seem to be the best one comparing with the other two methods stated above. In fact, this method does not consume so much time to execute fuzzy programs and besides can generate characters fairly freely, because the machine instruction with a higher grade of membership than the previous one is to be selected compulsory when a "Back-up" occurred.

5.3 Character Generation with Learning Process

Incorrect portions of a character generated by a fuzzy program will usually be corrected also by "fuzzy correcting instructions." If there remain still incorrect portions in the character thus corrected, the same procedure will be repeated again and again. In this case, however, it should be suggested that the fuzzy program will be able to generate the more correct character faster than usual through a learning process which is encountered always in human study of writing.

5.3.1 Fuzzy Instruction for Correction & Learning Algorithm

Let the fuzzy instructions for correcting bad portions of a character be composed of the following instructions (a) - (h) and the following adjectives (i) - (k).

(a) LONG, (b) SHORT, (c) ANTICLOCKWISE, (d) CLOCKWISE,

(e) RIGHT, (f) LEFT, (g) UP, (h) DOWN

(i) VERY, (j) LITTLE, (k) MUCH

The learning algorithm used in this simulation is a linear reinforcement rule given by Equation (22).

$$w_{n+1}(x) = \lambda w_n(x) + (1 - \lambda)\chi_n \tag{22}$$

where w_n denotes the grade of a parameter x at the n-th learning stage which is involved in the membership function specifying a fuzzy instruction and $0 \leq \lambda \leq 1$.

$$\chi_n = \begin{cases} 1, & \text{if } x \text{ is adequate,} \\ 0, & \text{if } x \text{ is not adequate,} \end{cases}$$

that is, if the parameter x (such as, say, a stroke length or a stroke inclination, etc) modified by fuzzy correcting instructions is assumed to be adequate by the teacher, $\chi_n = 1$ and if it is assumed to be not adequate, $\chi_n = 0$.

In practice, bad portions of a character displayed on a graphic unit are corrected by fuzzy correcting instructions through a light pen as shown in Fig. 16. After cleaning out all of bad portions, the character thus corrected is displayed again and the instruction named "GOOD" is pointed out by a light pen if there is no bad portion.

It should be noticed that, in the present case, the initial fuzzy instructions for character generation are modified through learning process under supervision of the fuzzy instructions which is given to correct bad portions of a character.

5.3.2 Simulation of Learning Process

The fuzzy program to generate a character "B" is as shown in Fig. 15. Fig. 16 exemplifies the character "B" generated by the execution of this program by making use of the PROB-Method with Reinforced Modification.

Let us now correct bad portions of the character "B" thus displayed. This is simply performed as follows. By applying the fuzzy correcting instructions, the grade of membership of the fuzzy instructions given originally can be updated so as to generate the more correct character. This updating procedure is just a learning process and its algorithm is based on a linear reinforcement rule. On the lefthand side of Fig. 16 are shown the fuzzy correcting instructions used and Fig. 17 demonstrates the corrected character "B" by learning correction mentioned above.

Of course, the simulation has been performed also in the two cases of MAX-Method and PROB-Method with Simple Modification other than PROB-Method with Reinforced Modification mentioned above.

NO.	FUZZY INSTRUCTIONS
1	START FROM A POINT NEARBY (-20,70).
2	TURN BY ABOUT (-180) DEGREES.
3	DRAW ABOUT (150) STEPS.
4	START FROM A POINT NEARBY ((-20,70).
5	TURN BY ABOUT (90) DEGREES.
6	DRAW ABOUT (60) STEPS.
7	TURN BY ABOUT (-45) DEGREES.
8	DRAW ABOUT (30) STEPS.
9	TURN BY ABOUT (-45) DEGREES.
10	DRAW ABOUT (30) STEPS.
11	TURN BY ABOUT (-45) DEGREES.
12	DRAW ABOUT (30) STEPS.
13	TURN BY ABOUT (-45) DEGREES.
14	DRAW ABOUT (60) STEPS.
15	IF THE POINT IS NOT CLOSE TO (-20,-2) THEN BACK UP.
16	TURN BY ABOUT (-180) DEGREES.
17	DRAW ABOUT (60) STEPS.
18	TURN BY ABOUT (-45) DEGREES.
19	DRAW ABOUT (30) STEPS.
20	TURN BY ABOUT (-45) DEGREES.
21	DRAW ABOUT (40) STEPS.
22	TURN BY ABOUT (-45) DEGREES.
23	DRAW ABOUT (30) STEPS.
24	TURN BY ABOUT (-45) DEGREES.
25	DRAW ABOUT (60) STEPS.
26	IF THE POINT IS NOT CLOSE TO (-20,-80) THEN BACK UP.
27	END.

Figure 15 Fuzzy Instructions for Generation of "B"

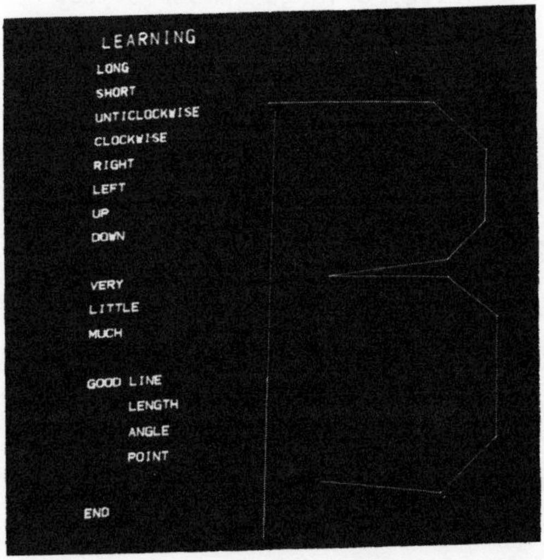

Figure 16 Character "B" Displayed Originally

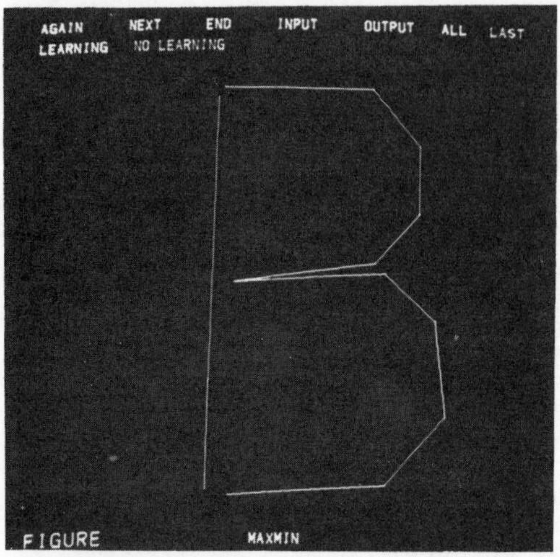

Figure 17 Character "B" Displayed after Learning Correction

Summarizing those simulation results, the following comparison can be obtained. The MAX-Method is not adequate to generate a character written freely while the number of correction procedures is small and it can generate a character as directed by the program. The PROB-Method with Simple Modification has a merit to be able to generate a freely-written character excepting that it requires a large number of correction procedures and not so enough learning effect can be expected. Contrary to the above two methods, the PROB-Method with Reinforced Modification may seem to be the best way because it can generate a fairly free character and none the less enough learning effect can be expected by a fairly few correction procedures.

6. CONCLUSION

In this paper, as an abstract model of a machine for the execution of fuzzy programs, a generalized fuzzy machine has been formulated from which a variety of fuzzy machines have been introduced.

Several methods of execution of fuzzy programs have been investigated by making use of the fuzzy machines introduced above. It has been pointed out that there exist three ways such as fuzzy- , probabilistic- and nondeterministic way depending on the specific character of the respective fuzzy machine with repsect to the way of selecting a machine instruction to a given fuzzy instruction and the way of state transition. Thereby, a unified survey for various execution methods of fuzzy programs can be obtained.

In addition, as some application examples using the presented methods for execution of fuzzy programs, the two simulation experiments such as human driver's behavior and character generation with learning process have been discussed. As the conclusion of this simulation, it has been found that in case of human driver's behavior the MAX-Method is best and in case of character generation the PROB-Method with Reinforced Modification is most favorable.

As a topic for further discussion, there remains an investigation of interpretation and execution methods of the more complicated fuzzy programs by making use of the concept of fuzzy semantics [4] as well as that of fuzzy algorithm [5].

REFERENCES

[1] S. K. Chang: "On the Execution of Fuzzy Programs Using Finite-State Machines," IEEE Trans. on Comp. Vol. C-21, No. 3, March 1972.

[2] E. S. Santos & W. G. Wee: "General Formulation of Sequential Machines," Inf. & Cont., Vol. 12, pp. 5-10, 1968.
E. S. Santos: "Maximin, Minimax and Composite Sequential Machines," J. Math. Annals & Appl. Vol. 24, pp. 246-259, 1968.

[3] R. Jakubowski & A. Kasprzak: "Application of Fuzzy Programs to the Design of Matching Technology," Bulletin de L'Academie, Polonaise des Sciences, Vol. XXI, No. 1,1973.

[4] L. A. Zadeh: "Quantitative Fuzzy Semantics," Inf. Sci. Vol. 3, pp. 159-176, 1971.
[5] L. A. Zadeh: "Fuzzy Algorithms," Inf. & Cont., Vol. 12, pp. 94-102, 1968.
[6] M. Mizumoto & K. Tanaka: "Various Kinds of Automata with Weights," J.CSS (to appear).

FUZZY GRAPHS†

Azriel Rosenfeld
Computer Science Center
University of Maryland
College Park, Maryland 20742

ABSTRACT

Some basic properties of fuzzy relations are reviewed, and generalized to the case where the underlying set is a fuzzy set. Fuzzy analogues of several basic graph-theoretic concepts (e.g., bridges and trees) are defined, and some of their properties are established.

1. INTRODUCTION

Fuzzy relations on a set S -- i.e., mappings from SxS into [0,1] -- have been studied by Zadeh [1-2] and by Tamura, et al. [3]; they are also discussed in detail by Kaufmann [4]. In particular, fuzzy analogs have been defined for the (ir-) reflexivity, (anti-)symmetry, and transitivity properties of relations; in terms of these, fuzzy analogs of equivalence and order relations have been introduced.

In the first part of the present paper, some of these ideas are reviewed, and it is shown how one can generalize them to fuzzy relations "on" a fuzzy subset of the given set S, rather than on S itself.

In the second part of the paper, graph terminology is introduced, and fuzzy analogs of several basic graph-theoretic

†The support of the Office of Computing Activities, National Science Foundation, under Grant GJ-32258X, is gratefully acknowledged, as is the help of Shelly Rowe in preparing this paper.

concepts are defined, including subgraphs; paths and connectedness; cliques; bridges and cutnodes; forests and trees. Much more could be done along these lines; it is hoped that this paper will serve to stimulate further work on fuzzified graph theory.

2. FUZZY RELATIONS ON FUZZY SETS

Let S be a set. We recall that a <u>fuzzy subset</u> of S is a mapping $\sigma: S \to [0,1]$ which assigns to each element $x \in S$ a degree of membership, $0 \leq \sigma(x) \leq 1$. Similarly, a <u>fuzzy relation</u> on S is a fuzzy subset of SxS, i.e., a mapping $\mu: S\times S \to [0,1]$ which assigns to each ordered pair of elements (x,y) a degree of membership, $0 \leq \mu(x,y) \leq 1$. In the special cases where σ and μ can only take on the values 0 and 1, they become the characteristic functions of an ordinary subset of S and an ordinary relation on S, respectively.

If $T \subseteq S$ is a subset of S, and $R \subseteq S\times S$ a relation on S, then R is a relation on T provided that $(x,y) \in R$ implies $x \in T$ and $y \in T$ for all x,y. Let τ and ρ be the characteristic functions of T and R, respectively; then this condition can be restated as

$$\rho(x,y) = 1 \text{ implies } \tau(x) = \tau(y) = 1 \qquad \text{for all } x,y \text{ in } S$$

This is readily equivalent to

$$\rho(x,y) \leq \tau(x) \wedge \tau(y) \qquad \text{for all } x,y \text{ in } S,$$

where $\wedge$ means "inf".

Returning to the general case where σ is a fuzzy subset of S and μ a fuzzy relation on S, we shall say that <u>μ is a fuzzy relation on σ</u> if

$$\mu(x,y) \leq \sigma(x) \wedge \sigma(y) \qquad \text{for all } x,y \text{ in } S.$$

In other words, for μ to be a fuzzy relation on σ, we require that the degree of membership of a pair of elements never exceed the degree of membership of either of the elements themselves. (If we think of the elements as nodes in a graph, and of the pairs as arcs (see Section 6), this amounts to requiring that the "strength" of an arc can never exceed the strengths of its end nodes.)

<u>Proposition 2.1.</u> For a given fuzzy subset σ of S, the strongest fuzzy relation on S that is a fuzzy relation on σ is μ_σ, defined by

$$\mu_\sigma(x,y) = \sigma(x) \wedge \sigma(y) \qquad \text{for all } x,y \text{ in S.} \qquad //$$

<u>Proposition 2.2.</u> For a given fuzzy relation μ on S, the weakest fuzzy subset of S on which μ is a fuzzy relation is σ_μ, defined by

$$\sigma_\mu(x) = \sup_{y \in S}[\mu(x,y) \vee \mu(y,x)] \qquad \text{for all } x \text{ in S,}$$

where $\vee$ means "sup".//

For any t, $0 \leq t \leq 1$, the set

$$\sigma_t = \{x \in S \mid \sigma(x) \geq t\}$$

is a subset of S, and the set

$$\mu_t = \{(x,y) \in S \times S \mid \mu(x,y) \geq t\}$$

is a relation on S. Using this notation, we can state

<u>Proposition 2.3.</u> Let μ be a fuzzy relation on σ, and let $0 \leq t \leq 1$; then μ_t is a relation on σ_t.

<u>Proof:</u> For any pair $(x,y) \in \mu_t$ we have $t \leq \mu(x,y) \leq \sigma(x) \wedge \sigma(y)$. Thus $\sigma(x)$ and $\sigma(y)$ each $\geq t$, and so are in σ_t.//

3. COMPOSITION OF FUZZY RELATIONS

Let μ and ν be fuzzy relations on σ. By the <u>composite</u>

of μ and ν is meant the fuzzy set $\mu\circ\nu$ defined by

$$(\mu\circ\nu)(x,z) = \sup_{y\varepsilon S}[\mu(x,y) \wedge \nu(y,z)] \quad \text{for all } x,z \text{ in } S.$$

(Other definitions of composition of fuzzy relations will be discussed below; the present definition is called <u>max-min composition</u>.)

<u>Proposition 3.1.</u> $\mu\circ\nu$ is a fuzzy relation on σ.

<u>Proof</u>: For all x,y,z we have $\mu(x,y) \leq \sigma(x) \wedge \sigma(y)$ and $\nu(y,z) \leq \sigma(y) \wedge \sigma(z)$. Thus

$$\mu(x,y) \wedge \nu(y,z) \leq \sigma(x) \wedge \sigma(y) \wedge \sigma(z) \leq \sigma(x) \wedge \sigma(z)$$

for every y, so that

$$(\mu\circ\nu)(x,z) = \sup_{y\varepsilon S}[\mu(x,y) \wedge \nu(y,z)] \leq \sigma(x) \wedge \sigma(z)$$

for all x,z.//

It is well known that composition of fuzzy relations is <u>associative</u>, i.e., for all μ,ν,ρ we have $\mu\circ(\nu\circ\rho) = (\mu\circ\nu)\circ\rho$. We can thus uniquely define the <u>powers</u> of a fuzzy relation as $\mu^1 = \mu$, $\mu^2 = \mu\circ\mu$, $\mu^3 = \mu\circ\mu^2 = \mu\circ\mu\circ\mu$, and so on. We shall also define

$$\mu^\infty = \sup_{k=1,2,\ldots} \mu^k.$$

Finally, it is convenient to define

$$\left.\begin{array}{l} \mu^0(x,y) = 0 \text{ if } x \neq y \\ \mu^0(x,x) = \sigma(x) \end{array}\right\} \quad \text{for all } x,y \text{ in } S.$$

<u>Proposition 3.2.</u> For all t, $0 \leq t \leq 1$, we have $(\mu\circ\nu)_t = \mu_t\circ\nu_t$.

<u>Proof</u>: The following statements are all equivalent:

a) $(x,z) \varepsilon (\mu\circ\nu)_t$

b) $(\mu\circ\nu)(x,z) \geq t$

c) $\mu(x,y) \wedge \nu(y,z) \geq t$ for some $y \varepsilon S$

d) $\mu(x,y) \geq t$ and $\nu(y,z) \geq t$

e) $(x,y) \in \mu_t$ and $(y,z) \in \nu_t$

f) $(x,z) \in \mu_t \circ \nu_t$ (by definition of composition of ordinary relations).//

Proposition 3.3. If $\mu \leq \nu$ and $\lambda \leq \rho$, then $\mu\circ\lambda \leq \nu\circ\rho$.

Proof: $(\mu\circ\lambda)(x,z) = \sup_{y\in S}[\mu(x,y) \wedge \lambda(y,z)]$ $\leq \sup_{y\in S}[\nu(x,y) \wedge \rho(y,z)] = (\nu\circ\rho)(x,z)$ for all x,z in S.//

Other definitions of composition of fuzzy relations are sometimes used. For example, we can define the max-prod and max-av composites of μ and ν, respectively, by

$$(\mu \underset{\bullet}{\circ} \nu)(x,z) + \sup_{y\in S}[\mu(x,y) \cdot \nu(y,z)]$$

$$(\mu \underset{+}{\circ} \nu)(x,z) = \frac{1}{2} \sup_{y\in S}[\mu(x,y) + (\nu(y,z)]$$

Note that

$$\mu \underset{\bullet}{\circ} \nu \leq \mu\circ\nu \leq \mu \underset{+}{\circ} \nu \qquad \text{for any } \mu,\nu.$$

Such alternative definitions seem, in many cases, to be more intuitively appealing than the max-min definition. For example, with the max-min composite, if we can find a y that is strongly μ-related to x and strongly ν-related to z, e.g., $\mu(x,y) = \nu(y,z) = \frac{1}{2}$, we have $(\mu\circ\nu)(x,z) \geq \frac{1}{2} \wedge \frac{1}{2} = \frac{1}{2}$, i.e., x and z are just as strongly $(\mu\circ\nu)$-related to each other. This may seem counterintuitive; just because there is a y that is closely similar to both x and z, it should not follow that x and z are just as similar to each other. Here the max-prod composite would give a more plausible result, namely, $(\mu \underset{\bullet}{\circ} \nu)(x,z)$ $\frac{1}{2} \cdot \frac{1}{2} = \frac{1}{4}$.

On the other hand, many useful properties of the max-min composite fail to hold for the alternative definitions. For example, Propositions 3.1-2 are true for max-prod, but not for max-av, though Proposition 3.3 is true for both. As we shall see below, however, many other properties of the max-min composite hold for max-av but not for max-prod. We shall continue to use max-min as our basic definition. It would be of interest to formulate conditions on a definition of composition that would be both necessary and sufficient for the various properties to hold -- in particular, to determine the greatest (or least) composition that satisfies a given property.

4. REFLEXIVITY AND SYMMETRY

Let μ be a fuzzy relation on σ. We call μ reflexive if

$$\mu(x,x) = \sigma(x) \qquad \text{for all } x \in S.$$

(This generalizes the usual definition, which requires $\mu(x,y) = 1$ for all x.)

Proposition 4.1. If μ is reflexive, then $\mu(x,y) \leq \mu(x,x)$ and $\mu(y,x) \leq \mu(x,x)$ for all x,y in S.

Proof: $\mu(x,y) \leq \sigma(x) \wedge \sigma(y) \leq \sigma(x) = \mu(x,x)$.//

Proposition 4.2. If μ is a reflexive fuzzy relation on σ, then for any $0 \leq t \leq 1$, μ_t is a reflexive relation on σ_t.

Proof: For all $x \in \sigma_t$ we have $t \leq \sigma(x) = \mu(x,x)$, so that $(x,x) \in \mu_t$.//

Proposition 4.3. If μ is reflexive, then for any ν we have

$$\mu \circ \nu \geq \nu \text{ and } \nu \circ \mu \geq \nu.$$

Proof: $(\mu \circ \nu)(x,z) = \sup_{y \in S}[\mu(x,y) \wedge \nu(y,z)] \geq \mu(x,x) \wedge \nu(x,z) = \sigma(x) \wedge \nu(x,z)$. But $\nu(x,z) \leq \sigma(x) \wedge \sigma(x)$; hence $\sigma(x) \wedge \nu(x,z) = \nu(x,z)$, so that we have proved $(\mu \circ \nu)(x,z) \geq \nu(x,z)$ for all x,z in S.//

Corollary 4.4. If μ is reflexive, $\mu \leq \mu \circ \mu$.//

Corollary 4.5. If μ is reflexive, $\mu^0 \leq \mu^1 \leq \mu^2 \leq \cdots \leq \mu^\infty$.//

Corollary 4.6. If μ is reflexive, $\mu^0(x,x) = \mu^1(x,x) = \mu^2(x,x) = \cdots = \mu^\infty(x,x) = \sigma(x)$.//

Proposition 4.7. If μ and ν are reflexive, so is $\mu\circ\nu$.

Proof: $(\mu\circ\nu)(x,x) = \sup_{y\in S}[\mu(x,y) \wedge \nu(y,x)] \geq \mu(x,x) \wedge \nu(x,x) = \sigma(x) \wedge \sigma(x) = \sigma(x)$.//

Propositions 4.3 and 4.7 do not hold for the max-prod composition, but they do hold for the max-av composition.

We call μ symmetric if $\mu(x,y) = \mu(y,x)$ for all x,y in S. It is clear that if μ is symmetric, so is μ_t for any threshold t. Note that the symmetry property does not depend on the choice of fuzzy subset σ, unlike reflexivity.

Proposition 4.8. If μ and ν are symmetric, then $\mu\circ\nu$ is symmetric if and only if $\mu\circ\nu = \nu\circ\mu$.

Proof: $\mu\circ\nu$ symmetric means

$$\sup_{y\in S}[\mu(x,y) \wedge \nu(y,z)] = \sup_{y\in S}[\mu(z,y) \wedge \nu(y,x)]$$

for all x,z, while $\mu\circ\nu = \mu\circ\nu$ means

$$\sup_{y\in S}[\mu(x,y) \wedge \nu(y,z)] = \sup_{y\in S}[\nu(x,y) \wedge \mu(y,z)]$$

for all x,z.

If μ and ν are symmetric, the two right-hand sides are equal.//

Corollary 4.9. If μ is symmetric, so is every power of μ.//

These last results would hold for any definition of composition that is based on a commutative operation on [0,1].

5. TRANSITIVITY

We call μ transitive if $\mu\circ\mu \leq \mu$. Note that, like symmetry, this property does not depend on σ. Readily, transitivity implies $\mu^k \leq \mu$ for all k, so that $\mu^\infty \leq \mu$. It is also easily seen that, for any μ, μ^∞ is transitive.

Proposition 5.1. If μ is symmetric and transitive, then $\mu(x,y) \leq \mu(x,x)$ for all x,y in S.

Proof: $\mu(x,x) \geq (\mu\circ\mu)(x,x) = \sup_{y\varepsilon S}[\mu(x,y) \wedge \mu(y,x)] = \sup_{y\varepsilon S}\mu(x,y)$.//

Proposition 5.2. If μ is a transitive relation on σ, then for any $0 \leq t \leq 1$, μ_t is a transitive relation on σ_t.

Proof: $\mu(x,z) \geq (\mu\circ\mu)(x,z) \geq \mu(x,y) \wedge \mu(y,z)$ for any x,y and z; hence $\mu(x,y) \geq t$ and $\mu(y,z) \geq t$ imply $\mu(x,z) \geq t$.//

Proposition 5.3. If μ is transitive and ν, ρ each $\leq \mu$, then $\nu\circ\rho \leq \mu$.

Proof: $(\nu\circ\rho)(x,z) = \sup_{y\varepsilon S}[\nu(x,y) \wedge \rho(y,z)] \leq \sup_{y\varepsilon S}[\mu(x,y) \wedge \mu(y,z)] = (\mu\circ\mu)(x,z) \leq \mu(x,z)$ for all x,z.//

Corollary 5.4. If μ is transitive, ν is reflexive, and $\nu \leq \mu$, then $\nu\circ\mu = \nu\circ\mu = \mu$.

Proof: Propositions 4.3 and 5.3.//

Corollary 5.5. If μ is reflexive and transitive, then $\mu\circ\mu = \mu$.//

Corollary 5.6. If μ is reflexive and transitive, then $\mu^0 \leq \mu^1 = \mu^2 = \cdots = \mu^\infty$.//

Proposition 5.7. If μ and ν are transitive, and $\mu\circ\nu = \mu\circ\nu$, then $\mu\circ\nu$ is transitive.

Proof: By associativity and by the fact that μ and ν commute, we have $(\mu\circ\nu) \circ (\mu\circ\nu) = (\mu\circ\mu) \circ (\nu\circ\nu) \leq \mu\circ\nu$ (the last step uses Proposition 3.3).//

Propositions 5.1 and 5.2 hold for the max-av composition, but not for max-prod; Propositions 5.3 and 5.7 hold for both (but the corollaries hold only for max-av). On the other hand, the max-av (and max-min) definitions of transitivity impose severe restrictions on the values that the fuzzy relation can take on. In particular, suppose that μ is symmetric and (max-min) transitive; let x,y,z be any three elements of S, and suppose -- without loss of generality --

that $\mu(x,z) \leq \mu(x,y) \leq \mu(y,z)$. Then by transitivity we have

$$\mu(x,z) \geq (\mu\circ\mu)(x,z) \geq \mu(x,y) \wedge \mu(y,z) = \mu(x,y).$$

Thus $\mu(x,z) = \mu(x,y)$ -- i.e., for any three elements of S, the degrees of relatedness of the two less related pairs must be equal. The situation is even worse for the max-av definition, where we have

$$\mu(x,z) \geq (\mu\circ\mu)(x,z) \geq \frac{1}{2}[\mu(x,y) + \mu(y,z)],$$

which together with $\mu(x,z) \leq \mu(x,y) \leq \mu(y,z)$ implies that all three must be equal! For max-prod, on the other hand, we have only

$$\mu(x,z) \geq \mu(x,y) \cdot \mu(y,z)$$

and no two of the μ's need be equal; however, the smallest must be at least equal to the square of the next smallest.

6. FUZZY GRAPHS

Any relation $R \subseteq S\times S$ on a set S can be regarded as defining a graph with node set S and arc set R. Similarly, any fuzzy relation $\mu: S\times S \to [0,1]$ can be regarded as defining a weighted graph, or fuzzy graph, where the arc $(x,y) \in S\times S$ has weight $\mu(x,y) \in [0,1]$. In this and the following sections we shall use graph terminology, and introduce fuzzy analogs of several basic graph-theoretic concepts. For simplicity, we will consider only undirected graphs -- i.e., we assume that our fuzzy relation is symmetric, so that all arcs can be regarded as unordered pairs of nodes. (We will never need to consider loops, i.e., arcs of the form (x,x); we can assume, if we wish, that our fuzzy relation is reflexive.)

Formally, a fuzzy graph $G = (\sigma,\mu)$ is a pair of functions $\sigma: S \to [0,1]$ and $\mu: S\times S \to [0,1]$, where for all x,y in S we have $\mu(x,y) \leq \sigma(x) \wedge \sigma(y)$. The fuzzy graph $H = (\tau,\nu)$ is called a (partial) fuzzy subgraph of G if

$$\tau(x) \leq \sigma(x) \qquad \text{for all } x \in S$$

and $\nu(x,y) \leq \mu(x,y)$ for all $x,y \in S$.

For any threshold t, $0 \leq t \leq 1$, if we let $\sigma_t = \{x \in S | \sigma(x) \geq t\}$, and $\mu_t = \{(x,y) \in S \times S | \mu(x,y) \geq t\}$, then as seen earlier, we have $\mu_t \subseteq \sigma_t \times \sigma_t$, so that (σ_t,μ_t) is a graph with the node set σ_t and arc set μ_t.

Proposition 6.1. If $0 \leq u \leq v \leq 1$, then (σ_v,μ_v) is a subgraph of (σ_u,μ_u).//

Proposition 6.2. If (τ,ν) is a fuzzy subgraph of (σ,μ), then for any threshold t, $0 \leq t \leq 1$, (τ_t,ν_t) is a subgraph of (σ_t,μ_t).//

We say that the fuzzy subgraph (τ,ν) spans the fuzzy graph (σ,μ) if $\tau(x) = \sigma(x)$ for all x. In this case, the two graphs have the same fuzzy node set; they differ only in the arc weights.

For any fuzzy subset τ of σ, i.e., such that $\tau(x) \leq \sigma(x)$ for all x, the fuzzy subgraph of (σ,μ) induced by τ is the maximal fuzzy subgraph of (σ,μ) that has fuzzy node set τ. Evidently, this is just the fuzzy graph (τ,ν), where

$$\nu(x,y) = \tau(x) \wedge \tau(y) \wedge \sigma(x,y) \qquad \text{for all } x,y \in S.$$

We will assume from now on that the underlying set S of a fuzzy graph is always finite.

7. PATHS AND CONNECTEDNESS

A path ρ in a fuzzy graph is a sequence of distinct nodes $x_0,x_1,\ldots,x_n$ such that $\mu(x_{i-1},x_i) > 0$, $1 \leq i \leq n$; here $n \geq 0$ is called the length of ρ. The consecutive pairs (x_{i-1},x_i) are called the arcs of the path. The strength of

ρ is defined as $\wedge_{i=1}^{n}\mu(x_{i-1},x_i)$. In other words, the strength of a path is defined to be the weight of the weakest arc of the path. If the path has length 0, it is convenient to define its strength to be $\sigma(x_0)$. We call ρ a <u>cycle</u> if $x_0 = x_n$ and $n \geq 3$.

Two nodes that are joined by (i.e., are the first and last nodes of) a path are said to be <u>connected</u>. It is evident that "connected" is a reflexive, symmetric relation, and it is readily seen to be transitive also. In fact, clearly x and y are connected if and only if $\mu^\infty(x,y) > 0$. The equivalence classes of nodes under this relation are called <u>connected components</u> of the given fuzzy graph; they are just its maximal connected fuzzy subgraphs. A <u>strongest path</u> joining any two nodes x,y has strength $\mu^\infty(x,y)$; we shall sometimes refer to this as the strength of connectedness between the nodes.

<u>Proposition 7.1.</u> If (τ,ν) is a fuzzy subgraph of (σ,μ), then for all x,y in S we have $\nu^\infty(x,y) \leq \mu^\infty(x,y)$.//

One could attempt to define the "distance" between x and y as the length of the shortest strongest path between them. This "distance" is, in fact, symmetric and positive definite (by our definition of a fuzzy graph, no path from x to x can have strength greater than $\sigma(x)$, which is the strength of the path of length 0). However, it is not triangular, as we see from the example

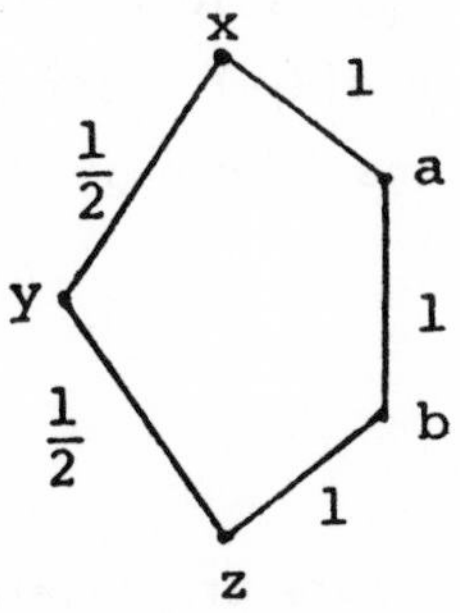

Here any path from x to y or from y to z has strength $\leq \frac{1}{2}$, since it must involve either arc (x,y) or arc (y,z); thus the shortest strongest paths between them have length 1. On the other hand, there is a path from x to z, through a and b, that has length 3 and strength 1. Thus $d(x,z) = 3 > 1+1 = d(x,y) + d(y,z)$ in this case.

A better notion of distance in a fuzzy graph can be defined as follows: For any path $\rho = x_0, \ldots, x_n$, define the <u>μ-length</u> of ρ as the sum of the reciprocals of ρ's arc weights, i.e.,

$$\ell(\rho) = \sum_{i=1}^{n} \frac{1}{\mu(x_{i-1}, x_i)}$$

If n = 0, we define $\ell(\rho) = 0$. Clearly, for $n \geq 1$ we have $\ell(\rho) \geq 1$. For any two nodes x,y, we can now define their <u>μ-distance</u> $\delta(x,y)$ as the smallest μ-length of any path from x to y.

<u>Proposition 7.2.</u> $\delta(x,y)$ is a metric.

<u>Proof</u>: a) $\delta(x,y) = 0$ if and only if x = y, since $\ell(\rho) = 0$ if and only if ρ has length 0.

(b) $\delta(x,y) = \delta(y,x)$, since the reversal of a path is a path, and μ is symmetric.

(c) $\delta(x,z) \leq \delta(x,y) + \delta(y,z)$, since the concatenation of a path from x to y and a path from y to z is a path from x to z, and ℓ is additive for concatenation of paths.//

In the non-fuzzy case, $\ell(\rho)$ is just the length n of ρ, since all the μ's are 1. Hence $\delta(x,y)$ becomes the usual definition of distance, i.e., the length of the shortest path between x and y.

8. CLUSTERS

In ordinary graphs, there are several ways of defining

"clusters" of nodes. One approach is to call the set C of nodes a <u>cluster</u> of order k if

a) For all nodes x,y in C we have $d(x,y) \leq k$

b) For all nodes $z \notin C$ we have $d(z,w) > k$ for some $w \in C$.

where d(a,b) is the length of a shortest path between a and b. In other words, in a k-cluster C, every pair of nodes are within distance k of each other, and C is maximal with respect to this property -- i.e., no node outside C is within distance k of every node in C.

When k=1, a k-cluster is called a <u>clique</u>; it is a maximal complete subgraph -- i.e., a maximal subgraph in which each pair of nodes is joined by an arc. At the other extreme, if we let $k \to \infty$, a k-cluster becomes a connected component -- i.e., a maximal subgraph in which each pair of nodes is joined by a path (of any length).

These ideas can be generalized to fuzzy graphs as follows: In $G = (\sigma,\mu)$, we can call $C \subseteq S$ a <u>fuzzy cluster</u> of order k if

$$\inf_{x,y\in C} \mu^k(x,y) > \sup_{z\notin C}(\inf_{w\in C}\mu^k(w,z))$$

Note that C is an ordinary subset of S, not a fuzzy subset. If G is an ordinary graph, we have $\mu^k(a,b) = 0$ or 1 for all a and b; hence this definition reduces to

a) $\mu^k(x,y) = 1$ for all x,y in C

b) $\mu^k(w,z) = 0$ for all $z\notin C$ and some $w\in C$

which is the same as the definition above.

In fact, the k-clusters obtained using this definition are just ordinary cliques in graphs obtained by thresholding the kth power of the given fuzzy graph. Indeed, let C be a fuzzy k-cluster, and let $\inf_{x,y\in C} \mu^k(x,y) = t$. If we threshold μ^k (and σ) at t, we obtain an ordinary graph in which C is now an ordinary clique.

9. BRIDGES AND CUTNODES

Let $G = (\sigma,\mu)$ be a fuzzy graph, let x,y be any two distinct nodes, and let G' be the fuzzy subgraph of G obtained by deleting the arc (x,y); i.e., $G' = (\sigma,\mu')$ where

$$\mu'(x,y) = 0; \ \mu' = \mu \quad \text{for all other pairs.}$$

We say that (x,y) is a <u>bridge</u> in G if $\mu'^{\infty}(u,v) < \mu^{\infty}(u,v)$ for some u,v -- in other words, if deleting the arc (x,y) reduces the strength of connectedness between some pair of nodes. Evidently, (x,y) is a bridge if and only if there exist u,v such that (x,y) is an arc of every strongest path from u to v.

<u>Theorem 9.1.</u> The following statements are equivalent:

a) (x,y) is a bridge

b) $\mu'^{\infty}(x,y) < \mu(x,y)$

c) (x,y) is not the weakest arc of any cycle

<u>Proof</u>: If (x,y) is not a bridge, we must have $\mu'^{\infty}(x,y) = \mu^{\infty}(x,y) \geq \mu(x,y)$; thus (b) implies (a). If (x,y) is a weakest arc of a cycle, then any path involving arc (x,y) can be converted into a path not involving (x,y) but at least as strong, by using the rest of the cycle as a path from x to y; thus (x,y) cannot be a bridge, so that (a) implies (c). If $\mu'^{\infty}(x,y) \geq \mu(x,y)$, there is a path from x to y, not involving (x,y), that has strength $\geq \mu(x,y)$, and this path together with (x,y) forms a cycle of which (x,y) is a weakest arc; thus (c) implies (b).//

Let w be any node, and let G^* be the fuzzy subgraph of G obtained by deleting the node w; i.e., G^* is the fuzzy subgraph induced by σ^*, where

$$\sigma^*(w) = 0; \ \sigma^* = \sigma \text{ for all other nodes.}$$

Note that in $G^* = (\sigma^*, \mu^*)$ we must have $\mu^*(w,z) = 0$ for all z. We say that w is a <u>cutnode</u> in G if $\mu^{*\infty}(u,v) < \mu^{\infty}(u,v)$ for some u,v (other than w) -- in other words, if deleting the node w reduces the strength of connectedness between some other pair of nodes. Evidently, w is a cutnode if and only if there exist u,v, distinct from w, such that w is on every strongest path from u to v.

G is called <u>nonseparable</u> (or sometimes: a <u>block</u>) if it has no cutnodes. It should be pointed out that a block may have bridges (this cannot happen for non-fuzzy graphs). For example, in

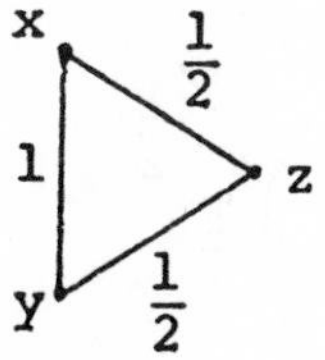

arc (x,y) is a bridge, since its deletion reduces the strength of connectedness between x and y from 1 to $\frac{1}{2}$. However, it is easily verified that no node of this fuzzy graph is a cutnode.

If between every two nodes x,y of G there exist two strongest paths that are disjoint (except for x,y themselves), G is evidently a block. This is analogous to the "if" of the non-fuzzy graph theorem that G is a block (with at least three nodes) if and only if every two nodes of G lie on a common cycle. The "only if", on the other hand, does not hold in the fuzzy case, as the example just given shows.

10. FORESTS AND TREES

We recall that a (non-fuzzy) graph that has no cycles is called <u>acyclic</u>, or a <u>forest</u>; and a connected forest is called a <u>tree</u>. We shall call a fuzzy graph a forest if the graph consisting of its nonzero arcs is a forest, and a tree if this graph is also connected.

More generally, we call the fuzzy graph $G = (\sigma,\mu)$ a <u>fuzzy forest</u> if it has a fuzzy spanning subgraph $F = (\sigma,\nu)$ which is a forest, where for all arcs (x,y) not in F (i.e., such that $\nu(x,y) = 0$), we have $\mu(x,y) < \nu^{\infty}(x,y)$. In other words, if $(x,y) \in G$ but $\notin F$, there is a path in F between x and y whose strength is greater that $\mu(x,y)$. It is clear that a forest is a fuzzy forest.

For example, the following are fuzzy forests:

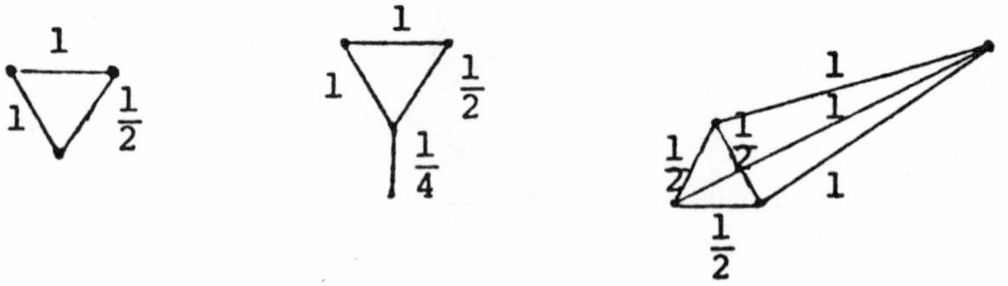

but the following are not:

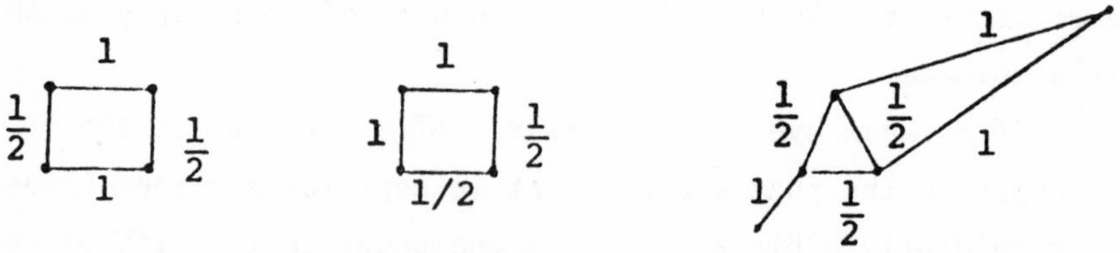

If G is connected, readily so is F (any arc of a path in G is either in F, or can be diverted through F). In this case we call G a <u>fuzzy tree</u>. The examples of fuzzy forests given above are all fuzzy trees.

Note that if we replaced $<$ by $\leq$ in the definition, then even

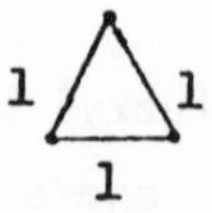

would be a fuzzy forest, since it has subgraphs such as

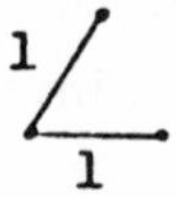

<u>Theorem 10.1.</u> G is a fuzzy forest if and only if, in any cycle of G, there is an arc (x,y) such that $\mu(x,y) < \mu'^{\infty}(x,y)$, where the prime denotes deletion of the arc (x,y) from G.

<u>Proof</u>: To see "if", let (x,y) be an arc, belonging to a cycle, which has the property of the theorem and for which $\mu(x,y)$ is smallest. (If there are no cycles, G is a forest and we are done.) If we delete (x,y), the resulting fuzzy subgraph satisfies the path property of a fuzzy forest. If there are still cycles in this graph, we can repeat the process. Note that at each stage, no previously deleted arc is stronger than the arc being currently deleted; hence the path guaranteed by the property of the theorem involves only arcs that have not yet been deleted. When no cycles remain, the resulting fuzzy subgraph is a forest F. Let (x,y) not be an arc of F; thus (x,y) is one of the arcs that we deleted in the process of constructing F, and there is a path from x to y that is stronger than $\mu(x,y)$ and that does not involve (x,y) nor any of the arcs deleted prior to it. If this path involves arcs that were deleted later, it can be diverted around them using a path of still stronger arcs; if any of these were deleted later, the path can be further diverted; and so on. This process eventually stabilizes with a path consisting entirely of arcs of F. Thus G is a fuzzy forest.

Conversely, if G is a fuzzy forest, let ρ be any cycle; then some arc (x,y) of ρ is not in F. Thus by definition of a fuzzy forest we have $\mu(x,y) < \nu^{\infty}(x,y) \le \mu'^{\infty}(x,y)$, which proves "only if".//

Note that if G is connected, so is the constructed F in the first part of the proof, since no step of the construction disconnects.

<u>Proposition 10.2.</u> If there is at most one strongest path between any two nodes of G, then G must be a fuzzy forest.

<u>Proof</u>: If not, by Theorem 10.1 there would be a cycle ρ in G such that, for all arcs (x,y) of ρ, we have $\mu(x,y) \geq \mu'^{\infty}(x,y)$; thus (x,y) itself constitutes a strongest path from x to y. If we choose (x,y) to be a weakest arc of ρ, it follows that the rest of ρ is also a strongest path from x to y, contradiction.//

The converse of Proposition 10.2 is false; G can be a fuzzy forest and still have multiple strongest paths between nodes. This is because the strength of a path is that of its weakest arc, and as long as this arc lies in F, there is little constraint on the other arcs. For example, the fuzzy graph

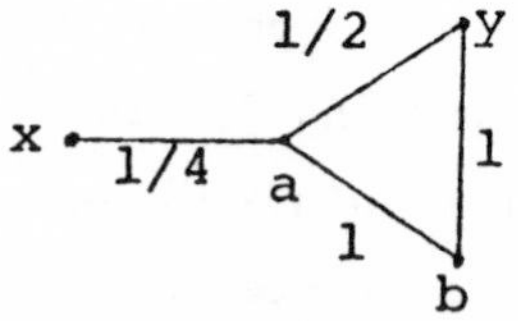

is a fuzzy forest; here F consists of all arcs except (a,y). The strongest paths between x and y have strength 1/4, due to the arc (x,a); both x, a, b, y and x, a, y are such paths, where the former lies in F but the latter does not.

<u>Proposition 10.3.</u> If G is a fuzzy forest, the arcs of F are just the bridges of G.

<u>Proof</u>: An arc (x,y) not in F is certainly not a bridge, since $\mu(x,y) < \nu^{\infty}(x,y) \leq \mu'^{\infty}(x,y)$. Conversely, let (x,y)

be an arc in F. If it were not a bridge, we would have a path ρ from x to y, not involving (x,y), of strength $\geq \mu(x,y)$. This path must involve arcs not in F, since F is a forest and has no cycles. However, by definition, any such arc (u_i, v_i) can be replaced by a path ρ_i in F of strength $> \mu(u,v)$. Now ρ_i cannot involve (x,y), since all its arcs are strictly stronger than $\mu(u,v) \geq \mu(x,y)$. Thus by replacing each (u_i, v_i) by ρ_i, we can construct a path in F from x to y that does not involve (x,y), giving us a cycle in F, contradiction.//

By this last proposition, if G is a fuzzy forest, its spanning forest F is unique.

REFERENCES

1. L. A. Zadeh, "Fuzzy Sets," Info. Control 8, 1965, 338-353.
2. L. A. Zadeh, "Similarity Relations and Fuzzy Orderings," Info. Science 3, 1971, 177-200.
3. S. Tamura, S. Higuchi, and K. Tanaka, "Pattern Classification Based on Fuzzy Relations," IEEE Trans. SMC-1, 1971, 61-66.
4. A. Kaufmann, Introduction a la Theorie des Sous-Ensembles Flous, (Vol. 1), Masson, Paris, 1973, ch. 2 (pp. 41-189).

FUZZINESS IN INFORMATIVE LOGICS

Tosio Kitagawa
International Institute for
Advanced Study of Social Information Science
and Kyushu University
Fukuoka, Japan

1. INTRODUCTION

In order to make clear the logical foundation of information science approaches, we have introduced three coordinate systems in our previous monograph [25] and paper [29]. The first coordination is concerned with three logical aspects of information structure: (a) objectivity, (b) subjectivity and (c) practices, while the second coordination is concerned with three aspects of information function: (α) cognition, (β) direction and (γ) evaluation. By picking up and rearranging the eighteen notions explained in the first and the second coordinations, we define the third coordination regarding the three principal aspects of subjective attitudes of information usage for feasibility of existence: (III_1) control aspect, (III_2) evolution (eizon) aspect, and (III_3) creation aspect. In three coordinations there are eighteen fundamental notions such as (a_1) pattern, (a_2) chaos, (a_3) transformation; (b_1) operation, (b_2) adaptation, (b_3) strategy; (c_1) optimalization, (c_2) stability, (c_3) learning; (α_1) deduction, (α_2) induction, (α_3) abduction; (β_1) control, (β_2) eizon, (β_3) creation; (γ_1) efficiency, (γ_2) reliability, (γ_3) plasticity, as shown in Table 1.

So far as shown in Table 1, these fundamental notions are given in linguistic expressions. Nevertheless it ought to be noted that we have illustrated these notions in connection with various mathematical formulations which can be

(III)	Control	Eizon	Creation
(I) Objectivity	(a_1) Pattern	(a_2) Chaos	(a_3) Transformation
(I) Subjectivity	(b_1) Operation	(b_2) Adaptation	(b_3) Strategy
(I) Practices	(c_1) Optimalization	(c_2) Stability	(c_3) Learning
(II) Cognition	(α_1) Deduction	(α_2) Induction	(α_3) Abduction
(II) Direction	(β_1) Control	(β_2) Eizon	(β_3) Creation
(II) Evaluation	(γ_1) Efficiency	(γ_2) Reliability	(γ_3) Plasticity

Table I. *Three coordination systems for information science approaches*

found in the area of cybernetics, information theory, control theories, automata theories, and documentation science, as shown in our papers and monographs [24], [25], [28], [31], [32]. In fact the eighteen fundamental notions as well as nine coordinate aspects mentioned above have originated from these areas which are closely connected with mathematical science approaches based upon the various mathematical models. At the same time there is a certain advantage for appealing to some literal and hence non-mathematical expressions, which, at the same time involve a certain disadvantage in the sense that some sort of vagueness could be inherited in these literal expressions. The situation should be recognized to be worth while to deep considerations on the real reasons why some kinds of vagueness are being in fact involved. Moreover, we should appreciate the merits which some sort of vagueness and fuzziness could bring with themselves.

This paper aims to analyze some aspects of vagueness which have been appearing in the realm of generalized logics which we propose to call informative logics, and to make some assertions as our proposal to investigate a new direction for fuzziness concept in developing information science approaches.

2. INFORMATIVE LOGICS

The purpose of this section is to explain the reason why we want to introduce the notion of informative logics and to show some possible advantages that can be expected by preparing a general framework of the logics covering three coordinate systems shown in Section 1. Let us start with a few observations to be emphasized in preparing our information science approaches.

Observation 2.1. Information science approaches aim at establishing a unified scientific discipline which is valid for various phenomena connected with information processing; that is, (i) production of information, (ii) transformation of information, (iii) storage and retrieval of information, and (iv) circulation and usage of information.

Observation 2.2. These information processing phenomena can be found in biological existences, social lives of human beings, as well as in machines manufactured by human beings.

Observation 2.3. Information science has been developed on the base of technological innovations on information processing technologies including (c_1) computer technology, (c_2) communication technology, and (c_3) control technology in each of which the notion of information (I_1) has been shown to work as the indispensable common currency through which they can be organized into a system S as their integration (I_3).

Observation 2.4. From the view point of information functions, there are three aspects which we have shown in the second coordination enunciated in Section 1, namely (C) cognition, (D) direction, and (E) evaluation. These aspects of information function can be designated as (I_2) intelligence in a broad sense. In combining Observations 2.3 and 2.4, we may present the following graphic representations shown in Figure 1.

$$I_2 \;\text{———}\; \begin{Bmatrix} C \\ D \\ E \end{Bmatrix} \;\text{———}\; I_1 \;\text{———}\; S \;\text{———}\; I_3$$

Fig. 1. (I_1)*Information*-(I_2)*Intelligence*-(I_3)*Integration through System*

In conclusion we can assert that these four observations just made lead us to the recognition that the notions of (I_2) intelligence, (I_3) integration are indispensable in our information science approaches, besides the two notions (I_1) information and (S) system.

2.1. Intelligence Aspects and Informative Logics

We would like to point out some substantial achievements in the specific areas of scientific researches with which the present author has been intimately concerned as a mathematician and/or as a statistician, who has been led into the investigations on information science in views of these achievements [17-24] in the specific areas of scientific researches.

(1) The logic of deduction. It is well known that the predominant direction of developments in this connection has been realized in symbolic logics, which, in fact, has been recently formulated in the realm of automata theory.

(2) The logic of induction. It is one of the most remarkable features of this century that the logic of induction has been formulated as testing hypothesis in a strict mathematical formulation in statistical sciences, due to the outstanding contributions by R. A. Fisher [8], J. Neyman [45], and A. Wald [64], among others.

(3) The logic of abduction. We are now so much

conscious of the fact that the principle (α_3) has not yet fully developed in contrast with the former two (α_1) and (α_2), except achievements by a few authors such as Moles [41], beyond the current theories of logics. Nevertheless, at least so far as Japanese scientific circles are concerned, there have been a few remarkable achievements given by several scholars such as K. Ichikawa [12] in equivalent transformational thinking theory and techniques known as KJ-method, field science [14] and problem-solving methodologies [15] in general, and also M. Nakayama [41], [43] referring to neuronic models and recently to "Zen" philosophy. The author of the present paper refers to the process of system formation in his monograph, Kitagawa [25], in order to understand the abduction procedure advocated by Kawakita [13]. We also show its connection with the logic of design [26], with particular reference to Japanese classical architecture principles. There is also a systematic study of heuristics which has been intensively done by Nakamura group in Nagoya University and which has a profound implication to abduction logics, as explained in Oda [46].

(4) The logic of dialectics. In our previous monograph, Kitagawa [27], on long range scientific planning in Japan, we have referred to the notion of feedback principle to be inherited in promoting scientific researches. This idea of ours is fundamental to our own studies in statistical inference processes [17] as well as in statistical control processes [18], which are strictly belonging to statistical sciences with special reference to statistical quality controls, response surface analysis, sample surveys and designs of experiments.

The reason why we have hitherto mentioned some specific and concrete achievements belonging to certain areas of individual sciences as having some connection with the two types of the logics (3) and (4) is based upon our intention to promote an investigation of these two logics in the following way:

(i) to formulate a certain type of logical space which yields us a general framework to discuss the logical implications of not only the above mentioned achievements but also of those to be expected in the future in connection with information science approaches.

(ii) to locate in our logical space all the fundamental notions in Table 1 which we have found to be indispensable in information science approaches.

(iii) to keep in our logical space the germs from which further development of information science can be expected to realize.

In fact, according to our intension we have introduced a new terminology called informative space in our monograph [25]. The three coordinate systems explained in Section 1 of the present paper are embedded in this informative space. Our principal attitutdes for introducing the informative logics can be explained in the following two points.

Assertion 2.1. We shall proceed step by step to the rigorous formulation of the fundamental notions of informative logics, on the basis of definite achievements obtained in various areas of information science approaches.

Assertion 2.2. We shall pick up the vagueness that are inherited in the literal expressions of the fundamental notions of informative logics, and we shall make an analysis of the vagueness in order to find out the procedures of how to manage them.

2.2. Integration Aspects in System Formation

In order to explain the integration aspects in system formation, which are crucial in information science approaches, we shall enunciate the five fundamental principles being adopted for this purpose, which we have enunciated in our monograph [25].

1. The principle of cutting. This principle is so fundamental in our logic because without any application of this principle one can not pick up any object of his concerns from the totality of the world existence. In fact, the principle of cutting is used to classify and to analyze the totality of the world existence.

2. The principle of self-conservation. The parts separated from the choastic totality of world existence, which have been obtained in virtue of application of the principle of cutting, should be stable existences, that is, they should satisfy the principle of self-conservation, because otherwise no existence can be secured either as real existences or as concepts. It is to be noted that these two principles of cutting and self-conservation are the basis for defining the notion of set.

3. The principle of construction. In general, there are a set of transformations $\{T_\sigma; \sigma \in \Sigma\}$ which can be applied to each of the following arrays and to their combinations.

(a) Sequential arrays (b) Parallel arrays.

4. The principle of integration. Besides the principle of construction which serves to an analysis of the system, there is another type of the principle called the principle of integration which is complementary to an analysis of the system and which serves to organize the system in an integrative way. The following two methods are basically important

in realizing the principle of integration in system formation.

(c) Black box method (d) The feedback principle

5. The principle of multitude. The multitude to be observed in a system derives fundamentally from the following three sources (e)-(f)-(g) and manifest itself in the output set (h).

(e) Multitude of inputs

(f) Variability of each transformation

(g) Multitude of transformation set

(h) Multitude of outputs

3. MECHANICAL INTELLIGENCE AND GENETIC EPISTEMOLOGY

In order to make clearer the key ideas underlying in our introduction of informative logics, it may be adequate to refer to some of our previous works [25], [26], [29], [30], [31]. We shall summarize our viewpoints in the following five observations.

Observation 3.1. We share the same views with J. Piaget [47], [48], to the effect that (a) intelligence constitutes the state of equilibrium towards which tend all the successive adaptations of a sensori-motor and congnitive nature, as well as all the assimilatory and accomodatory interactions between the organism and the environment, that (b) intelligence itself does not consist of an isolated and sharply differentiated class of cognitive processes, and that (c) the machanism of transition from one level of intelligence to another one is an equilibration process.

Observation 3.2. The general framework of genetic epistemology due to J. Piaget [47] has its correspondence to our formulation of logic of information science in the following sense:

(i) Each of the three key notions in setting up the intelligence of the lower level such as (a') environment, (b') organisms, and (c') action is a prototype of each of the three constituent axes of the first coordination, (a) objectivity, (b) subjectivity, and (c) practices, respectively.

(ii) Each of the two main features of organisms, cognition and affections, is the origin of structuration and valuation. In our formulation of the logics of information, the information structure is formulated in the framework of the first coordination, and it can be recognized as an objective realization of the three cognitive functions of information, namely (α) cognition given in the second coordination. On the other hand, valuation based upon affective life of organism is objectively described by means of two information function, namely, (β) direction and (γ) evaluation in the second coordination.

(iii) The notion of biological adaptation explained by Piaget [47], [48] in connection with development of intelligence, which implies three features (i) assimulation, (ii) accomodation and (iii) equilibration, can be considered as a prototype of each of three fundamental notions of feasibility of existence as the usages of informtion, namely (i) control, (ii) eizon, and (iii) creation respectively.

Observation 3.3. In view of Observation 3.1, the notion of biological intelligence should be introduced besides human intelligence and mechanical one. The mathematical models of biological intelligence are given by biorobots in ecosphere, whose intelligence aspects can be discussed in connection of the informative logics, because of the facts enunciated in Observation 3.2.

Observation 3.4. The mathematical models of biological

intelligence called biorobots in ecosphere can be expected to be most adequately discussed in the general framework of informative logics whose fundamental notions are being derived from mathematical science approaches in various cybernetical fields of control, communication and computation. Genetic epistemological considerations accompanied by informative logics are expected to contribute to an establishment of link between biological and mechanical intelligence.

Observation 3.5. The achievements due to artificial intelligence in sciences and technologies can be evaluated with respect to the second coordinate system of informative logics, that is, information functions. The essential procedures adopted in realization of artificial intelligence lie in their mechanization procedures of some information functions enunciated in the second cordinate systems. In this sense we use the terminology "mechanical intelligence" instead of artificial intelligence. The more informative logics are being formulated as informative machine (informative organism), the more human intelligence can be formulated in the framework of mechanical intelligence. In combination of these five Observations 3.1-3.5 just made, we may endeavor to mechanize some information functions of human intelligence as much as possible, in order to establish an intelligent and integrated system of information processings in the era of information oriented society. Nevertheless, there is one thing which we ought not to forget. In fact, there does exist an intrinsic vagueness in our formulation of informative logics, due to the fact that all the fundamental notions given in three coordinate systems mentioned in Table 1 are being given in literal terms which involve more or less an intrinsic vagueness. It is therefore our problem how to overcome such a vagueness in informative logics. We shall discuss the problem in the following Section 4.

In summing up Observations 3.1 ∿ 3.5 we obtain the following Table 2.

Logical space	Biopsychological space
(I_1) Objectivity (I_2) Subjectivity (I_3) Practices	(1_1) Environment (1_2) Organism (1_3) Action
(II_1) Cognition (II_2) Direction (II_3) Evaluation	(2_1) Structuration (2_2), (2_3) } Valuation
(III_1) Control (III_2) Eizon (III_3) Creation	(3_1) Assimulation (3_2) Accomodation (3_3) Equilibration

Table 2. *The genetic epistemological correspondence*

4. VAGUENESS AND FUZZINESS IN INFORMATIVE LOGICS

In Subsection 4.1.1 and 4.1.2, we shall make a general survey on the possible features of vagueness in connection with the fundamental notions in the first and the second coordination, respectively. In Subsection 4.1.3 we turn to a discussion on vagueness inherited in integration aspects of system formulation enunciated in Subsection 2.2. It is pointed out that the fundamental mathematical concepts such as set, function, relation, topological space, automata, and formal languages should be reconsidered in order that they can work adequately in the realm of information science approaches where vagueness will be always involved both in information structures and in information functions. The usefulness and the merit of the mathematical notion called fuzziness in the sense of Zadeh [68] can be recognized as a remedy to overcome the difficulties encountered in this connection. It is the purpose of Subsection 4.2 to point out the characteristic features of the fuzziness in the sense of Zadeh with respect to the fundamental notions given in Table 1.

4.1.1. Vagueness in information functions. There are three fundamental axes in the second coordinate system regarding information functions, namely, (C) cognition, (D) direction, and (E) evaluation.

In fact, the successive processes of cognition-direction-evaluation schemes can be formulated in a system formulation where the flows of information can be traced in the system. Now there are many sources of vagueness inherited in the notions belonging to the second coordinate system, because the system formation as well as its component functions designated as either of cognition, direction and evaluation can accompany with their intrinsic vagueness. The following Observations are important to our specific discussions.

Observation 4.1.1. The vagueness of the subsystems may not necessarily be hereditary. In fact, a simple system C may be a vague cognition, but D can be a definitive direction without any vagueness by which action can be taken.

Observation 4.1.2. Each of the functions (α_1) deduction, (α_2) induction, and (α_3) abduction has been formulated in such a way that each mechanical procedure of deduction, induction and abduction has been formulated in a certain sense respectively. Nevertheless there remains a broad area of mechanical intelligence where biological intelligence should be formulated in view of both ethological achievements and computer technolgoy developments.

4.1.2. Vagueness in information structures. The first coordinate system in Table 1 is concerned with information structure. The three coordinate axes are given by (a) objectivity, (b) subjectivity, and (c) practices. So far as mathematical programming formulations are concerned, (a), (b) and (c) can be formulated in exact mathematical expressions, and they can be handled with brainware notions in the sense defined in our recent work [32]. The formulation of cybernetics due to N. Wiener [66] can be also given in terms of mathematical expressions through which we can understand some of the fundamental notions such as operations, optimizations, chaos, stability, transformation, strategy and learning given in the first coordinate system of Table 1 in an exact and precise way. The situations are also favourable in the case of deterministic and statistical control processes, evolutionary operation programs as shown in [31]. Nevertheless there are certainly crucial problems where neither the deterministic nor the probabilistic formulation can be adopted, as shown and discussed by Bellman-Zadeh [3] and Asai-Kitajima [1] and Tanaka-Okuda-Asai [58].

4.1.3. Vagueness in system formation. The principles of system formation enunciated in Subsection 2.2 are realized by means of mathematical concepts and procedures which are defined in a rigorous way without any ambiguity and any vagueness. Thus, the notion of set is a realization of two principles of cutting and self-conservation, while the notion of transformation has been formulated in the mathematical theories of dynamical systems, automata and formal languages. The principles of construction and integration illustrated in Subsection 2.2 can be recognized to have been realized in a general way so far as the mathematical approaches are concerned. The application of the principle of multitude can be observed in the modern stochastic formulation by which to overcome the chaostic multitudes inherited in inputs, outputs, and flow of informations again by a rigorous mathematical formulation. In spite of the remarkable achievements given by these modern mathematical theories which have been so intimately connected with some aspect of information science approaches we have to point out that there remains a certain set of uncultivated research areas for which the present formulations do not seem to be completely adequate. Thus, any theory of large system does require us to introduce some system technique by which we can manage the vagueness inherited in large system, as discussed in our paper [31]. There is also another aspect to be considered. There are in fact various lower levels of intelligence equilibriums for which the current authentic mathematical models are not suited to give any adequate description. For instance, a certain stage of intelligence development of a child, J. Piaget [48] introduced the notion of groupment in the place of group in order to describe the pre-operational stage of cognition. In order to establish a broad and profound formulation of the key aspects of information science approaches, we ought to take into our serious consideration the areas of topics which have not been adequately formulated in the present mathematical formulations. This implies the needs for (i) locating any discrepancy that may exist between real situations and their mathematical models, (ii) pointing out any vagueness that may be admitted as a compromise to manage such a discrepancy, and then (iii) analyzing the states of vagueness in order to be able to attain some notion suited for mathematical treatment. The notion of fuzziness due to Zadeh [68] can be considered as one solution to satisfy our present needs. It ought to be pointed out that there may be another kind of solution to our present problem, as we shall discuss in the following Subsection 4.2.

4.2. Fuzziness Concept Used in the Logics of Information Science Approaches

In what follows a few observations on the extensive uses of fuzziness concepts are made in order to review them from the standpoints of informative logics, thereby leading us to find out where further developments are being required.

Observation 4.2.1. The essential aspect of fuzziness concept is to appeal to the membership function of a set instead of the classical characteristic function of a set which shows a distinction of the set from the surrounding environment. Here we can observe a fuzzy application of cutting principle explained in Subsection 2.2. Another fundamental principle of self-conservation does not seem so apparent to be adopted. This is due to the fact that the uses of fuzziness have been restricted to one equilibrium level of intelligence and that any transition from one level to another, that is, equilibration, has not been considered so much, except a certain aspect in learning fuzzy automata by Wee-Fu [65] and in connection with learning languages by Tamura-Tanaka [55]. On the other hand, the principle of transformation in system formation has been the serious topics of fuzziness, as we can observe in various types of fuzzy automata. Moreover, the principle of multitude explained in Subsection 2.2 may be said to have been considered in fuzzy environment in Bellman-Zadeh [3].

Observation 4.2.2. Regarding information structure aspects, the following researches have been done.

(i) The basic notions of equivalence, similarity, ordering, classification and abstraction, which are concerned with (a_1) pattern aspect in the first coordinate system, have been discussed by several authors such as Zadeh [72], Bellman-Kalaba-Zadeh [2], Ruspini [49], Gitman-Levine [9], Shimura [52], [53], and Tamura-Higuchi-Tanaka [56], with respect to the fuzziness, that is to say, the membership functions. It is to be noted that, although most of these approaches are concerned with one equilibrium level of intelligence, some of them can be considered with equilibration process of learning in connection with control processes, however without any clear formulation of abduction process.

(ii) In fact one of the essential merits of the fuzziness notion lies in its contribution to an introduction of a mathematical formulation for certain type of chaostic vagueness which is to be distinguished with the mathematical concept of probability. At the same time it ought to be noted that any transitory feature from fuzziness to probability has not been fully discussed, although some interesting works have been by Nasu-Honda [44], Zadeh [70], and Sugano [54], [55], in another direction.

Observation 4.2.3. Regarding information function aspects, the following points are worth while to be mentioned.

(i) With respect to the cognition axis in the second coordination, deduction has been discussed in the name of fuzzy logics by several authors such as Marinos [36], Lee and Chang [34], Lee [33], and De Luca and Termini [7] which refers to Brouwerian logics. Fuzzy algorithms discussed by Zadeh [71] and Santos [49], [50] are also concerned with deduction. None of these works have not been concerned with the other two aspects of cognition, namely induction and abduction, where one may expect various features of vagueness and ambiguity to be encountered with.

(ii) With respect to the evaluation axis in the second coordination, nothing has been done in any literature on the fuzziness concept in the direction to look into the discussion on (γ_1) efficiency, (γ_2) reliability, and (γ_3) plasticity. It seems to us that no systematic approach has been developed which will contribute to the positive advantage of the fuzziness concept in connection with the two aspects called (II) eizon, and (III) creation. The recent work of Zadeh [75] regarding a new approach to the analysis of complex systems and decision processes is also concerned with control aspect, so far as fuzzy decisional algorithms are concerned, although it does provide us an approximate yet effective means of describing the behavior of systems which are too complex or too ill-defined to admit precise mathematical analysis.

Observation 4.2.4. In connection with linguistics, which should have a predominant position in information approaches as shown in Zadeh [73], the concept of linguistic variable has recently been introduced by Zadeh [74] ∿ [76] in the general framework of fuzziness approaches to serve to quantitative fuzzy semantics, which has its substantial contributions to approximate reasoning and the analysis of complex systems and decision processes, as illustrated in Zadeh [74], [75]. A series of contributions on various fuzzy grammars have been given by Tanaka group of Osaka University, which, so far as can be observed from Mizumoto-Toyoda-Tanaka [38], [40], Kim-Mizumoto-Toyoda-Tanaka [16], and Tamura-Tanaka [57], the induction of membership functions to the classical formulations of mathematical linguistics along the line of automata theory seems to us to be their essential contribution. These tendencies can be recognized to follow to the authentic development line of the concept of fuzziness to enlarge its application fields. Nevertheless, if we return back to the general relationship between linguistics and information science approaches, it does seem to us that there remain a lot of vagueness aspect which is worth while to a serious reconsideration.

Observation 4.2.5. With regard to the third coordination which is concerned with the three principal aspects of information usage for feasibility of existence, (III_1) control, (III_2) evolution (eizon), and (III_3) creation, the uses of fuzziness concept have been entirely concerned with the (III_1) control aspect, as we can observe in the contributions by Zadeh [70] in optimization under fuzzy constraints, by Bellman Zadeh [3] in decision making in a fuzzy environment, by Chang and Zadeh [6] in fuzzy mapping and control, by Asai and Kitaj Kitajima [1] in optimizing control using fuzzy automata, by H. Tanaka-Okuda-Asai [58] in fuzzy mathematical programming.

5. A NEW DIRECTION FOR A GENERALIZATION OF FUZZINESS CONCEPT TO BE USED IN DEVELOPING INFORMATION SCIENCE APPROACHES

Besides general observations so far given in the previous Sections, there are several specific topics belonging to information science approaches, which have been the chief concerns of the present author and which have turned out to give an impulse for the author to seek for a new direction for a generalization of fuzziness concept which may be useful for our purposes.

These topics are (1) previous knowledge in successive processes of statistical inferences and controls in [17] $\sim$ [20], (2) relativistic logic in mutual specification in [21], (3) control processes in large systems in [31], (4) system approaches in ecosphere in [28], and (5) general linguistics and system formation in [25]. Instead of analysing each of these backgrounds and their implications to the concepts of vagueness and/or fuzziness, we shall be content here with an enunciation of three assertions which will point out clearly the essential aspects of our proposal to find out a new direction for a generalization of fuzziness concept.

Assertion 5.1. An application of the principles of system formation enunciation in Subsection 2.2 to a theoretical analysis of information processing precedures in language communications from the standpoint of informative logics leads us to the needs for introducing a fuzziness concept in our information structure formulations which are not confined to linguistics, in such a general way that

(i) Besides static model, we are concerned with dynamic model.

(ii) Fuzzy structural stability is defined in a fuzzy topological space where dynamic model as well as static one is introduced.

(iii) Fuzzy catastrophes and fuzzy attracters are defined along the line of R. Thom [60], with a view covering the ideas suggested by him in [61] $\sim$ [63] with reference to Tesniere [59] and Greenberg [11].

(iv) Topological, lattice theoretic and time-dependent approaches in connection with fuzziness, which were given by Brown [4], Chang [5], Goguen [10] and Lientz [35], can be appreciated for further investigation.

Assertion 5.2. Creation process as one of the three cognition aspects in information functions is analysed in view of the principles of system formation, and the analysis leads us to a recognition of the roles of three evaluation aspects (γ_1) efficiency, (γ_2) reliability, and (γ_3) plasticity in this connection. Furthermore the positive roles of the fuzziness concept in its generalized formulation given in Assertion 5.1 can be evaluated. In order to establish a sound theoretical foundation for any large system approach involving their ecological features, the positive merits of fuzziness should be more emphasised than it has been considered in the current literatures.

Assertion 5.3. An integrated system of three information functions, namely, (C) cognition, (D) direction and (E) evaluation, which can be observed in the real usages of information functions, is worth while to our systematic investigation in the general framework of informative logics. In view of our investigations of successive processes of statistical inferences and controls developed in a series of papers [17] [23], which can be recognized as the special integrated systems of information functions, two directions do seem to us to be indispensable. The first direction is to proceed to various formulations of integrated system of three information functions which are automatically controlled, that is, machanized in the sense which we have explained in Section 3 and which was precisely defined in our paper [23] with regard to successive processes of statistical inferences and controls. The second direction is to deal with the real situations which involve sophisticated considerations on the topics such as we have discussed:

(i) Mutual specification in relativistic logics in incompletely specified model in [21].

(ii) Combined usages of informations whose sources may be neither exactly identified nor precisely formulated in [17].

In the consequence there arises the problem whether and how far the mathematical models based upon automated system of information functions can be used in real applications by introducing fuzziness concept in every part of the systems where it is required.

Here we have to show the most essential aspects of the Thom theory of structural stability and morphogenesis which we need in the sense explained in Assertion 4.1. This may be done by citing his own conclusions explained in Subsection 13.7 as conclusion of his monograph [60] p. 321, which may be translated as follows:

(1) Every object or every physical form can be represented by an attracter C of dynamical system on a space M of internal variables.

(2) An object does neither represent its stability nor be perceptible, unless its corresponding attracter is structurally stable.

(3) Every creation or destruction of the formes, that is, every morphogenesis, can be described by the extinction of the attracters representing the initial formes and their replacement by the attracters representing the final forms. Such a process which is called catastrophe can be described on a space of external variables.

(4) Every structurally stable process is described by one (or a system of) catastrophe(s) which is (are) stable on the space P.

(5) Every natural process can be decomposed into structurally stable small islands called chreods. The set of these chreods and the multidimensional syntax regulating their respective positions constitute a semantic model.

(6) If one considers a chreod C as a word of the multidimensional language, the signification of the word is nothing but the global topology of one (or plural) catastrophe(s) associated with it and the catastrophes to which it is subject. In particular, for a given attracter, its signification is

definitely the geometry of its existence domain in the space P and the topology of catastrpohes of regulation which restrict this domain.

In short, semantic topology in which the noun is described by a potential wall in the dynamics of mental activities and the verb is described by an oscillator in the unfolding space of a spatial catastrophe does require to introduce a systematic use of fuzziness concept in its formulation. We are now in a position to show our research programs of generalizing usages of fuzziness concept in accordance with three Assertions 5.1 ∿ 5.3.

Program 1. A fuzzy topological space T, according to the definition of Chang [5], for instance, is used to define fuzzy structural stability in the space T. The set $\mathcal{B}$ of all the fuzzy structural stable points in the space T is fuzzy open in this space, and the set T - $\mathcal{B}$ is called the set of catastrophes. Our Program 1 is to make a systematic survey along the line which will be similar to that developed by Thom [60].

Program 2. In accordance with the applications of the catastrophe theory due to Thom [60] ∿ [63], to general linguistics, the principles of system formation are applied to topological analysis and synthesis of natural languages. For these applications theoretical models can be provided by fuzzy topological spaces which will be developed in Program 1.

Program 3. In accordance with progress of implementation of Program 2, conventional usages of natural languages in various scientific approaches including social sciences and humanities are expected to become more intelligible and better organized from the standpoint of informative logics than they have been. This assertion implies in particular the following crucial advantages:

(i) Structural stability aspects of various phenomena can be identified and traced in spite of their superflucious appearance of foggy vagueness and chaostic disorder which have been believed to be inherited in these fields.

(ii) Catastrophic phenomena can be mathematically formulated so as to admit a systematic approach based upon the informative logics explained in Section 2.

Program 4. In accordance with the progress of implementation of Programs 2 and 3, by which to confirm systematic usages of linguistic variables in quantitative semantics according to the sense of Zadeh [74] with the brainware regarding the world structure description, we can expect to have scientific methodologies for treating with some sort of vagueness concepts in the context of generalized aspects of information usages for feasibility of our existence including (III_2) eizon aspect and (III_3) creation aspect besides (III_1) control aspect.

Program 5. Throughout the whole area of scientific researches and surveys where the usage of data banks are required in supplementing our own experiments and/or surveys, there is an urgent need to establish a methodology of integrating various informations with each vague domain of validity. It is our program to have a systematic investigation of fuzziness concept in connection with Programs 1 $\sim$ 4 with an intension to promote the direction suggested by our previous papers [17] $\sim$ [24] on statistical inference and control processes.

Remark. After the Seminar and before the preparation of the final draft for publication, the author has the opportunity of having the recent papers on fuzzy topological spaces by a participant of the Seminar C. K. Wong and also a Japanese monograph by H. Nakai on Information Communication, which suggests an application of Thom catastrophe theory on documentation science.

REFERENCES

[1] Asai, K., and Kitajima, S.: "Optimizing Control Using Fuzzy Automata," Automatica, 8(1972), 101-104.
[2] Bellman, R., Kalaba, R., and Zadeh, L.: "Abstraction and Pattern Classification," Journ. Math. Analysis and Applications, 13(1966), 1-7.
[3] Bellman, R., and Zadeh, L. A.,: "Decision-Making in a Fuzzy Environment," Management Science, 17(1970), 141-164.
[4] Brown, J. G.: "A Note on Fuzzy Sets," Information and Control 18, (1971), 32-39.
[5] Chang, C. L.: "Fuzzy Topological Spaces," Journ. Math. Analysis and Applications, 24 (1968), 182-190.
[6] Chank, S. S. L., and Zadeh, L. A.: "On Fuzzy Mapping and Control," IEEE Transactions on Systems, Man, and Cybernetics, Vol. SMC-2, No. 1 (1972), 30-34.
[7] De Luca, A. and Termini, S.: "A Definition of a Non-probabilistic Entropy in the Setting of Fuzzy Sets Theory," Information and Control 20, (1972), 301-312.
[8] Fisher, R. A.: Contributions to Mathematical Statistics, John Wiley & Sons/Chapman & Hall, (1950).
[9] Gitman, I., and Levine, M.: "An Algorithm for Detecting Unimodal Fuzzy Sets and Its Application as a Clustering Technique," IEEE Transactions on Computer, Vol. C-19, No. 7, (1970), 583-593.
[10] Goguen, J. A.: "L-Fuzzy Sets," Journ. Math. Analysis and Applications, 18 (1967), 145-174.
[11] Greenberg, J.: "Some Universals of Grammar with Particular Reference to the Order of Elements," Universal of Language, pp. 58-90, Ed. J. Greenberg, M.I.T., (1966).
[12] Ichikawa, K.; "Science of Creativity-Illustrations and Introduction to the Equivalent Transformational Thinking (In Japanese)," Nippon Hososhuppan Kyokai, Tokyo, (1970).
[13] Kawakita, J.: "Abduction Methods (In Japanese)," Chuko-Shinsho 136, Chuo Koronsha, (1967).
[14] Kawakita, J.: "Methodologies of Field Sciences," (In Japanese), Chuko-Shinsho, 332, Chuo Koronsha, (1973).
[15] Kawakita, J. and Makishima, S.: "Science of Problem Solving-Working Book for KJ Method (In Japanese)," Kodansha, (1970).
[16] Kim, H. H., Mizumoto, M., Toyada, J., and Tanaka, K.:

"L-Fuzzy Grammars," Information Sciences, 1974 (to appear).
[17] Kitagawa, T.: "Successive Process of Statistical Inferences," (1) Mem. Fac. Sci. Kyushu University, Ser. A, 6 (1950), 139-180; (2) ibid., Ser. A, 6 (1951), 54-95; (3) ibid., Ser. A, 6 (1952), 54-95; (5) ibid., A, 7 (1953), 81-106; (6) ibid., A, 8 (1953), 1-29; Bull. Math. Statistics, 5 (1951), 35-50.
[18] Kitagawa, T.: "Successive Process of Statistical Controls," (1) Mem. Fac. Sci., Kyushu Univ., Ser. A (1952), 13-26; (2) ibid., Ser. A (1959), 1-16; (3) ibid., 8 (1959), 80-114.
[19] Kitagawa, T.: "The Logical Aspect of Successive Processes of Statistical Inferences and Controls," Bull. Math. Intern. Statist. Inst., 38 (1961), 151-164.
[20] Kitagawa, T.: "Automatically Controlled Sequence of Statistical Procedures in Data Analysis," Mem. Fac. Sci. Kyushu Univ., A, 17 (1963), 106-129.
[21] Kitagawa, T.: "The Relativistic Logic of Mutual Specification in Statistics," Mem. Fac. Sci. Kyushu Univ., Ser. A, 17 (1963), 76-105.
[22] Kitagawa, T.: "Estimation After Preliminary Test of Significance," Univ. Calif. Publ. Statist. 3 (1963), 147-186.
[23] Kitagawa, T.: Automatically Controlled Sequence of Statistical Procedures, Springer-Verlag, Berlin, Heidelberg, New York (1965), 146-178.
[24] Kitagawa, T.: "Information Science and Its Connection with Statistics," Proc. 5th Berkeley Symposium on Math. Statist. and Probability, I (1967), 491-530.
[25] Kitagawa, T.: The Logic of Information Science (In Japanese), Tokyo, Kodansha, Contemporary Series, Vol. 200, (1969).
[26] Kitagawa, T.: "The Logic of Design (In Japanese)," Design Series in Architecture, Vol. 4, Creation in Design, Shokokusha, Tokyo, (1970).
[27] Kitagawa, T.: Science Planning-Seventeen Years in Science Council of Japan (In Japanese), Kyoritsu Shuppan, (1969).
[28] Kitagawa, T.: "Environments and Eizon-Sphere (In Japanese)," Social Information Science Series, Vol. 10, Environmental Science and Human Beings, Gakkuken (1973), pp. 132-199.
[29] Kitagawa, T.: "Three Coordinate Systems for Information Science Approaches," Information Sciences, 5 (1973), 157-169.
[30] Kitagawa, T.: "Biorobots for Simulation Studies of

Learning and Intelligent Controls," U. S.-Japan Seminar of Learning Control and Intelligent Control, Gainesville. October 22-25, (1973).
[31] Kitagawa, T.: "The Logic of Information Sciences, and Its Implications for Control Processes in Large Systems, The Second 'FORMATOR'," Symposium on Mathematical Methods for the Analysis of Large-Scale Systems, Prague, June, 18-21, (1974).
[32] Kitagawa, T.: "Brainware Concept in Intelligent and Integrated System of Information," FID/RI Committee Meeting "Information Science, Its Scope, Objects and Problems" Moscow, April 24-26, (1974) (to be published).
[33] Lee, R.C.T.: "Fuzzy Logic and the Resolution Principle," Journ. Assoc. Comp. Machinery, 19 (1972), 109-129.
[34] Lee, R. C. T., and Chang, C. L.: "Some Properties of Fuzzy Logic," Information and Control 9, (1971), 413-431.
[35] Lientz, B. P.: "On Time Dependent Fuzzy Sets," Information Sciences 4 (1972), 367-376.
[36] Marinos, P. N.: "Fuzzy Logic and Its Application to Switching Systems," IEEE Transactions on Computers, Vol. C-18, No. 4 (1969), 343-348.
[37] Mizumoto, M., Toyoda, J., and Tanaka, K.: "Some Considerations on Fuzzy Automata," Journ. Computer and System Sciences, 3 (1969), 409-422.
[38] Mizumoto, M., Toyoda, J., and Tanaka, K.: "General Formulation of Formal Grammars," Information Sciences, 4 (1972), 87-100.
[39] Mizumoto, M., Toyoda, J., and Tanaka, K.: "Examples of Formal Grammars with Weights," Information Processing Letters 2 (1973), 74-78.
[40] Mizumoto, M., Toyoda, J., and Tanaka, K.: "N-Fold Fuzzy Grammars," Information Sciences,5, (1973), 25-43.
[41] Moles, A.: "La Creation Scientifique," Editions Rene Kister, Geneve, (1957).
[42] Nakayama, M.: "Techniques of Creative Thinking (In Japanese)," Contemporary Series, 231, Kodansha (1970.
[43] Nakayama, M.: "Introduction to Creativity Engineering," (In Japanese), Sangyo Noritsu Tandai Shuppanbu, (1972).
[44] Nasu, M. and Honda, N.: "Fuzzy Events Realized by Finite Probabilistic Automata," Information and Control, 12 (1968), 284-303.
[45] Neyman, J.: A Selection of Early Statistical Papers of J. Neyman, Univ. Calif. Press, Berkeley and Los Angeles, (1967).
[46] Oda, M.: "Heuristics-General Theory and Experimental Considerations," (In Japanese), Symposium on Information

Science and Mathematical Programming, Research Institute of Fundamental Information Science, Kyushu Univ., January 18-19, (1973).

[47] Piaget, J.: The Psychology of Intelligence, (Translated from the French by Piercy, M. and Berklyne, D. E.), Putledge & Kagan Paul LTD, London, (1950).

[48] Piaget, J.: Genetic Epistemology (Translated by Duckwork, E.), Columbia Univ. Press, New York, (1970).

[49] Ruspini, E. H.: "Numerical Methods for Fuzzy Clustering," Information Sciences, 2 (1970), 319-350.

[50] Santos, E. S.: "Maximum Automata," Information and Control, 13 (1968), 363-377.

[51] Santos, E. S.: "Max-Product Machines," Journ. Math. Analysis and Applications, 37 (1972), 677-686.

[52] Shimura, M.: "Fuzzy Sets Concept in Rank-Ordering Objects," Journ. Math. Analysis and Applications, 43 (1973), 717-733.

[53] Shimura, M.: "Application of Fuzzy Functions to Pattern Classification," (In Japanese), Journ. IECE Japan Shin Gaku Ron (D), 55-D, (1972), 218-225.

[54] Sugano, M.: "Fuzzy Measure and Fuzzy Integral," (In Japanese), Transaction of SICE (The Society of Instrument and Control Engineers), Collection of Papers, 8 (1972), 218-226.

[55] Sugano, M.: "Constructing Fuzzy Measure and Grading Similarity Patterns by Fuzzy Integral," (In Japanese), Society of Instrument and Control Engineers, 9 (1973), 361-368. "Constructing of Fuzzy Measure and Grading Similarity of Patterns by Fuzzy Integral," (In Japanese), Transactions of SICE (The Society of Instrument and Control Engineers), 9 (1973), 361-368.

[56] Tamura, S., Higuchi S., and Tanaka, K.: "Pattern Classification Based on Fuzzy Relations," IEEE Transactions on Systems Man, and Cybernetics, Vol. SMC-1, No. 1, (1971), 61-66.

[57] Tamura, S., and Tanaka, K.:"Learning of Fuzzy Formal Language," IEEE Transactions on Systems, Man, and Cybernetics, Vol. SMC-3, No. 1 (1973), 98-102.

[58] Tanaka, H., Okuda, T., and Asai, K.: "On Fuzzy-Mathematical Programming," Journal of Cybernetics, (to appear).

[59] Tesnière, T.: Eléments de Syntaxe Structurelle, Klincksieck, Paris (1968).

[60] Thom, R.: Stabilité Structurelle et Morphogenèse, Benjamin, (1973).

[61] Thom, R.: "Topologie et Lingustique," Essays on Topology and Related Topics, Springer, (1970).

[62] Thom, R.: "Sur la Typologie des Langues Naturelles," in The Formal Analysis of Natural Languages, Mouton (1973).
[63] Thom, R.: "Sur la Typologie des Langues Naturelles," Conference à la Maison Franco-Japonaise, la 19 avril (1973).
[64] Wald, A.: Statistical Decision Functions, John Wiley and Sons, (1950).
[65] Wee, W. and Fu, K. S.: "A Formulation of Fuzzy Automata and Its Application as a Model of Learning Systems," IEEE Transactions on Systems and Cybernetics, Vol. SSC-5, No. 3 (1969), 215-223.
[66] Wiener, N.: Cybernetics-Control and Communication in the Animal and the Machine, MIT Press, Second Edition John Wiley and Sons, (1961).
[67] Yukawa, H.: Jamp into Creation (in Japanese), Kodansha, (1968); specially Part II, 139-176.
[68] Zadeh, L. A.: "Fuzzy Sets," Information and Control, 8, (1965) 338-353.
[69] Zadeh, L. A.: "Fuzzy Sets and Systems," Proc. of the Symposium on System Theory, Polytechnic Institute of Brooklyn, New York, (1965) 29-39.
[70] Zadeh, L. A.: "Probability Measures of Fuzzy Events," Journal Math. Analysis and Applications, 23 (1968), 421-427.
[71] Zadeh, L. A.: "Fuzzy Algorithms," Information and Control 12, (1968), 94-102.
[72] Zadeh, L. A.: "Similarity Relations and Fuzzy Orderings," Information Sciences, 3 (1971), 177-200.
[73] Zadeh, L. A.: "Note on Fuzzy Languages," Information Sciences, 1 (1969), 421-434.
[74] Zadeh, L. A.: "Quantitative Fuzzy Semantics," Information Sciences, 3 (1971), 159-176.
[75] Zadeh, L. A.: "Outline of a New Approach to the Analysis of Complex Systems and Decision Processes," IEEE Transactions on Systems, Man, and Cybernetics, Vol. SMC-3, No. 1, (1973), 28-44.
[76] Zadeh, L. A.: "The Concept of a Linguistic Variable and Its Application to Approximate Reasoning," Learning Systems and Intelligent Robots, Plenum Publishing Co., New York, 1974, pp. 1-10.

FUZZY RELATIONS, FUZZY GRAPHS, AND THEIR APPLICATIONS TO CLUSTERING ANALYSIS*

Raymond T. Yeh and S. Y. Bang
Department of Computer Science
University of Texas
Austin, Texas U.S.A.

I. INTRODUCTION

Techniques for pattern classification and clustering analysis are well studied in the literature. Most of the techniques are relational or graph-theoretical. The usual approaches, given a data matrix or graph, are to obtain threshold graphs and then apply various connectivity properties of the graph to form clusters. These kinds of approaches have a common weakness in that weights in the graphs are not treated evenly. This paper is partly motivated from the desire to extend the existing graph theoretical techniques to work on data graphs directly rather than on the threshold graphs. In order to do this, it is necessary to extend many standard graph theoretical concepts.

In Section III, an algebra of fuzzy relations is developed which parallels closely that of the binary relations. Sections IV and V involve extensions of various connectedness concepts of digraphs and undirected graphs into fuzzy graphs. Section VI provides evidence that techniques for clustering analysis is extendible and desirable.

II. PRELIMINARIES

We will provide in this section basic terminologies and notations which are necessary for the understanding of

*The research reported here is supported in part by the National Science Foundation under Grant No. GJ-31528.

subsequent results.

A fuzzy (binary) relation R from a set X to a set Y is a fuzzy subset of X x Y characterized by a membership function μ_R: $X \times Y \to [0,1]$. For each $x \in X$ and $y \in Y$, $\mu_R(x,y)$ is referred to as the strength of the relation between x and y. If X = Y, then we say R is a fuzzy relation on X. As in the case of nonfuzzy binary relations, every fuzzy relation R on X can be represented by a directed fuzzy graph consisting of a set of vertices X such that there is a weighted arc connecting each pair of vertices x_i, x_j and the weight on the arc (x_i, x_j) is $\mu_R(x_i, x_j)$. Equivalently, R can be represented by a fuzzy matrix, M_R, whose $(i,j)^{th}$ entry is $\mu_R(x_i, x_j)$.

In the following definitions, the symbols $\vee$ and $\wedge$ stand for max and min, respectively.

Let R and S be two fuzzy relations from X to Y. R is said to be contained in S, in symbols, $R \subseteq S$, if $\mu_R(x,y) \leq \mu_S(x,y)$, for all $(x,y) \in X \times Y$. The union of R and S, denoted by $R \cup S$, is defined by $\mu_{R \cup S} = \mu_R \vee \mu_S$. The intersection of R and S, denoted by $R \cap S$, is defined by $\mu_{R \cap S} = \mu_R \wedge \mu_S$. The complement of R, denoted by $\overline{R}$, is defined by $\mu_{\overline{R}} = 1 - \mu_R$. The inverse of R, denoted by R^{-1}, is a fuzzy relation from Y to X defined by $\mu_{R^{-1}} = \mu_R$.

If R and S are fuzzy relations from X to Y and from Y to Z, respectively, then the composition of R and S, denoted by R o S (or simply by RS), is a fuzzy relation from X to Z defined by

$$\mu_{R \circ S}(x,z) = \bigvee_y [\mu_R(x,y) \wedge \mu_S(y,z)], \quad x \in X, \ z \in Z.$$

The n-fold composition $\underbrace{R \circ R \circ \ldots \circ R}_{n \text{ times}}$ is denoted by R^n.

We assume here that basic operations for fuzzy matrices are Max and Min respectively. Note that $M_{RS} = M_R M_S$.

III. AN ALGEBRA OF FUZZY RELATIONS

In this section we will show that the concept of similarity relation introduced by Zadeh [18] is derivable in much the same way as equivalence relation. Furthermore, through this derivation, the resolution identity [15] is brought out quite naturally.

Definition 1: Let R be a fuzzy relation on a set X. We define the following notions:

1. R is ε-reflexive iff $(\forall x \in X)\ [\mu_R(x,x) \geq \varepsilon]$. A 1-reflexive relation will simply be referred to as a reflexive relation.

2. R is irreflexive iff $(\forall x \in X)\ [\mu_R(x,x) = 0]$.

3. R is symmetric iff $\mu_R(x,y) = \mu_R(y,x)$ for all x,y in X.

4. R is weakly reflexive iff for x,y in X, $\mu_R(x,y) = \varepsilon \rightarrow \mu_R(x,x) \geq \varepsilon$.

5. R is transitive iff $R \supseteq R \circ R$.

Lemma 1. If R is a fuzzy relation from X to Y, then the relation RR^{-1} is weakly reflexive, and symmetric.

Proof: i)
$$\mu_{RR^{-1}}(x,x') = \bigvee_y [\mu_R(x,y) \wedge \mu_{R^{-1}}(y,x')]$$
$$\leq \bigvee_y [\mu_R(x,y) \wedge \mu_R(x,y)]$$
$$= \mu_{RR^{-1}}(x,x)$$

Hence, RR^{-1} is weakly reflexive.

ii) $(RR^{-1})^{-1} = RR^{-1}$. Hence, RR^{-1} is symmetric.

Let R now be a weakly reflexive and symmetric relation on X. Define a family of non-fuzzy sets F^R as follows:

$$F^R = \{K \subseteq X \mid (\exists\, 0 < \varepsilon \leq 1)[\forall x \varepsilon X)[x \varepsilon K \leftrightarrows (\forall x' \varepsilon K)[\mu_R(x,x') = \varepsilon]]\} \quad (1)$$

We note that if we let

$$F^R_\varepsilon = \{K \subseteq X \mid (\forall x \varepsilon X)[X \varepsilon K \leftrightarrows (\forall x' \varepsilon K)[\mu_R(x,x') \geq \varepsilon]]\}^1 \quad (2)$$

then we see that $\varepsilon_1 \leq \varepsilon_2 \Rightarrow F^R_{\varepsilon_2} \preceq F^R_{\varepsilon_1}$, where "$\preceq$" denotes covering relation, i.e. every element in $F^R_{\varepsilon_2}$ is a subset of an element in $F^R_{\varepsilon_1}$.

A subset J of X is called <u>ε-complete</u> with respect to R iff $(\forall x,x' \varepsilon J)[\mu_R(x,x') \geq \varepsilon]$. A <u>maximal</u> ε-complete set is one which is not properly contained in any other ε-complete set.

<u>Lemma 2</u>. F^R is the family of all maximal ε-complete sets with respect to R for $0 \leq \varepsilon \leq 1$.

Proof: Let $K \varepsilon F^R$ and $x \varepsilon K$. Then there exists $0 < \varepsilon \leq 1$ such that $x' \varepsilon K$, $\mu_R(x,x') \geq \varepsilon$ by (1). Hence K is complete. Next, consider an ε-complete set J which is not maximal. This means that there exists a maximal ε-complete set J' such that $J \subset J'$, which implies that there exists $x' \varepsilon J' - J$. But since J' is ε-complete, we conclude that for each x in J, $\mu_R(x',x) \geq \varepsilon$. Hence, by (1) we must conclude that $x' \varepsilon J$, is a contradiction. Hence, J must also be maximal.

<u>Lemma 3</u>. Whenever $\mu_R(x,x') > 0$, there is some ε-complete set $K \varepsilon F^R$ such that $\{x,x'\} \subseteq K$.

Proof: If $x = x'$, then $\{x\}$ is certainly ε-complete for $\varepsilon = \mu_R(x,x)$. Otherwise, if $x \neq x'$, then since $\mu_R(x,x') = \mu_R(x',x)$ by symmetry, and $\mu_R(x,x) \geq \mu_R(x,x')$ and $\mu_R(x',x')$

---

[1] Note that each element in F^R defines an ε-level-set of Zadeh [18].

$\geq \mu_R(x,x')$ by weak reflexivity, we see that $\{x,x'\}$ is ε-complete, where $\varepsilon = \mu_R(x,x')$. Thus, $\{x,x'\}$ is contained in some ε-complete set C. Denote by C_ε the family of all ε-complete sets C' which include C. Then C is partially ordered under set inclusion and hence satisfies the condition of Zorn's lemma. Therefore, we conclude from Zorn's lemma that C_ε has a maximal element K. This element is also maximal in the family of all ε-complete sets since any sets including K must also include C. Hence $K \varepsilon F^R$ by lemma 2, and the proof is completed.

It should be remarked here that sometimes a subclass of F^R, satisfying the condition of lemma 3, will cover the set X. For example, let R be the fuzzy relation on X = {a,b,c, d,e,f} given by the following matrix.

$$\begin{bmatrix} 1 & .3 & .4 & 0 & .4 & .3 \\ .3 & 1 & .2 & .3 & 0 & .3 \\ .4 & .2 & 1 & .3 & .5 & 0 \\ 0 & .3 & .3 & 1 & 0 & 0 \\ .4 & 0 & .5 & 0 & 1 & 0 \\ .3 & .4 & 0 & 0 & 0 & 1 \end{bmatrix}$$

We see that the family three maximal complete set {a,b,f}, {b,c,d} and {a,c,e} satisfy the condition of lemma 3 but it does not contain the maximal complete set {a,b,c}.

Let 0 and I denote two special relations on a set X such that for all x,x' in X,

$$\mu_0(x,x') = 0, \quad \mu_I(x,x') = 1.$$

<u>Lemma 4</u>. If $R \neq 0$ is a weakly reflexive and symmetric relation on X, then there exists a set Y and a fuzzy relation S form X to Y such that $R = SS^{-1}$.

Proof: Denote by Y the set $\{K^* | K \varepsilon F^R\}$, we define a fuzzy relation S from X to Y as follows:

$$\mu_S(x,K^*) = \begin{cases} \alpha, \text{ if } x \in K \text{ and } \alpha \text{ is the largest number} \\ \text{for which } K \in F^R_\alpha. \\ 0 \text{ otherwise.} \end{cases} \quad (3)$$

If $\mu_R(x,x') = \alpha > 0$, then by lemma 3, there is an α-complete set $K \in F^R$ such that $\{x,x'\} \subseteq K$. Since

$$\mu_{SS^{-1}}(x,x') = \bigvee_{K^*} [\mu_S(x,K^*) \wedge \mu_S(x',K^*)] \geq \alpha = \mu_R(x,x'),$$

we conclude that $R \subseteq SS^{-1}$.

Suppose now that $\mu_{SS^{-1}}(x,x') = \beta$. Then there exists $K^* \in F_\beta$ such that $\mu_S(x,K^*) = \mu_S(x',K^*)$. This means that $\{x,x'\} \subseteq K$ and hence $\mu_R(x,x') \geq \beta$. Therefore, $SS^{-1} \subseteq R$.

Combining lemmas 1 and 4, we have the following result.

Theorem 1. A fuzzy relation $R \neq 0$ on a set X is weakly reflexive and symmetric iff there is a set Y and a fuzzy relation S from X to Y such that $R = SS^{-1}$.

In the sequel, we shall use the notation ϕ_R to denote the relation S defined in (3).

Definition 2: A cover C on a set X is a family of subsets X_i, $i \in I$, of X such that $\bigcup_{i \in I} X_i = X$.

Definition 3: Let R be a fuzzy relation from X to Y, we define the following notations:

R is ε-determinate iff for each $x \in X$, there exists at most one $y \in Y$ such that $\mu_R(x,y) \geq \varepsilon$.

R is ε-productive iff for each $x \in X$ there exists at least one $y \in Y$ such that $\mu_R(x,y) \geq \varepsilon$.

R is an ε-function iff it is both ε-determinate and ε-productive.

Lemma 5. If R is an ε-reflexive relation on X then ϕ_R is ε-productive and for each $\varepsilon' \leq \varepsilon$, $F^R_{\varepsilon'}$ is a cover of X.

Proof: Since for each $x \in X$, $\mu_R(x,x) \geq \varepsilon$, and because $\{x\}$ is ε-complete, there is some K in $F^R_{\varepsilon'}$ $(\varepsilon' \leq \varepsilon)$ such that $x \in K$. Hence, $F^R_{\varepsilon'}$ is a cover. Also, by definition of ϕ_R, $x \in K$ implies that $\mu_{\phi_R}(x,K^*) \geq \varepsilon$ which implies that ϕ_R is ε-productive.

In the sequel, we shall use the term productive for 1-productive, etc.

Corollary 1. If R is reflexive, then ϕ_R is productive and each F^R_ε $(0 < \varepsilon \leq 1)$ is a cover of X.

The following result is a consequence of lemma 4 and corollary 1.

Corollary 2. R is reflexive and symmetric relation on X iff there is a set Y and a productive fuzzy relation S from X to Y such that $R = SS^{-1}$.

Lemma 6. Let R be a weakly reflexive, symmetric and transitive relation on X, and let ϕ^ε_R denote the relation ϕ_R whose range is restricted to F^R_ε. Then for each $0 < \varepsilon \leq 1$, ϕ^ε_R is -determinate and elements of F^R_ε are pairwise disjoint.

Proof: Let K and K' be two not necessarily distinct elements of F^R_ε and assume that $k \cap K' \neq \emptyset$. For any $q_1 \in K \cap K'$, we have $\mu_R(q,q_1) \geq \varepsilon$, for all q in K and $\mu_R(q_1,q') \geq \varepsilon$, for all q' in K'. Since R is transitive, we see that $\mu_R(q,q') \geq \varepsilon$, for $q \in K$, and $q' \in K'$. Since R is weakly reflexive and symmetric, we conclude that $K \cup K'$ is ε-complete. However, since K and K' are maximal ε-complete, we must conclude that $K = K'$. Hence, $K \neq K' \rightarrow K \cap K' = \emptyset$. Now since $\mu_{\phi^\varepsilon_R}(x,K^*) = \varepsilon$, and since x cannot belong to any other sets in F^R, ϕ^ε_R is determinate.

Definition 4. A similarity relation R on X is a fuzzy relation on X which is reflexive, symmetric and transitive. R is called an <u>ε-similarity</u> relation if it is ε-reflexive for some $0 < \varepsilon \leq 1$.

Since clearly reflexivity implies weak reflexivity, we have the following consequence of lemmas 5 and 6.

Corollary 3. If R is a similarity relation on X, then for each $0 < \varepsilon \leq 1$, F_ε^R is a partition on X.

Note that corollary 3 says that every similarity relation R admits a resolution $\bigcup_\alpha \alpha R_\alpha$, where R_α is the equivalence relation induced by the partition F_α^R. Indeed, it was pointed out by Zadeh [18] that if the R_α, $0 < \alpha \leq 1$, are a nested sequence of distinct equivalence relations on X, with $\alpha_1 > \alpha_2 \Longleftrightarrow R_{\alpha_1} \subseteq R_{\alpha_2}$, R_1 is nonempty and domain of R_1 is equal to domain of R_2, then $R = \bigcup_\alpha R_\alpha$ is a similarity relation on X.

The following result, which is a straightforward consequence of theorem 1 and corollary 3, offers another characterization of similarity relation.

Theorem 2. A relation R is an ε-similarity $(0 < \varepsilon \leq 1)$ relation on a set X iff there is another set Y and an ε-function f from X to Y such that $R = ff^{-1}$.

VI. FUZZY GRAPHS

In this section, fuzzy graphs will be analyzed from the connectedness viewpoint. The results will be applied to clustering analysis and modelling of information networks.

Definition 5: A fuzzy graph G is a pair [V,R], where V is a set of vertices, and R is a fuzzy relation on V.

Following the usual convention between binary relations and boolean matrices, we denote by M_G the corresponding fuzzy matrix of a fuzzy graph G. In other words, $(M_G)_{ij} = \mu_R(v_i, v_j)$.

The first part of the following result is due to Tamura,

Higucchi and Tanaka [15]. The second part is quite straightforward and hence is given without proof.

Theorem 3. Let $G = [V,R]$ be a given finite fuzzy graph, consisting of n vertices.

(i) If R is reflexive, then there exists $k \leq n$ such that

$$M_G < M_G^2 < \ldots < M_G^k = M_G^{k+1}$$

(ii) If R is irreflexive, then the sequence $\{M_G^i\}_{i=1}$ is eventually periodic.

In the following, we will only consider finite fuzzy graph G whose characterizing fuzzy relation R is reflexive.

Definition 6: Let $G = [V,R]$ be a fuzzy graph. A vertex v is said to be ε-reachable from another vertex u, for some $0 < \varepsilon \leq 1$, iff there exists a positive integer k such $\mu_{R^k}(u,v) \geq \varepsilon$. The reachability matrix of G, denoted by R_G, is the matrix M_G^k, where k is the smallest integer such that $M_G^k = M_G^{k+1}$. The ε-reachability matrix of G, denoted by R_G^ε, is obtained from R_G such that $R_G^\varepsilon(u,v) = 1$ iff $R_G(u,v) \geq \varepsilon$.

The following algorithm can determine the reachability between any pair of nodes in a fuzzy graph G.

Algorithm 1: Determination of R_G (i,j)

1. Let $R_i = (a_{i_1},\ldots,a_{i_n})$ denotes the ith row.
2. Obtain the new R_i by the following procedure:

$$a_{ij}(\text{new}) = \max_j \{\max_k \{\min \{a_{kj}, a_{ik}(\text{old})\}\}, a_{ij}(\text{old})\}.$$

3. Repeat 2 until no further changes occur.
4. $R_G(i,j) = a_{ij}(\text{new})$.

Note that a similar algorithm can be constructed for the determination of R_G^ε, $0 \leq \varepsilon \leq 1$.

Definition 7: Let G be a fuzzy graph. The connectivity of a pair of vertices u and v, denoted by c(u,v) is defined to be min $\{R_G(u,v), R_G(v,u)\}$. The connectivity matrix of G, denoted by C_G, is defined such that $C_G(u,v) = C(u,v)$. For $0 \leq \varepsilon \leq 1$, the ε-connectivity matrix of G, denoted by C_G^ε, is obtained from C_G such that $C_G^\varepsilon(u,v) = 1$ iff $C(u,v) \geq \varepsilon$.

Algorithm 2: Determination of C_G.

1. Construct R_G.
2. Construct R_{G^T}, where G^T is the transpose of G.
3. $C_G = \min(R_G, R_{G^T})$.

An algorithm for determining C_G^ε is similar to algorithm 2.

Definition 8: Let G be a fuzzy graph. G is called strongly ε-connected iff every pair of vertices are mutually ε-reachable. G is said to be initial ε-connected iff there exists $v \in V$ such that every vertex u in G is ε-reachable from v. A maximal strongly ε-connected subgraph (MSε-CS) of G is a strongly ε-connected subgraph not propertly contained in any other MSε-CS.

It is easily seen that strongly ε-connectedness implies initial ε-connectedness. Also, the following result is straightforward.

Theorem 4. A fuzzy graph G is strongly ε-connected iff there exists a vertex u such that for any other vertex v in G, $R_G(u,v) \geq \varepsilon$ and $R_G(v,u) \geq \varepsilon$.

Algorithm 3. Determination of all MSε-CS in G.

1. Construct C_G^ε.
2. The number of MSε-CS in G is given by the number of distinct row vectors in C_G^ε. For each row vector α in C_G^ε, the vertices contained in the corresponding MSε-CS are the nonzero element of the corresponding columns of α.

Example 1: Let G be a graph whose corresponding fuzzy matrix is given in figure 1a, and $C_G^{.5}$ is given in figure 1b. We see that the MS.5-CS's of G contain the following vertex sets, {1}, {2,4}, {3,5}, respectively.

$$M_G = \begin{bmatrix} 1.0 & .6 & .4 & 0 & 0 \\ 0 & 1.0 & .2 & .6 & .3 \\ 0 & .8 & 1.0 & 0 & .9 \\ .2 & .7 & .3 & 1.0 & .2 \\ .4 & 0 & .5 & .3 & 1.0 \end{bmatrix} \qquad C_G^{.5} = \begin{bmatrix} 1 & 0 & 0 & 0 & 0 \\ 0 & 1 & 0 & 1 & 0 \\ 0 & 0 & 1 & 0 & 1 \\ 0 & 1 & 0 & 1 & 0 \\ 0 & 0 & 1 & 0 & 1 \end{bmatrix}$$

(a) (b)

Figure 1: Fuzzy Matrix and Connectivity Matrix of a Fuzzy Graph

The previous result is now applied to clustering analysis. We assume that a data graph $G = [V,R]$ is given, where V is a set of data and $\mu_R(u,v)$ is a quantitive measure of the similarity of the two data items u and v. For $0 < \varepsilon \leq 1$, an <u>ε-cluster</u> in V is a maximal subset W of V such that each pair of elements in W is mutually ε-reachable. Therefore, the construction of ε-clusters of V is tantamount of finding all maximal strongly ε-connected subgraphs of G.

<u>Algorithm 4</u>. Construction of ε-clusters

1. Compute $R, R^2, \ldots, R^k$, where k is the smallest integer such that $R^k = R^{k+1}$;
2. Let $S = \sum_{i=1}^{k} R^i$. Note that S is a similarity relation;
3. Construct F_ε^S.

Then, each element in F_ε^S is an ε-cluster.

We may also define an ε-cluster in V as a maximal subset W of V such that every element of W is ε-reachable from a special element v in W. In this case, the construction of ε-clusters is equivalent to finding all maximal initial

ε-connected subgraphs of G. Note, however, that the relation induced by initial ε-connected subgraphs is not in general a similarity relation.

Another application is the use of fuzzy graphs to model information networks. Such a model was proposed by Nance, Karfhage and Bhat [12] utilizing the concepts of a directed graph. The most significant result of their work is the establishment of a measure of flexibility of a network. More specifically, let N be a network with m edges and n nodes, then the measure of flexibility of N, denoted by Z(N), is defined as follows:

$$Z(N) = \frac{m-n}{n(n-2)} \tag{3}$$

While equation (3) is quite useful in classifying certain graph structures related to information network, it also has some drawbacks in that it is insensitive to certain classes of graphs. It seems that the use of fuzzy graphs is a more desirable model for information network. The weights in each arc could be used as parameters such as number of channels between stations, costs for sending messages, etc. Thus, we propose here the use of a fuzzy graph to model an information network. Let N have n nodes; we define two measures of N: flexibility and balancedness, denoted by Z(N) and B(N) respectively, in the following:

$$Z(N) = \frac{\sum_{i=1}^{n} \sum_{j \neq i} \mu_R(v_i, v_j)}{n(n-1)} \tag{4}$$

$$B(N) = \frac{\sum_{i} \left| \sum_{j} \mu_R(v_i, v_j) - \sum_{k} \mu_R(v_k, v_i) \right|}{n(n-1)} \tag{5}$$

It is readily seen that the proposed two measures given

in (4) and (5) are much more sensitive to the structure of graphs than the one given in (3).

V. SYMMETRIC FUZZY GRAPHS

In this section, connectivity of a special class of fuzzy graphs will be investigated.

Definition 9: A fuzzy graph $G = [V,R]$ is called symmetric iff R is symmetric.

Following the usual convention of graph theory, a symmetric graph is simply a weighted undirected graph such that the edge <u,v> corresponding to a pair of symmetric arcs (u,v) and (v,u), and the weight between two nodes u and v is $\mu_R(u,v)$.

Definition 10: Let $G = [V,R]$ be a symmetric fuzzy graph. Define degree of a vertex v to be $d(v) = \sum_{u \neq v} \mu_R(v,u)$. The minimum degree of G is $\delta(G) = \min_v \{d(v)\}$, and the maximum degree of G is $\Delta(G) = \max_v \{d(v)\}$.

Definition 11: Let $G_i = [V_i,R_i]$, i=1,2 be two fuzzy graphs such that $V_1 \cap V_2 = \phi$. The sum of G_1 and G_2, denoted by $G_1 + G_2$, is a graph [V,R], where $V = V_1 \cup V_2$ and

$$\mu_R(u,v) = \begin{cases} \mu_{R_1}(u,v), & u,v \in V_1 \\ \mu_{R_2}(u,v), & \mu,v \in V_2 \\ 0, & \text{otherwise} \end{cases}$$

Definition 12: Let $G_i = [V_i,R_i]$, i = 1,2 be two fuzzy graphs. G_1 is said to be a subgraph of G_2 if $V_1 \subseteq V_2$ and R_1 is R_2 restricted to V_1.

The following result is a straightforward consequence from the definitions.

Lemma 7: Let $G = [V,R]$ be a symmetric fuzzy graph and $G_i = [V_i, R_i]$, $i = 1,\ldots,n$, be disjoint subgraphs of G such that $\sum_{i=1}^{n} G_i$ is connected. Then

i) $\delta(\sum_i G_i) \geq \min \{\delta(G_i)\}$

ii) $\Delta(\sum_i G_i) \geq \max \{\delta(G_i)\}$.

Definition 13: Let $G = [V,R]$ be a symmetric fuzzy graph. G is said to be connected iff for each pair of vertices u and v in V, there exists a $k > 0$ such that $\mu_R^k(u,v) > 0$.

G is called τ-degree connected, for some $\tau \geq 0$, if $\delta(G) \geq \tau$ and G is connected. A τ-degree component of G is a maximal τ-degree connected subgraph of G.

Theorem 5. For any $\tau > 0$, the τ-degree components of a fuzzy graph are disjoint.

Proof: Let G_1 and G_2 be two τ-degree components of G such that their vertex sets have at least one common element. Since $\delta(G_1 + G_2) \geq \min_{1=1,2} \{\delta(G_i)\}$ by lemma 7, hence, $G_1 + G_2$ is τ-degree connected. Since G_1 and G_2 are maximal with respect to τ-degree connectedness, we must conclude that $G_1 = G_2$.

Algorithm 5. Determination of τ-degree components of a finite symmetric fuzzy graph G.

1. Calculate the row sums of M_G.
2. If there are rows whose sums are less than τ, then obtain a new reduced matrix by eliminating those nodes, and go to 1.
3. If there is no such row, then stop.
4. Each disjoint subgraph of the graph induced by the nodes in the last matrix as well as each eliminated node is a maximal τ-degree connected subgraph.

Example 2: Let G be a finite symmetric fuzzy graph such that

$$M_G = \begin{array}{c} a \\ b \\ c \\ d \\ e \end{array} \begin{bmatrix} 0 & .6 & 0 & .3 & 0 \\ .6 & 0 & .4 & .7 & 0 \\ 0 & .4 & 0 & .7 & .6 \\ .3 & .7 & .7 & 0 & .4 \\ 0 & 0 & .6 & .4 & 0 \end{bmatrix} \begin{array}{c} .9 \\ 1.7 \\ 1.7 \\ 2.1 \\ 1.0 \end{array}$$

Applying algorithm 5, we obtain a reduced matrix

$$\begin{array}{c} b \\ c \\ d \end{array} \begin{bmatrix} 0 & .4 & .7 \\ .4 & 0 & .7 \\ .7 & .7 & 0 \end{bmatrix} \begin{array}{c} 1.1 \\ 1.1 \\ 1.4 \end{array}$$

Hence, the maximal 1.1-degree components of G are {a}, {b,c,d} and {d}.

Definition 14: Let G be a symmetric fuzzy graph, and $\{V_1,V_2\}$ be a dichotomy of its vertex set V. The set of edges joining vertices of V_1 and vertices of V_2 is called a cut-set of G, denoted by (V_1,V_2), relative to the dichotomy $\{V_1,V_2\}$. The weight of the cut-set (V_1,V_2) is defined to be $\sum_{\substack{u\varepsilon V_1 \\ v\varepsilon V_2}} \mu_R(u,v)$.

Definition 15: Let G be a symmetric fuzzy graph. The edge connectivity of G, denoted by $\lambda(G)$, is defined to be the minimum weight of cut-sets of G. G is called τ-edge connected if G is connected and $\lambda(G) \geq \tau$. A τ-edge component of G is a maximal τ-edge connected subgraph of G.

The following results can be proved similar to that of lemma 7 and theorem 5.

Lemma 8: Let G be a symmetric fuzzy graph and G_i, i=1,...,n, be disjoint subgraphs of G such that $\sum_i G_i$ is connected, then

$$\lambda(\sum_i G_i) \geq \min_i(\lambda(G_i))$$

Theorem 6. For $\tau > 0$, the τ-edge components of a symmetric fuzzy are disjoint.

The following algorithm for determining τ-edge components is based on a result of Matula [10].

Algorithm 6. Determination of τ-edge component of G.

1. Obtain the cohesive matrix H of the M_G.
2. Obtain the τ-threshold graph of H.
3. Each component of the graph is a maximal τ-edge connected subgraph.

Note that in Step 1, one can determine the cohesive matrix by using Corollary 9.5 of [10] after finding a narrow slicing [10]. And in order to obtain a narrow slicing, Matula's algorithm [10] can be used without modification.

Example 3: Let G be a symmetric fuzzy graph such that

$$M_G = \begin{array}{c} \\ \\ \\ \\ \\ \end{array}\begin{array}{c} \begin{matrix} a & b & c & d & e \end{matrix} \\ \begin{bmatrix} 0 & .8 & .2 & 0 & 0 \\ .8 & 0 & .4 & 0 & .4 \\ .2 & .4 & 0 & .8 & .3 \\ 0 & 0 & .8 & 0 & .8 \\ 0 & .4 & .3 & .8 & 0 \end{bmatrix} \end{array}$$

A narrow slicing and cohesive matrix of G are given in the following:

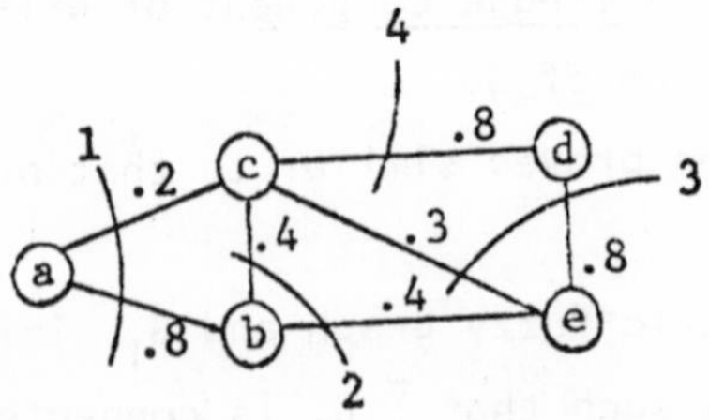

A narrow slicing of G

$$\begin{array}{c} a \\ b \\ c \\ d \\ e \end{array} \begin{bmatrix} 0 & 1.0 & 1.0 & 1.0 & 1.0 \\ 1.0 & 0 & 1.0 & 1.0 & 1.0 \\ 1.0 & 1.0 & 0 & 1.1 & 1.1 \\ 1.0 & 1.1 & 1.1 & 0 & 1.1 \\ 1.0 & 1.0 & 1.1 & 1.1 & 0 \end{bmatrix}$$

Cohesive matrix of M_G

It then follows from the algorithm that the 1.1-edge connected components are {a}, {b}, {c,d,e}.

Definition 16: A disconnection of a symmetric fuzzy graph G is a vertex set D whose removal results in a disconnected or a single vertex graph. The weight of D is defined to be $\sum_{V \varepsilon D} \{\min \mu_R(v,u) | \mu_R(v,u) \neq 0\}$.

Definition 17: The vertex connectivity of a symmetric fuzzy graph G, denoted by $\Omega(G)$, is defined to be the minimum weight of disconnection in G. G is said to be τ-vertex connected if $\Omega(G) \geq \tau$. A τ-vertex component is a maximal τ-vertex connected subgraph of G.

Note that τ-vertex components need not be disjoint as do τ-degree and τ-edge components. The following result is straightforward.

Theorem 7. Let G be a symmetric fuzzy graph, then

$$\Omega(G) \leq \lambda(G) \leq \delta(G)$$

Theorem 8. For any real numbers a, b and c such that $0 < a \leq b \leq c$, there exists a fuzzy graph G with $\Omega(G) = a$, $\lambda(G) = b$, and $\delta(G) = c$.

Proof: Let n be the smallest integer such that $c/n \leq 1$, and let $a' = a/n$, $b' = b/n$, and $c' = c/n$, then $0 < a' \leq b' \leq c' \leq 1$.

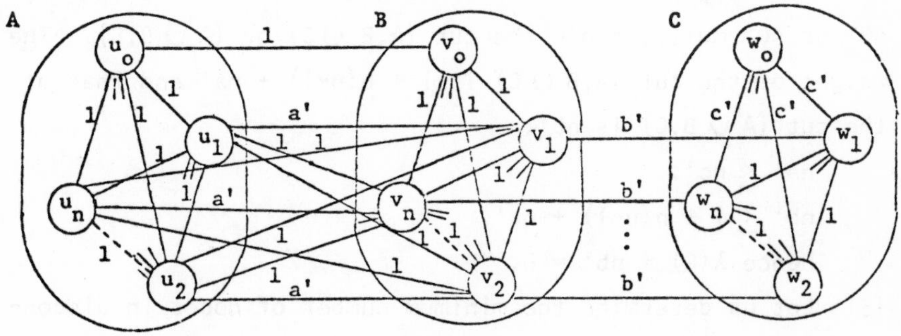

Figure 2

Let the fuzzy graph G in Figure 2 be constructed as follows. Each of subsets A, B, and C contains n+1 nodes. Let <A> denote the subgraph specified by the set A. In C $d(w_0) = nc'$ and $d(w_i) = (n-1) + c' + b'$ for $1 \leq i \leq n$. In other words <C - w_0> is 1.0-complete and <C> is c'-complete. In B $d(v_0) = n+1$ and $d(v_i) = n + (n-1) + a' + b'$ for $1 \leq i \leq n$. <B> is 1.0-complete. In A $d(u_0) = n+1$ and $d(u_i) = n + (n-1) + a'$ for $1 \leq i \leq n$. <A> is 1.0-complete. To make clearer the connections between subsets should be mentioned. Each w_i is connected to v_i with fuzzy value b' for $1 \leq i \leq n$. And each $u_i (i \neq 0)$ is connected to v_i with fuzzy value a' and to v_j's ($j \neq i$), 0) with fuzzy value 1.0. Finally u_0 is connected to v_0 with fuzzy value 1. All other fuzzy values in the fuzzy graph G are set to 0.

Now we will show that G thus constructed satisfies the conditions imposed.

(1) From the process of the construction described above it is clear that $\delta(G) = d(w_0) = nc' = c$.

(2) The number of edges in any cut of subgraphs <A>, <B> or <C> is greater than or equal to n since <A>, <B> and <C> are c'-complete. Therefore the weight of a cut is greater than or equal to nc', which means that the weight of any cut which contains a cut of <A>, <B> or <C> is greater than or equal to nc'. Only other cuts which do not contain a cut of <A>, <B> or <C> must contain the cut $(A, B \cup C)$ or $(A \cup B, C)$. The weight of the cut $(A, B \cup C)$ is $1 + n(n-1) + na'$ and that of the cut $(A \cup B, C)$ is nb'.

$nb' \leq nc'$.

$nb' \leq 1 + n(n-1) + na'$.

Hence $\lambda(G) = nb' = b$.

(3) Let us determine the minimum number of nodes in disconnection of G. Since <A>, <B> and <C> are at least

p

c'-complete, they can be disconnected or become a single node by removing at least n nodes. Only other possible ways to disconnect G are disconnections between A, B and C. Since $<(A - u_0) \cup (b - v_0)>$ is a'-complete and u_0 and v_0 are connected to each other and to $<(A - u_0) \cup (B - v_0)>$, any disconnection must contain at least n+1 nodes. On the other hand since <B> and <C> are connected by n edges, at least n nodes have to be removed to disconnect $<A \cup B>$ and <C>. But since nodes on both sides of edges are all different, at least n nodes have to be removed. Therefore at least n nodes have to be removed to disconnect the graph G. Then since $\min_{v \in V} \{f(v)\} = a'$ and actually $\{v_1, v_2, \ldots, v_n\}$ is a disconnection of G, the weight of the disconnection $\{v_1, v_2, \ldots, v_n\}$ specifies the node connectivity of the graph G, namely,

$$\Omega(G) = na' = a.$$

V. APPLICATION TO CLUSTERING ANALYSIS

The usual graph-theoretical approahces to clustering analysis involve first obtaining a threshold graph from a fuzzy graph, and then applying various techniques to obtain clusters as maximal components under different connectivity considerations. These methods have a common weakness, namely, the weight of edges are not treated fairly in that any weight greater (less) than the threshold is treated as 1(0). In this section, we will extend these techniques to fuzzy graphs. It will be shown that the fuzzy graph approach is more powerful.

In table 1 in the following, we provide a summary of various graph theoretical techniques for clustering analysis. This table is a modification of table II in Matula [10].

Cluster Procedure	Graph Theoretical Interpretation of Clusters	Cluster Independence	Extent of Chaining
Single Linkage	Maximal Connected Subgraphs	Disjoint	High
K-linkage	Maximal Connected Subgraphs of Minimum Degree k	Disjoint	Moderate
k-edge Connectivity	Maximal k-edge Connected Subgraph	Disjoint	Low
k-vertex Connectivity	Maximal k-vertex Connected Subgraph and Cliques on k or Less Vertices	Limited Overlap	Low
Complete Linkage	Cliques	Considerable Overlap	None

Table 1 - Cluster Procedures and Their Corresponding Graph Theoretic Characterizations and Properties

In the following definition, clusters will be defined based on various connectivities of a fuzzy graph.

Definition 18: Let $G = [V,R]$ be a symmetric fuzzy graph. A cluster of type i is defined by condition (i), $i = 1,2,3,4$, in the following.

(1) maximal ε-connected subgraphs, for some $0 < \varepsilon \leq 1$.

(2) maximal τ-degree connected subgraphs.

(3) maximal τ-edge connected subgraphs.

(4) maximal τ-vertex connected subgraphs.

It follows from the previous definition that clusters of type (1), (2) and (3) are hierarchical with different ε and τ, whereas clusters of type (4) are not due to the fact τ-vertex compoenents need not be disjoint.

It is also easily seen that all clusters of type (1) can be obtained by the single-linkage procedure. The difference between the two procedures lies in the fact that ε-connected subgraphs can be obtained difectly from R_G by at most n-1 matrix multiplication (where n is the rank of M_G), whereas in the single-linkage procedure, it is necessary to obtain as many threshold graphs as the number of distinct fuzzy values in the graph.

In the following, we will show that not all clusters of types 2, 3 and 4 are obtainable by procedures of k-linkage, k-edge connectivity, and k-vertex connectivity, respectively.

Example 4: Let G be a symmetric fuzzy graph given in figure 3a. The dendrogram in figure 3b indicates all the clusters of type 2. It is easily seen from the threshold graphs of G that the same dendrogram cannot be obtained by the k-linkage procedure. Those for k=1 and 2 are given in figures 4a and 4b, respectively.

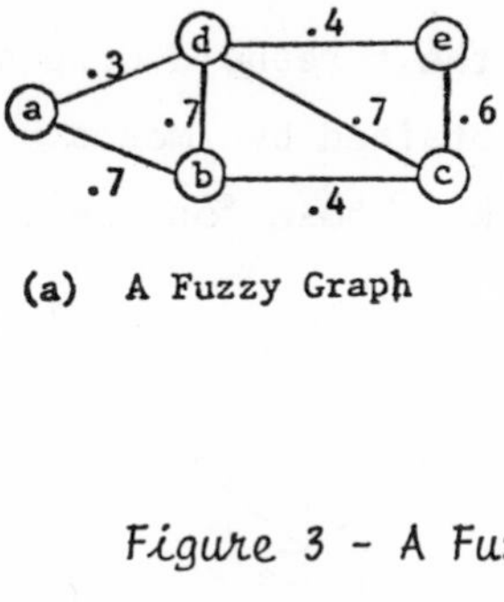

(a) A Fuzzy Graph

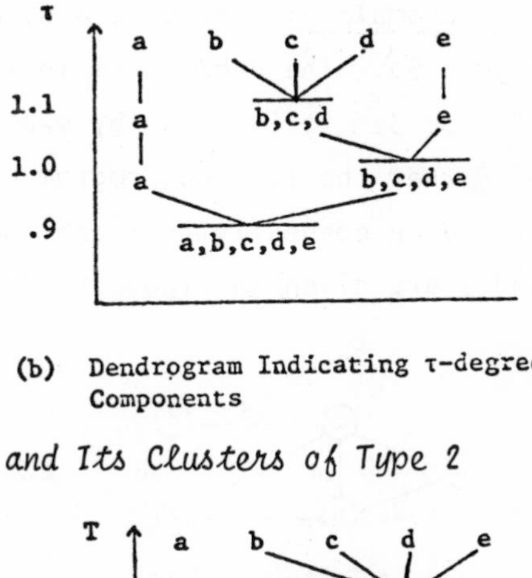

(b) Dendrogram Indicating τ-degree Components

Figure 3 - A Fuzzy Graph and Its Clusters of Type 2

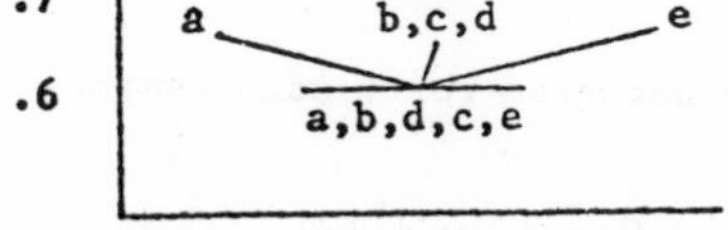

(a) k = 1

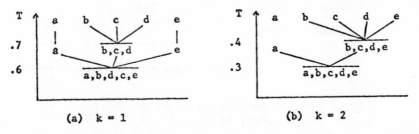

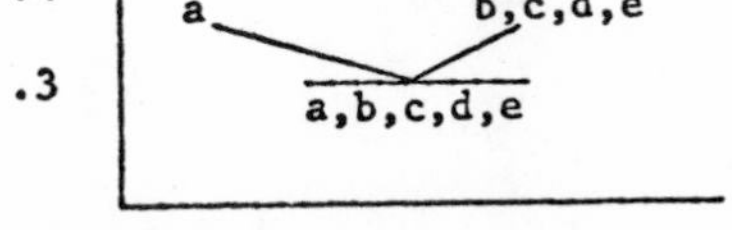

(b) k = 2

Figure 4 - Dendrograms for Clusters Obtained by k-linkage Method for k = 1 and 2

Theorem 9: The τ-degree connectivity procedure for the construction of clusters is more powerful than the k-linkage procedure.

Proof: In light of example 4 above, it is sufficient to show that all clusters obtainable by k-linkage procedure is also obtainable by τ-degree connectivity procedure for some τ.

Let G be a symmetric fuzzy graph. For $0 < \varepsilon \leq 1$, let G' be a graph obtained from G by replacing these weights less than ε in G by 0. For any k used in k-linkage procedure, set $\tau = k\varepsilon$. It is easily seen that a set is a cluster obtained by applying k-linkage procedure to G iff it is a cluster obtained by applying the τ-degree connectivity procedure to G'.

Example 5: Let G be a symmetric fuzzy graph given in figure 5a. The dendrogram in figure 5b gives all clusters of type 3. It is clear by examining all the threshold graphs of G that the same dendrogram cannot be obtained by means of k-edge connectivity technique for any k. Those for k = 1 and 2 are given in figure 6.

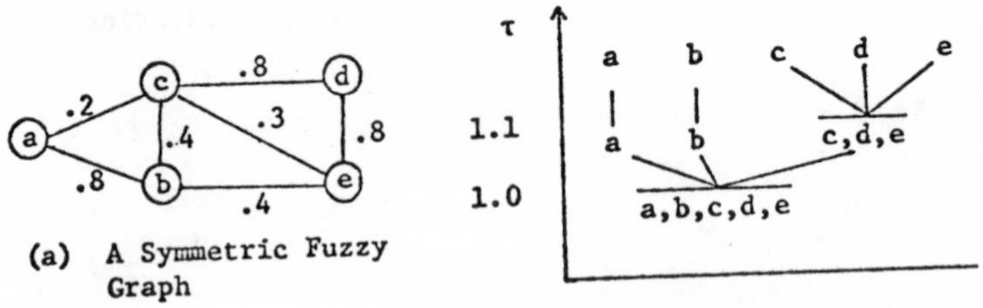

(a) A Symmetric Fuzzy Graph

(b) Dendrogram for τ-edge Component

Figure 5 - A Symmetric Fuzzy Graph and Its Clusters of Type 3

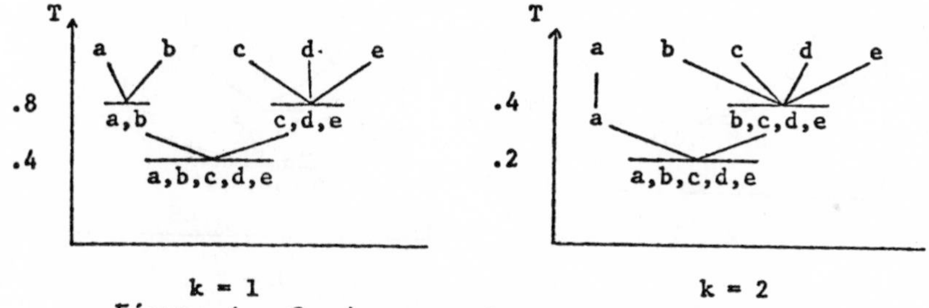

Figure 6 - Dendrograms for Clusters Obtained from k-edge method for k = ℓ and 2.

By example 5 and following the same proof procedure as in theorem 8, we have the following result.

Theorem 10. The τ-edge connectivity procedure for the construction of clusters is more powerful than the k-edge connectivity procedure.

Example 6: Let G be a symmetric fuzzy graph given in figure 7a. The dendrogram in figure 7b provides all clusters of type 4. It is easily seen that the same dendrogram cannot be obtained by means of k-vertex connectivity technique for any k. Those for k = 1 and 2 are given in figure 8.

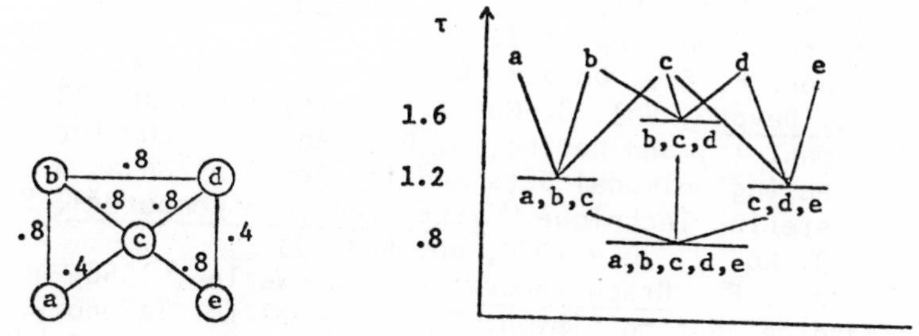

(a) A Fuzzy Graph

(b) Dendrogram for τ-vertex Components

Figure 7 - A Symmetric Fuzzy Graph and Its Clusters of Type 4

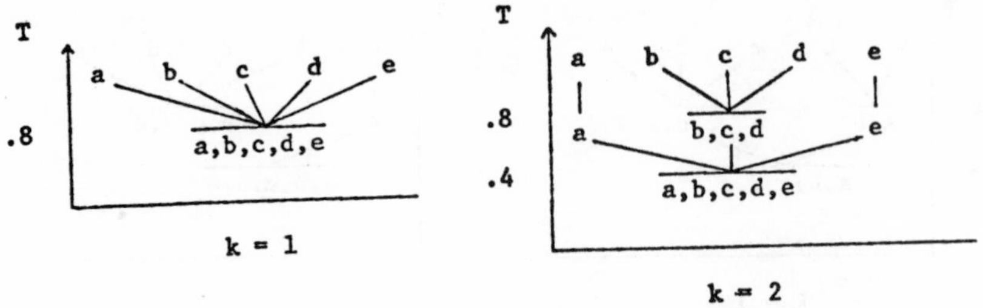

Figure 8 - Dendrograms for Clusters Obtained from k-vertex Method for k = 1 and 2.

Following the same proof procedure as in theorems 8 and 9, we conclude with the result below.

Theorem 11. The τ-vertex connectivity procedure for the construction of clusters is more powerful than the k-vertex connectivity procedure.

REFERENCES

1. Augustson, J. G., and Minker, J., "An Analysis of Some Graph Theoretical Cluster Techniques," JACM, Vol. 17, No. 4 , October 1970, pp. 571-588.
2. Bellman, R., Kalaba, R., and Zadeh, L., "Abstraction and Pattern Classification," J. Math. Anal. Appl., Vol. 13 January 1966, pp. 1-7.
3. Bonner, R. E., "On Some Clustering Techniques," IBM J. Res. Devpt., Vol. 8, No. 1, January 1964, pp. 22 - .
4. Gitman, I., and Levine, M. D., "An Algorithm for Detecting Unimodal SEts and Its Application as a Clustering Technique," IEEE Trans. on Computers, Vol. C-19, No. 7, July 1970, pp. 583-593.
5. Harary, F., Graph Theory, Addison-Wesley, 1968.
6. Jardine, N., and Sibson, R., "A Model for Taxonomy," Mathematical Biosciences, Vol. 2, 1968, pp. 465-482.
7. Jardine, N., and Sibson, R., "The Construction of Hierarchic and Non-hierarchic Classifications," The Computer J., Vol. 11, 1968, pp. 177-184.
8. Lance, G. N., and Williams, W. T., "A General Theory of Classificatory Sorting Strategies, 1. Hierarchical Systems," The Computer J., Vol. 9, No. 4, 1967, pp. 373-380.

9. Ling, R. F., "On the Theory and Construction of k-cluster," The Computer J., Vol. 15, No. 4, 1972, pp. 326-332.
10. Matula, D. W., "Cluster Analysis via Graph Theoretic Techniques," Proc. of Louisiana Conf. on Combinatrics, Graph Theory, and Computing, March 1970, pp. 199-212.
11. Matula, D. W., "k-components, Clusters, and Slicings in Graphs," SIAM J. Appl. Math., Vol. 22, No. 3, May 1972, pp. 459-480.
12. Nance, R. E., Korfhage, R. R., and Bhat, U. N., "Information Networks: Definitions and Message Transfer Models," Tech. Report CP-710011, July 1971, Computer Science/Operation Research Center, SMU, Dallas, Texas.
13. Ramamoorthy, C. V., "Analysis of Graphs by Connectivity Considerations," JACM., Vol. 13, No. 2, April 1966, pp. 211-222.
14. Sibson, R., "Some Observations on a Paper by Lance and Williams," The Computer J., Vol. 14, No. 2, 1971, pp. 156-157.
15. Tamura, S., Higuchi, S., and Tanaka, K., "Pattern Classification Based on Fuzzy Relations," IEEE Trans. on Systems, Man, and Cybernetics, Vol. SMC-1, No. 1, January 1971, pp. 61-66.
16. Yeh, R. T., "Toward an Algebraic Theory of Fuzzy Relational System," Tech. REport 25, Dept. of Computer Science, The Univesrity of Texas, July 1973.
17. Zadeh, L. A., "Fuzzy SEts," Infor. Contr., Vol. 8, June 1965, pp. 338-353.
18. Zadeh, L. A., "Similarity Relations and Fuzzy Orderings," Information Sci., Vol. 3, 1971, pp. 177-200.

CONDITIONAL FUZZY MEASURES AND THEIR APPLICATIONS

T. Terano and M. Sugeno
Tokyo Institute of Technology
Tokyo, Japan

1. INTRODUCTION

The concept of fuzzy sets [1, 2, 3] suggested by Prof. Zadeh gave us a powerful tool by which we can treat many complicated problems of human behavior . Comparing a fuzzy set with an ordinary set, one of the authors defined a set function "fuzzy measure" and a functional "fuzzy integral" in the former papers [4, 5]. And they examined analytically and experimentally the effectiveness of these concepts for the macroscopic evaluation of some fuzzy phenomena with fuzzy criterion [6].

In this paper, they introduce a set function "conditional fuzzy measure" and a relation between "a priori and a posteriori fuzzy measures". These are very useful for describing any kind of transition of fuzzy phenomena such as communication of rumors, the reprint of color photograph, and the development of human abilities by education. The common characteristics of these phenomena are that the measures describing their status are very vague, and that their transitions are influenced greatly by the subjectivity of the people who are involved in such fuzzy phenomena.

It is well known that the uncertainty of human behavior is sometimes conveniently expressed by the subjective probability (judgemental probability) in Bayesian statistics. But we can not say that this is the best way of expressing the fuzziness unless the problem is directly related to decision making.

Since the fuzzy measure is an extension of the probability measure and its physical meaning is similar to the membership function of a fuzzy set, the models presented in this paper are more suitable for describing the transition of fuzzy phenomena generally.

NOMENCLATURE

X, Y	:	arbitrary set
ϕ	:	empty set
x, y	:	elements, $x \in X$, $y \in Y$
$\mathcal{B}$	:	Borel field
$g(\cdot)$	:	fuzzy measure of $(X,\mathcal{B})$
h	:	mapping from X to $[0,1]$
$P(\cdot)$	:	probability measure of $(X,\mathcal{B})$
$\sigma_Y(\cdot\|x)$	:	conditional fuzzy measure of Y with respect to x
$g_X(\cdot)$	:	fuzzy measure of X (a priori fuzzy measure)
$g_Y(\cdot)$	:	fuzzy measure of Y
$\sigma_X(\cdot\|y)$	:	a posteriori fuzzy measure of X
$\{\ \}$	:	set including only one element x.
g^i	:	fuzzy density
$F(\lambda)$	:	spectrum of light
λ	:	$-1 < \lambda < \infty$ or wave length
$u(\lambda)$, $v(\lambda)$, $w(\lambda)$	:	tristimulus values of a spectrum color
$F_R(\lambda)$, $F_G(\lambda)$, $F_B(\lambda)$	:	spectrum colors selected for standard colors Red, Green, Blue
$t(\lambda)$	:	absorption coefficient of filter.

2. FUZZY MEASURES AND INTEGRALS

2.1 FUZZY MEASURES

Ordinary measures in the theory of Lebesgue integrals have additivity. Here the authors consider a measure as a set function with monotonicity but not always with additivity. The idea of fuzzy measures can be shortly expressed in three statements. These are very similar to the axioms of probability measure. The concept of "grade" is used instead of probability in the statements. Since "grade" is also used in fuzzy sets theory, we will easily understand the connotation.

Let X be an arbitrary set and ϕ be an empty set. A small letter x denotes a member of X. The first statement is: "The grade of $x \in X$ equals unity and that of $x \in \phi$ equals 0". Let A and B be subsets of X. Then, the second statement is: "If $A \subset B$, then the grade of $x \in B$ is larger than that of $x \in A$"†. The last statement will be seen in the definition of fuzzy measures.

Let $\mathcal{B}$ be a Borel field of X.

[Definition 1] A set function $g(\cdot)$ with the following properties is called a fuzzy measure.

1) $g(\phi) = 0.\ g(X) = 1.$
2) If $A, B \in \mathcal{B}$ and $A \subset B$, then $g(A) \leq g(B)$.
3) If $F_n \in \mathcal{B}$ and F_n is a monotone sequence, then
$$\lim_{n\to\infty} g(F_n) = g(\lim_{n\to\infty} F_n)$$

In the above definition, 1) means boundedness and nonnegativity, 2) monotonicity, and 3) continuity. If X is a finite set, the condition 3) can be omitted.

---

† The concept of the grade of $x \in A$ never implies that A is a fuzzy set. It can be interpreted in comparison with the probability of $x \in A$.

Let $h: X \to [0,1]$ and $F_\alpha = \{x | h(x) \geq \alpha\}$. If $F_\alpha \subset \mathcal{B}$ for and $\alpha \in [0,1]$, then h is called a $\mathcal{B}$-measurable function. A triplet $(X, \mathcal{B}, g)$ is called a fuzzy measure space.

In an ordinary probability space $(X, \mathcal{B}, P)$, P has the following proerties.

1) $0 \leq P(E) \leq 1$ for all $E \in \mathcal{B}$, particularly, $P(X) = 1$.

2) If $E_n \in \mathcal{B}$ for $1 \leq n < \infty$ and $E_i \cap E_j = \phi$ for $i \neq j$

$$\text{then } P(\sum_{n=1}^{\infty} E_n) = \sum_{n=1}^{\infty} P(E_n)$$

Now, P is one of the fuzzy measures because the monotonicity and the continuity of P can be derived from the above properties. In other words, the fuzzy measure is a set function obtained by loosing some properties of the probability measures.

We may use a probability measure with additivity to measure "fuzzy phenomena". However, it is quite doubtful if there exists "additivity" behind "fuzziness" in the phenomena where human subjectivity plays an important role. We might as well expect that we may be able to measure "fuzziness", if we construct a theory of the measures without additivity.

Monotonicity is a quite natural condition. It is not necessary that $g(A \cup B) = g(A) + g(B)$ where $A \cap B = \phi$, if the additivity is thrown away. It would be considered that a fuzzy measure is a means for measuring "fuzziness", while a probability measure is one for measuring "randomness". In this sense, fuzzy measures can be regarded as subjective scales by which "grade of fuzziness" is measured.

2.2 FUZZY INTEGRALS

Now, by using fuzzy measures, let us define fuzzy integrals which are very similar to Lebesgue integrals. Let h be a $\mathcal{B}$-measurable function.

[Definition 2] The fuzzy integral of h over A with respect to g is defined as follows:

$$⨍_A h(x) \circ g(\cdot) = \sup_{\alpha \in [0.1]} [\alpha \wedge g(A \cap F_\alpha)], \ F_\alpha = \{x | h \geq \alpha\}. \qquad (2.1)$$

The symbol ⨍ is an integral with a small bar. It is also a symbol of letter f. The small circle is the symbol of the composition used in the fuzzy sets theory. For simplification, the fuzzy integral over X is written as $⨍ h \circ g$. The fuzzy integrals can be also called fuzzy expectations in comparison with the probabilistic counterparts. They have the following properties.

1) $0 \leq - h \circ g(\cdot) \leq 1$ (2.2)

Let $a \in [0,1]$, then

2) $⨍(a \vee h) \circ g(\cdot) = a \vee ⨍h \circ g(\cdot)$, (2.3)

3) $⨍(a \wedge h) \circ g(\cdot) = a \wedge ⨍h \circ g(\cdot)$. (2.4)

4) $⨍(h_1 \vee h_2) \circ g(\cdot) \geq ⨍h_1 \circ g(\cdot) \vee ⨍ h_2 \circ g(\cdot)$ (2.5)

5) $⨍(h_1 \wedge h_2) \circ g(\cdot) \leq ⨍ h_1 \circ g(\cdot) \wedge ⨍ h_2 \circ g(\cdot)$ (2.6)

6) $⨍_{E \cup F} h \circ g(\cdot) \geq ⨍_E h \circ g(\cdot) \vee ⨍_F h \circ g(\cdot)$ (2.7)

7) $⨍_{E \cap F} h \circ g(\cdot) \leq ⨍_E h \circ g(\cdot) \wedge ⨍_F h \circ g(\cdot)$ (2.8)

Let $h_A(x)$ be a membership function of a fuzzy set A. Then the fuzzy measure of A is defined as follows:

$$g(A) = ⨍ h_A(x) \circ g(\cdot). \qquad (2.9)$$

The fuzzy integral over A is defined as

$$⨍_A h(x) \circ g(\cdot) = ⨍[h_A(x) \wedge h(x)] \circ g(\cdot). \qquad (2.10)$$

Here, the properties 6) and 7) also hold.

Let $\{h_i\}$ be a monotonically decreasing sequence of $\mathcal{B}$-measurable function and $\{a_i\}$ a monotonically increasing sequence of real numbers, then $\bigvee_{i=1}^{n} [a_i \wedge h_i]$ is also $\mathcal{B}$-measurable and there holds

8) $⨍[\bigvee_{i=1}^{n} (a_i \wedge h_i)] \circ g(\cdot) = \bigvee_{i=1}^{n} [a_i \wedge ⨍h_i \circ g(\cdot)]$. (2.11)

If $\{h_n\}$ is a monotone sequence of $\mathcal{B}$-measurable functions and $\lim_{n\to\infty} h_n = h$, then h is $\mathcal{B}$-measurable and

9) $\lim_{n\to\infty} \int h_n \circ g(\cdot) = \int h \circ g(\cdot)$. (2.12)

Let P be a probability measure. Then both integrals, Lebesgue and fuzzy, are defined with respect to P.

10) $|\int h(x)\, dP - \int h(x) \circ P(\cdot)\ | \leq 1/4$ (2.13)

Since the operations of fuzzy integrals include only comparisons. This inequality means that without using addition, we can obtain, b a fuzzy integral, the value different by at most 1/4 from the probabilistic expectation.

2.3 CONDITIONAL FUZZY MEASURES

[Definition 3] Let $\sigma_Y(\cdot|x)$ be a fuzzy measure of Y for any fixed $x \in X$.† Then it is called a condition fuzzy measure of Y with respect to X.

A fuzzy measure $g_Y(\cdot)$ of Y is constructed by $\sigma_Y(\cdot|x)$ and $g_X(\cdot)$ as follows:

$$g_Y(F) = \int_X \sigma_Y(F|x) \circ g_X(\cdot) \tag{2.14}$$

We have

$$\int_Y h(y) \circ g_Y = \int_X [\int_Y h(y) \circ \sigma_Y(\cdot|x)] \circ g_X \tag{2.15}$$

Analogously, we can consider $\sigma_X(\cdot|y)$. There holds the next relation between $\sigma_Y(\cdot|x)$ and $\sigma_X(\cdot|y)$.

$$\int_F \sigma_X(E|y) \circ g_Y = \int_E \sigma_Y(F|x) \circ g_X \tag{2.16}$$

The above equation corresponds to Bayes' theorem. In this sense, g_X is called a priori fuzzy measure and $\sigma_X(\cdot|y)$ a posteriori fuzzy measure.

When X and Y are finite sets, a posteriori fuzzy measure can be easily calculated from Eq. (2.16). Let $F = \{y_j\}$ where

---

†Note: More precisely, $\sigma_Y(\cdot|x)$ is arbitrary on a set E such that $g_X(E) = 0$ where $E \subset X$ and g_X is a fuzzy measure of X.

$y_j \in Y$. Then we obtain

$$\sigma_X(E|y_j) \wedge g_Y(\{y_j\}) = f_E\, \sigma_Y(\{y_j\}|x) \circ g_X. \tag{2.17}$$

From this follows that

(1) if $g_Y(\{y_j\}) \geq f_E\, \sigma_Y(\{y_j\}|x) \circ g_X$

$$\sigma_X(E|y_j) = f_E\, \sigma_Y(\{y_j\}|x) \circ g_X, \tag{2.18}$$

(2) if $g_Y(\{y_j\}) = f_E\, \sigma_Y(\{y_j\}|x) \circ g_X$

$$\sigma_X(E|y_j) \geq f_E\, \sigma_Y(\{y_j\}|x) \circ g_X. \tag{2.19}$$

Note that there holds always $g_Y(\{y_j\}) \geq f_E\, \sigma_Y(\{y_j\}|x) \circ g_X$. In the case (2), $\sigma_X(E|y_j)$ is not uniquely determined. We can, for instance, let $\sigma_X(E|y_j)$ be unity.

2.4 CONSTRUCTION OF FUZZY MEASURES

For simplification, assume that X is a finite set $\{x_1, x_2, \ldots, x_n\}$. Let us consider how to construct a fuzzy measure of a fuzzy measure space $(X, 2^X, g)$.
Let $0 \leq g^i \leq 1$ for $1 \leq i \leq n$. Here g^i is called a fuzzy density. Next let

$$\frac{1}{\lambda}\left[\prod_{i=1}^{n} (1 + \lambda g^i) - 1\right] = 1, \text{ where } -1 < \lambda < \infty. \tag{2.20}$$

Define for $E \subset X$

$$g_\lambda(E) = \frac{1}{\lambda}\left[\prod_{x_i \in E} (1 + \lambda g^i) - 1\right] \tag{2.21}$$

Then g_λ satisfies all consitions of a fuzzy measure.
From the definition, we obtain that

$$g_\lambda(\{x_i\}) = g^i \text{ for } 1 \leq i \leq n \tag{2.22}$$

and that if $A \cap B \neq \phi$, then

$$g_\lambda(A \cup B) = g_\lambda(A) + g_\lambda(B) + \lambda g_\lambda(A)\, g_\lambda(B). \tag{2.23}$$

When $\lambda = 0$, g_λ becomes additive and hence, equal to a probability measure. It follows from Eq. (2.23) that if $\lambda \leq 0$, then

$$g_\lambda(A \cup B) \leq g_\lambda(A) + g_\lambda(B) \tag{2.24}$$

and if $\lambda > 0$, then

$$g_\lambda(A \cup B) > g_\lambda(A) + g_\lambda(B) \tag{2.25}$$

Now let us consider how to calculate a fuzzy integral in $(X, 2^X, g)$. Let $h : X \to [0,1]$. Assume $h(x_1) \leq h(x_2) \leq \cdots \leq h(x_n)$. If not, rearrange $h(x)$ in an increasing order.
Define

$$F_i = \{x_i, x_{i+1}, \cdots, x_n\}, \text{ for } 1 \leq i \leq n.$$

Then we obtain from Definition 2

$$\int_X h(x) \circ g(\cdot) = \bigvee_{i=1}^{n} [h(x_i) \wedge g(F_i)]. \tag{2.26}$$

There exists at least one j such that

$$h(x_{j-1}) \wedge g(F_{j-1}) \leq h(x_j) \wedge g(F_j)$$

$$h(x_j) \wedge g(F_j) \geq h(x_{j+1}) \wedge g(F_{j+1}).$$

Clearly, there holds for this j

$$\int_X h(x) \circ g(\cdot) = h(x_j) \wedge g(F_j).$$

Thus, we obtain the value of a fuzzy integral without evaluating $h(x_i) \wedge g(F_i)$ for all i's.

3. TRANSITION OF FUZZY PHENOMENA

Using the concept of the conditional fuzzy measure, two types of models are considered as for the transition of fuzzy phenomena. The first model corresponds to Markov process Eq. (2.14). Here, the input is a fuzzy measure g_X and the output g_Y is the fuzzy integration of a conditional fuzzy measure $\sigma_Y(\cdot|x)$ respect to g_X as shown in Fig. 1. For example, supposing g_X is the measure of a gossip told to a person, g_Y is the measure of the gossip modified by his subjectivity σ_Y.

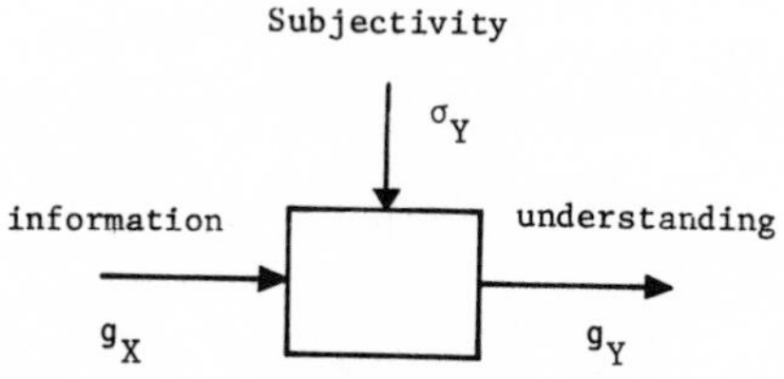

Figure 1. Model I

The second model corresponds to Bayes' theorem Eq. (2.16). The input is a priori fuzzy measure g_X as above, but the output is a posteriori fuzzy measure $\sigma_X(\cdot|y)$ as shown in Fig. 2. One of the explanations of this model is that one's preconception g_X is improved to σ_X by an information σ_Y. The example of this model is the ability of evaluation, judgement and diagnosis improved by experience, information or learning.

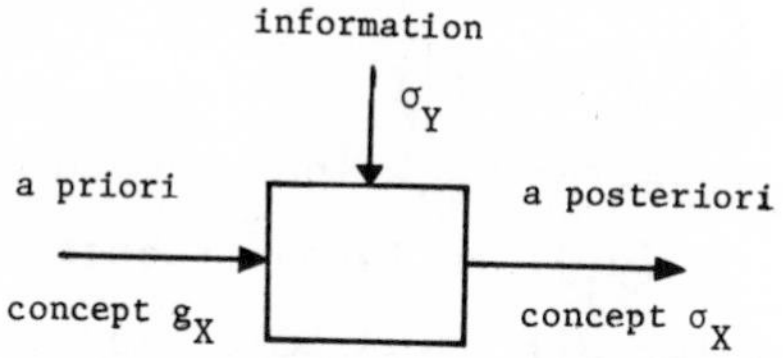

Figure 2. Model II

Now some numerical examples are shown. We assume that X and Y are finite sets each including only three elements. In the first model, if $\sigma_Y(\{\cdot\}|x)$ is independent of x, the output $g_Y(\cdot)$ is always equal to $\sigma_Y(\{\cdot\}|x)$. The result is almost the same when the trends of σ_Y are not so different with different x's, as shown in Fig. 3. This is analogous to the fact that the hardheaded people do not change their opinion by outside information.

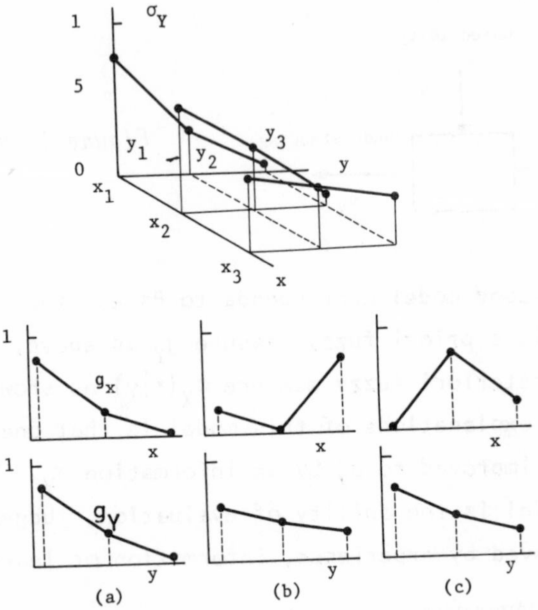

Figure 3. Example 1 of Model I

Since $g_Y(\{y\})$ is, generally speaking, a weighted mean of $\sigma_X(\{y\}|x)$ of which weight is g_X, $g_Y(\{y_j\})$ is large if both $g_X(\{x_i\})$ and $\sigma_Y(\{y_j\}|x_i)$ are large for all i's. Some examples are shown in Fig. 4.

We can see that g_Y becomes flat usually. Therefore, some amplifiers might be added when this model is applied to the actual transition problems.

A fuzzy transition phenomenon may be represented by a sequence of models or steps each of which is similar to one given in Fig. 1. The outputs of these steps converge rapidly to a certain value as the number of steps increase.

In the second model, the output is sometimes amplified as explained in Section 2.3. This corresponds to the case when a complete information is given in Bayesian statistics. In this case a posteriori probability becomes unity. In our model, however, $\sigma_X(\cdot|y_i)$ can take any values between $g_Y(\{y_j\})$

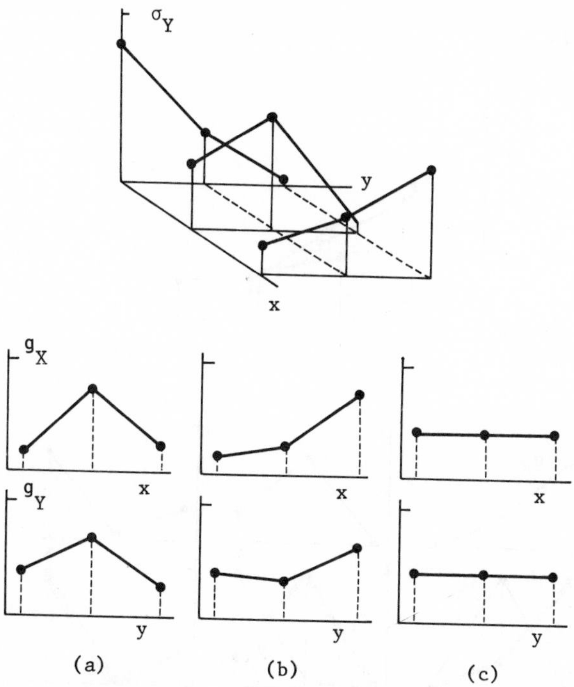

Figure 4. Example 2 of Model I

and 1 as shown in Eq. (2.19). Therefore, we can say that the information is not complete but comparatively reliable. When $\sigma_Y(\{y_j\}|x_i)$ does not change much with i as in Fig. 5, and y_1 is given as an information, the output σ_X is similar to the imput g_X. This is because $\sigma_Y(\{y_1\}|x_i)$ is large, or in other words, y_1 is an information telling us that something likely to happen has happened. This information is not valuable so that g_X is not modified. On the other hand, if y_3 is given, the input is affected greatly. This means that y_3 is very valuable because of its rareness.

Figs. 6 and 7 show other examples. In Fig. 6, the convex input is changed to a concave output and vice versa. This means that a priori fuzzy measure is strongly influenced by an effective information. The information is thought to be effective if $\sigma_Y(\{y_j\}|x_2)$ is large (small) when $g_X(\{x_2\})$ becomes small (large).

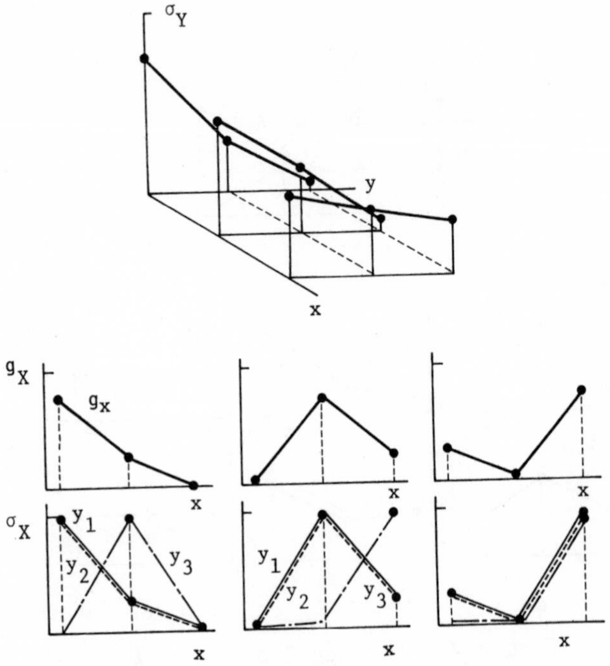

Figure 5. Example 1 of Model II

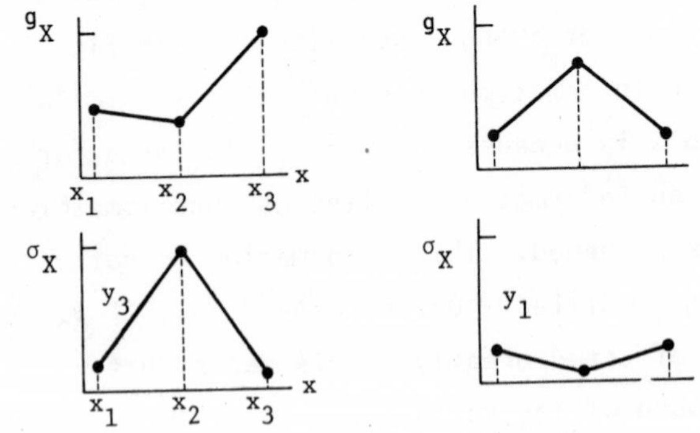

(a) (b)

Figure 6. Example 2 of Model II

In Fig. 7, some parts of the input are amplified. This can be considered as a case when one's preconceived opinion is intensified by same kind of external information.

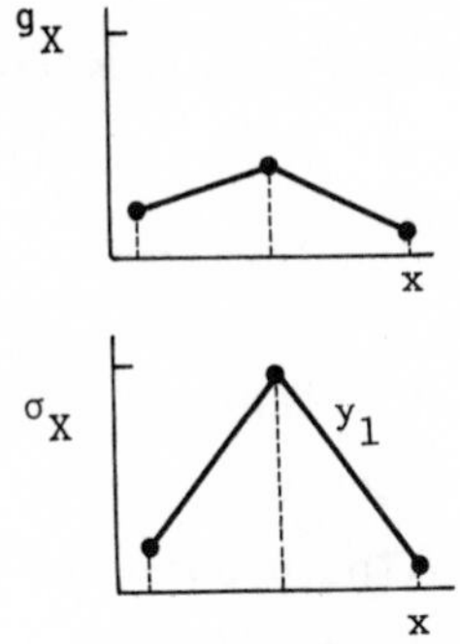

Figure 7. Example 3 of Model II

When g_X and σ_Y are not so small, σ_X becomes slightly flat but not so different from g_X. This corresponds to the case when a fair man gets an incomplete information.

4. APPLICATIONS

4.1 FILTERING OF COLOR

The color of light is changed when the light is filtered through a colored glass. The color of photograph by a photochemical reaction is also a kind of transition of color.

Let $F(\lambda)$ is the spectrum of a light source, where λ is a wave length. The color of $F(\lambda)$ is called "spectrum color". According to the theory of color [8], our feeling of color is said to be expressed by the trichromatic coefficients of light (u, v, w).

$$u = \frac{U}{U + V + W}, \quad v = \frac{V}{U + V + W}, \quad w = 1 - u - v, \tag{4.1}$$

where $U = \int \bar{u}(\lambda)F(\lambda)d\lambda$, $V = \int \bar{v}(\lambda)F(\lambda)d\lambda$, $W = \int \bar{w}(\lambda)F(\lambda)d\lambda$
Here, $\bar{u}(\lambda)$, $\bar{v}(\lambda)$ and $\bar{w}(\lambda)$ are the tristimulus values of a spetrum color defined by International Commission on Illumination. Consequently, a color is represented as a point on u-v plane. Since the characteristic of a filter is expressed by the

absorption coefficient $t(\lambda)$, $F(\lambda)$ is changed to $F'(\lambda) = t(\lambda) F(\lambda)$, when the light is filtered.

Now $F(\lambda)$ is formed by mixing three primary spectrum colors $F_R(\lambda)$, $F_G(\lambda)$ and $F_B(\lambda)$ which form day light if they are mixed equally.

$$F(\lambda) = aF_R(\lambda) + bF_G(\lambda) + cF_B(\lambda) \qquad (4.2)$$

When $F(\lambda)$ is filtered, the trichromatic coefficients u', v' w' of $F'(\lambda)$ are calculated from Eq. (4.1). Let us consider the problem of determining filtered color. The problem is solved, if a', b' and c' are identified so that $F''(\lambda)$ shows the same color as $F'(\lambda)$.

$$F''(\lambda) = a'F_R(\lambda) + b'F_G(\lambda) + c'F_B(\lambda) \qquad (4.3)$$

The basic spectra of colors by which we form an arbitrary light is shown in Fig. 8. Fig. 9 shows the characteristic of a filter used for the calculation. The values of a, b, c and a', b', c' are calculated as shown in Table 1. From this table, we can see that the filter emphasizes green.

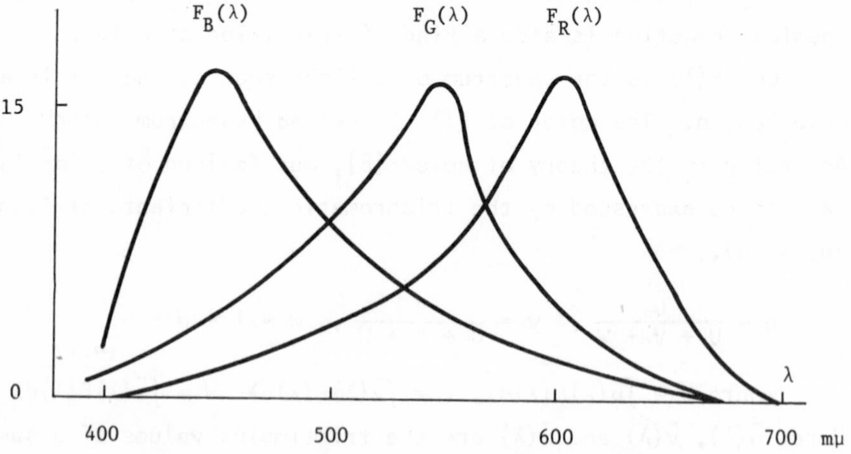

Figure 8. Spectrum Colors: Red, Green, Blue

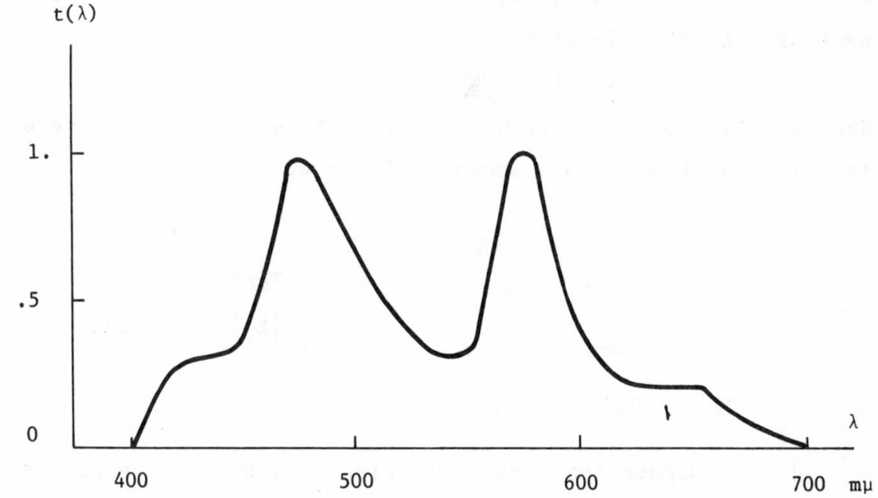

Figure 9. Filter Characteristics

a	b	c	a'	b'	c'
0.33	0.33	0.33	0.24	0.42	0.32
0.60	0.30	0.10	0.49	0.41	0.11
0.60	0.10	0.30	0.47	0.27	0.26
0.30	0.60	0.10	0.25	0.60	0.15
0.30	0.10	0.60	0.20	0.27	0.53
0.10	0.60	0.30	0.08	0.60	0.32
0.10	0.30	0.60	0.04	0.40	0.56

Table 1 Input and Output Light of Filter

This transition can be expressed by the first model of conditional fuzzy measure. Let $g_X(\cdot)$ be a fuzzy measure which represents the mixing ratio of light sources, where X consists of x_1(red), x_2(green) and x_3(blue).

In this example, we assume that $g_X(\cdot)$ has additivity. Assuming that Y also consists of three axes of red, green, blue for the filtered light, the mixing ratio of three colors is expressed by the equation

$$g_Y(F) = \int_X \sigma_Y(F|x) \circ g_X.$$

The densities of $\sigma_Y(\cdot|x)$, which have the same filtering characteristics as Fig. 9, are shown in Table 2.

	y_1	y_2	y_3
x_1	0.50	0.40	0
x_2	0.10	0.60	0.10
x_3	0.05	0.30	0.55

Table 2
$\sigma_Y(\{y_j\}|x_i)$ *of Filter*

If we suppose that the coefficients a, b, c correspond to the densities of g_X, we can claculate the densities of g_Y as shown in Table 3. Comparing Table 1 with Table 3, we can see that the process of color change when light is filtered is well simulated by the first model. This example is rather physical, so the calculation can be carried out without fuzziness. But in analysing the transition of the more sensible variables, such as contrast and tone of color, the merit of this model will be seen more saliently.

g_X^1	g_X^2	g_X^3	g_Y^1	g_Y^2	g_Y^3
0.33	0.33	0.33	0.33	0.40	0.33
0.60	0.30	0.10	0.50	0.40	0.10
0.60	0.10	0.30	0.50	0.30	0.30
0.30	0.60	0.10	0.30	0.60	0.10
0.30	0.10	0.60	0.30	0.40	0.55
0.10	0.60	0.30	0.10	0.60	0.30
0.10	0.30	0.60	0.10	0.40	0.55

Table 3 Calculated Densities of g_X *and* g_Y *by Model II*

4.2 EVALUATION OF HOUSES

When a person evaluates a house, he considers many factors simultaneously such as equipment, space, cost, environment, shopping, etc. The weights of these factors may differ from person to person, and some parts of the factors may be redundant. Moreover, these weights are being refined according with his experience or learning.

The authors already pointed out [6,7] that the fuzzy integral is a powerful tool for the synthetic evaluation of fuzzy systems. Now, they apply the second model of the conditional fuzzy measure to this learning process.

As the factors of evaluation of houses, we choose five items: equipment, space, environment, transportation and shopping convenience. A volunteer student scores the above items $x_1 \sim x_5$ according to his a priori evaluation of houses generally. These scores are the densities of a priori fuzzy measure g_X.

Table 4 shows this experimental scores. The conditional fuzzy measure σ_Y, in this case, is calculated objectively from the data of four sample houses ($y_1 \sim y_4$) as shown in Table 5. Here, $\sigma_Y(\cdot|x_i)$ is considered as the scores of each house from the viewpoints of an item x_i. Substituting g_X and σ_Y into Eqs. (2.18) and (2.19), we calculate a posteriori fuzzy measure σ_X. Tables 6 and 7 show the densities of g_Y and σ_X, respectively.

	$g_X(\{x_i\})$
x_1	0.17
x_2	0.26
x_3	0.22
x_4	0.21
x_5	0.06

Table 4
A Priori Weights of x_i

	y_1	y_2	y_3	y_4
x_1	0.50	0.20	0.40	0.20
x_2	0.60	0.40	0.20	0.10
x_3	0.35	0.70	0.10	0.10
x_4	0.15	0.15	0.40	0.60
x_5	0.20	0.20	0.30	0.60

Table 5 Conditional Fuzzy Measures
$\sigma_y(\{y_j\}|x_i)$

Comparing Table 4 with Table 7, we can see that the weights of evaluation do not change remarkably when y_j is fixed. This is becuase he does not have any special preference as shown in Table 4 and all the samples (houses) are not remarkable different too. The items of which weights are changed are transportation for y_1, y_2, environment for y_3, space and environment for y_4.

The order of his synthetic preference is y_1, y_2, y_3, y_4, and coincides with the integration in Table 6. That is to say, he prefers the house for which he does not need to change his a priori weight. This result is quite natural.

	y_1	y_2	y_3	y_4
$g_Y(\{y_j\})$	0.42	0.40	0.39	0.28

Table 6 Densities of g_y

	x_1	x_2	x_3	x_4	x_5
y_1	0.17	0.26	0.22	0.15	0.06
y_2	0.17	0.26	0.22	0.15	0.06
y_3	0.17	0.20	0.10	0.21	0.06
y_4	0.17	0.10	0.10	0.21	0.06

Table 7 A Posteriori Weights of x_i Calculated by Model II

In this example, it is difficult to examine the accuracy of our model experimentally. We shall be able to check it, if we adopt the examples of diagnosis or cause-estimation.

5. CONCLUSION

The authors show in this paper some idea and the formulation of "fuzzy measure" and "fuzzy integrals". These are suitable for the synthesis of fuzzy elements because of their special constitution and their simplicity of calculation. Especially, "conditional fuzzy measures" and their related formula are quite effective for expressing the transition of fuzzy phenomena.

Only two examples are described here. But many other fields of applications, such as decision making, optimization, artificial intelligence and thinking models, can be considered. Some of them are being studied by the authors.

ACKNOWLEDGEMENTS

The authors wish to express their appreciation to Professor Y. Morita for his valuable discussion.

REFERENCES

[1] L. A. Zadeh: "Fuzzy Sets," Information and Control, Vol. 8, 338-353, 1965.

[2] L. A. Zadeh: "Fuzzy Algorithms," Information and Control, Vol. 12, 94-102, 1968.

[3] L. A. Zadeh: "Probability Measures of Fuzzy Events," Journal of Mathematical Analysis and Applications, Vol. 23, 421-427, 1972.

[4] M. Sugeno: "Fuzzy Measures and Fuzzy Integrals," Trans. S. I. C. E. (In Japanese), Vol. 8, No. 2, 1972.

[5] M. Sugeno: "Constructing Fuzzy Measures and Grading Similarity of Patterns by Fuzzy Integrals," Trans. S. I. C. E. (In Japanese), Vol. 9, 1973.

[6] M. Sugeno. T. Terano: "An Approach to the Identification

of Human Characteristics by Applying Fuzzy Integrals," Proc. 3rd IFAC Sympo. on Identification and System Parameter Estimation, Hauge, 1973.

[7] M. Sugeno, Y. Tsukamoto, T. Terano: "Subjective Evaluation of Fuzzy Objects," IFAC Sympo. on Stochastic Control, Budapest, 1974.

[8] F. W. Sears: Optics, Addison-Wesley.

FUZZY TOPOLOGY

C. K. Wong
Computer Sciences Department
IBM Thomas J. Watson Research Center
Yorktown Heights, New York 10598

ABSTRACT

The theory of general topology is based on the set operations of union, intersection and complementation. Fuzzy sets as introduced by Zadeh [1] have the same kind of operations. It is, therefore, natural to extend the concept of point set topology to fuzzy sets, resulting in a theory of fuzzy topology. Various results similar to those in general topology as well as some significant ramifications are discussed.

I. INTRODUCTION

In his classical paper [1], Zadeh first introduced the fundamental concept of fuzzy sets. An immediate application of this idea can be found in the theory of general topology since it is based mainly on the operations of union, intersection and complementation of sets [2,10]. In the development of a parallel theory based on fuzzy sets, many interesting phenomena have been observed [3-6]. Specifically, one notices many differences between the two theories. A good example is the Tychonoff Theorem in general topology: any product of compact spaces is compact. Its fuzzy counterpart holds only for finite products. A couterexample of infinite product can be easily constructed [4,6]. The concept of point can also be fuzzified and a local theory is therefore possible [5]. With the proper fuzzification of the unit interval, one can study the separation properties of fuzzy topological spaces, in particular, normality and uniformity [7,8].

All these constitute a rich body of theory which is largely parallel to that of general topology but has many significant ramifications as well. It can therefore be aptly termed the theory of <u>fuzzy topology</u>.

For simplicity and unity, we will follow the original definition of fuzzy sets as introduced by Zadeh [1], i.e. functions from X to the unit interval. Replacement of the unit interval by various lattices at different points is definitely possible and will lead to slightly more general results [6-9]. Proofs will be omitted but references will be provided at the end of each section.

II. BASIC DEFINITIONS AND PROPERTIES

Let $X = \{x\}$ be a point set. A fuzzy set A in X is characterized by a membership function $\mu_A(x)$ from X to the unit interval $I = [0,1]$.

<u>Definition 2.1</u>. Let A and B be fuzzy sets in X. Then:

$A = B \iff \mu_A(x) = \mu_B(x)$, for all $x \in X$.

$A \subset B \iff \mu_A(x) \leq \mu_B(x)$, for all $x \in X$.

$C = A \cup B \iff \mu_C(x) = \max[\mu_A(x), \mu_B(x)]$, for all $x \in X$.

$D = A \cap B \iff \mu_D(x) = \min[\mu_A(x), \mu_B(x)]$, for all $x \in X$.

$E = A' \iff \mu_E(x) = 1 - \mu_A(x)$, for all $x \in X$.

More generally, for a family of fuzzy sets, $A = \{A_i | i \in I\}$, the union $C = \bigcup_I A_i$ and the intersection $D = \bigcap_I A_i$ are defined by

$$\mu_C(x) = \sup_I\{\mu_{A_i}(x)\} \quad x \in X,$$

and

$$\mu_D(x) = \inf_I\{\mu_{A_i}(x)\} \quad x \in X.$$

The symbol Φ will be used to denote the empty fuzzy set ($\mu_\Phi(x) = 0$ for all x in X). For X, we have by definition $\mu_X(x) = 1$ for all x in X.

Definition 2.2. A fuzzy topology is a family T of fuzzy sets in X which satisfies the following conditions:

(a) $\Phi, X \in T$.

(b) If $A, B \in T$, then $A \cap B \in T$.

(c) If $A_i \in T$ for each $i \in I$, then $\bigcup_I A_i \in T$.

T is called a fuzzy topology for X, and the pair (X,T) is a fuzzy topological space, or fts for short. Every member of T is called a T-open fuzzy set (or simply open fuzzy set). A fuzzy set is T-closed (orsimply closed) iff its complement is T-open.

As in general topology, the indiscrete fuzzy topology contains only Φ and X, while the discrete fuzzy topology contains all fuzzy sets.

Definition 2.3. Let A be a fuzzy set in a fts (X,T). The largest open fuzzy set contained in A is called the interior of A and is denoted by $A°$. The smallest closed fuzzy set containing A is called the closure of A and is denoted A^-.

Definition 2.4. Let f be a function from X to Y. Let B be a fuzzy set in Y with membership function $\mu_B(y)$. Then the inverse of B, written as $f^{-1}[B]$, is a fuzzy set of X whose membership function is defined by

$$\mu_{f^{-1}[B]}(x) = \mu_B(f(x)) \text{ for all } x \text{ in } X.$$

Conversely, let A be a fuzzy set in X with membership function $\mu_A(x)$. The image of A, written as $f[A]$, is a fuzzy set in Y whose membership function is given by

$$\mu_{f[A]}(y) = \sup_{z \in f^{-1}[y]}\{\mu_A(z)\} \text{ if } f^{-1}[y] \text{ is not empty,}$$

$$= 0 \quad \text{otherwise,}$$

for all y in Y, where $f^{-1}[y] = \{x|f(x) = y\}$.

Theorem 1.1. Let f be a function from X to Y. Then:

(a) $f^{-1}[B'] = \{f^{-1}[B]\}'$ for any fuzzy set B in Y.

(b) $f[A'] \supset \{f[A]\}'$ for any fuzzy set A in X.

(c) $B_1 \subset B_2 \Rightarrow f^{-1}[B_1] \subset f^{-1}[B_2]$, where B_1, B_2 are fuzzy sets in Y.

(d) $A_1 \subset A_2 \Rightarrow f[A_1] \subset f[A_2]$, where A_1 and A_2 are fuzzy sets in X.

(e) $B \supset f[f^{-1}[B]]$ for any fuzzy set B in Y.

(f) $A \subset f^{-1}[f[A]]$ for any fuzzy set A in X.

(g) Let f be a function from X to Y and g a function from Y to Z. Then $(g \circ f)^{-1}[C] = f^{-1}[g^{-1}[C]]$ for any fuzzy set C in Z, where $g \circ f$ is the composition of g and f.

Definition 2.5. A function f from a fts$(X,\mathcal{T})$ to a fts $(Y,\mathcal{U})$ is F-continuous iff the inverse of each $\mathcal{U}$-open fuzzy sets is $\mathcal{T}$-open.

Theorem 1.2. A function f from a fts$(X,\mathcal{T})$ to a fts$(Y,\mathcal{U})$ is F-continuous iff the inverse of each $\mathcal{U}$-closed fuzzy set is $\mathcal{T}$-closed.

Definition 2.6. A function f from a fts$(X,\mathcal{T})$ to a fts $(Y,\mathcal{U})$ is said to be F-open (F-closed) iff it maps an open (closed) fuzzy set in $(X,\mathcal{T})$ onto an open (closed) fuzzy set in $(Y,\mathcal{U})$.

A fuzzy homeomorphism is an F-continuous one-to-one mapping of a fts X onto a fts Y such that the inverse of the mapping is also F-continuous. If there exists a fuzzy homeomorphism of one fuzzy space onto another, the two fuzzy spaces are said to be F-homeomorphic and each is a fuzzy homeomorph of the other. Two fts's are topologically F-equivalent iff they are F-homeomorphic.

For reference see [2].

III. COMPACTNESS AND COUNTABILITY

Definition 3.1. Let (X,T) be a fts. A family A of fuzzy sets is a cover of a fuzzy set B iff $B \subset \cup \{A | A \in A\}$. It is an open cover iff each member of A is an open fuzzy set. A subcover of A is a subfamily which is also a cover.

Definition 3.2. A fts is compact iff each open cover of the space has a finite subcover.

Definition 3.3. A fts is countably compact iff every coutable open cover of the space has a finite subcover.

Definition 3.4. Let T be a fuzzy topology. A subfamily B of T is a base for T iff each member of T can be expressed as the union of some members of B.

Definition 3.5. Let T be a fuzzy topology. A subfamily B of T is a subbase for T iff the family of finite intersections of members of S forms a base for T.

Definition 3.6. A fts(X,T) is said to be C_{II} if there exists a coutable base B for T.

Theorem 3.1. If a fts (X,T) is C_{II}, then compactness and countable compactness are equivalent.

Theorem 3.2. Let f be an F-continuous function from a compact (countably compact) fts X onto a fts Y. Then Y is compact (countably compact).

Next, we will have a characterization of compactness and countable compactness peculiar to fuzzy topological spaces.

Given a cover $A = A_i$, $i \in I$, it means that $\sup_{i \in I}\{\mu_{A_i}(x)\} = 1$ for all $x \in X$. Therefore, for any $0 < \varepsilon < 1$, and for any $x \in X$, there exists a fuzzy set A_i such that $\mu_{A_i}(x) \geq 1 - \varepsilon$. At each point $x \in X$, select one such A_i and group together all points x with the same A_i. Let $\Gamma_{i,\varepsilon}$ denote the set of all such x's. For a fixed $\varepsilon, \{\Gamma_{i,\varepsilon}\}$ form a partition of X, called an ε-partition by A. Note that the partition depends on the initial choice of A_i's.

If in addition, for any $x \in X$, there exits A_i such that $\mu_{A_i}(x) = 1$, then group all points x with the same A_i and denote it $\{\Gamma_{i,0}\}$. A is then said to have a 0-partition of X.

We have the following characterization theorem.

Theorem 3.3. A fts (X,T) is compact (coutably compact) iff each open cover (countable open cover) has a finite 0-partition of X.

Theorem 3.4. If there exists an open cover (countable open cover) A of X and a point $x \in X$ such that $\mu_{A_i}(x) < 1$ for all $A_i \in A$, then (X,T) is not compact (countably compact).

Definition 3.7. A fts(X,T) is Lindelöf iff every open cover of X has a countable subcover.

Theorem 3.5. If a fts (X,T) is C_{II}, it is also Lindelöf..

Theorem 3.6. Let f be an F-continuous function from a Lindelöf fts X onto a fts Y. Then Y is Lindelöf.

Theorem 3.7. A fts (X,T) is Lindelof iff each open cover has a countable ε-partition of X for all ε such that $0 < \varepsilon < 1$.

For reference see [3].

IV. PRODUCT AND QUOTIENT SPACES

Let $\{X_\alpha\}$, $\alpha \in I$, be a family of spaces. Let $X = \prod_{\alpha\in I} X_\alpha$ be the usual product space, and let P_α be the projection from X onto X_α.

Further assume that each X_α is a fts with fuzzy topology T_α. Let $B \in T_\alpha$; then by Definition 2.4, $P_\alpha^{-1}[B]$ is a fuzzy set in X. The family of fuzzy sets $S = \{P_\alpha^{-1}[B] | B \in T_\alpha, \alpha \in I\}$ is now used to generate a fuzzy topology T for X in the following manner: Let B be the family of all finite intersections of members of S. Let T be the family of all unions of members of B. It is clear that T is indeed a fuzzy topology for X, with B as a base and S a subbase.

Definition 4.1. Given a family of fts $\{(X_\alpha, T_\alpha)\}$, $\alpha \in I$, the fuzzy topology defined as above is called the product fuzzy topology for $X = \prod_{\alpha \in I} X_\alpha$ and (X, T) is called the product fts.

Some immediate consequences from this definition are listed here.

Theorem 4.1. Let (X, T) be the product fts of the family of fts's $\{(X_\alpha, T_\alpha)\}$, $\alpha \in I$.

(i) For each $\alpha \in I$, the projection P_α is F-continuous.

(ii) The product fuzzy topology is the smallest fuzzy topology for X such that (i) is true.

(iii) Let (Y, U) be a fts and let f be a function from Y to X. Then f is F-continuous iff for every $\alpha \in I$, $P_\alpha \circ f$ is F-continuous.

Next we state a product theorem for C_{II} spaces. One would notice the difference between fuzzy topology and general topology.

Theorem 4.2. Let $\{(X_\alpha, T_\alpha)\}$, $\alpha = 1, 2, \ldots$, be a countable family of C_{II} fts's. Then the product fts(X, T) is also C_{II}.

However, uncountable products of C_{II} spaces may not be C_{II}; hence, the above result is, in a sense, the best one can get.

Theorem 4.3. Let $\{(X_\alpha, T_\alpha)\}$, $\alpha \in I$, be an uncountable family of C_{II} spaces such that

(i) none is indiscrete, and

(ii) in each fts (X_α, T_α), for any $F \in T_\alpha$ and $F \neq \Phi$, there exists a point $x \in X_\alpha$ such that $\mu_F(x) = 1$, where μ_F is the membership function of F.

Then the product fts$(X,\mathcal{T})$ is not C_{II}.

After countability, we shall present a product theorem for compact spaces. Once again, departure from general topology is evident.

Theorem 4.4. Let $\{(X_\alpha,\mathcal{T}_\alpha)\}$, $\alpha = 1,2,\ldots,n$ be a finite family of compact (countably compact) fts's. Then the product fts $(X,\mathcal{T})$ is also compact (countably compact).

The following counterexample shows that, in a sense, Theorem 4.4 is the best one can hope for.

Let Y be any point set. Let n be any positive integer. Let A_n be the fuzzy set in Y with membership function $\mu_{A_n} = 1 - (1/n)$ for all $y \in Y$. Let $X_n = Y$ and let $\mathcal{T}_n = \{\Phi, A_n, Y\}$. Then $(X_n,\mathcal{T}_n)$ is a compact (countably compact) fts since X_n is the only open cover of X_n. However, the product fts of the countable family $\{(X_n,\mathcal{T}_n)\}$, $n = 1,2,\ldots$, is not compact (countably compact).

To see this, note that the fuzzy set $P_n^{-1}[A_n]$ has membership function

$$\mu_{P_n^{-1}[A_n]}(x) = 1 - (1/n) \text{ for all } x \in X = \prod_{n=1}^{\infty} X_n.$$

By definition of product fts, the family

$$\mathcal{S} = \{\Phi, X, P_n^{-1}[A_n],\ n = 1,2,\ldots\}$$

is used to generate the product fuzzy topology $\mathcal{T}$ by first taking the finite intersections and then the unions of these intersections. Clearly, the product fuzzy topology thus generated is exactly $\mathcal{S}$ itself. The family $\{P_n^{-1}[A_n]\}$, $n = 1, 2,\ldots$, is an open cover (countable open cover) of $(X,\mathcal{T})$ which has no finite subcover.

Next we discuss another method of constructing new fuzzy topology, which can be regarded as the dual of the product fuzzy topology.

Let X be a point set. Let R be an equivalence relation defined on X. Let X/R be the usual quotient set, and let P be the usual projection from X onto X/R.

If $(X,\mathcal{T})$ is a fts, one can define a fuzzy topology on X/R such that P is F-continuous as follows. Let $\mathcal{U}$ be the family of fuzzy sets in X/R defined by $\mathcal{U} = \{B \mid P^{-1}[B] \in \mathcal{T}\}$. Then $\mathcal{U}$ is a fuzzy topology, called the quotient topology for X/R, and $(X/R, \mathcal{U})$ is called the quotient fts.

We have results similar to Theorem 4.1.

Theorem 4.5. (i) The quotient fuzzy topology is the largest fuzzy topology such that P is F-continuous.

(ii) Let $(Y,\mathcal{V})$ be a fts. Let g be a function from the quotient fts $(X/R,\mathcal{U})$ to $(Y,\mathcal{V})$. Then g is F-continuous iff $g \circ P$ is F-continuous.

Theorem 4.6. Let f be a F-continuous function from a fts $(X,\mathcal{T})$ onto a fts $(Y,\mathcal{V})$ such that f is either F-open or F-closed, then there exists an equivalence relation R on X such that $(Y,\mathcal{V})$ is F-homeomorphic to the quotient fts $(X/R,\mathcal{U})$.

Definition 4.2. Let A be a fuzzy set in a fts $(X,\mathcal{T})$. Let R be an equivalence relation on X, which is therefore decomposed into disjoint subsets $\mathcal{D} = \{X_i\}$, $i \in I$ ($x,y \in X_i$ iff they are R related). Define two new fuzzy sets A_1, A_2 in X with membership functions as follows:

$$\mu_{A_1}(x) = \sup_{y \in X_i} \mu_A(y) \quad \text{for } x \in X_i$$

and

$$\mu_{A_2}(x) = \inf_{y \in X_i} \mu_A(y) \quad \text{for } x \in X_i.$$

They will be called upper and lower fuzzy sets of A, respectively.

Theorem 4.7. Let P be the projection from a fts $(X,\mathcal{T})$ onto its quotient fts $(X/R,\mathcal{U})$. Then the following statements are equivalent.

(i) P is F-open.

(ii) If A is an open fuzzy set in $(X,\mathcal{T})$, then its upper fuzzy set A_1 is open.

(iii) If A is a closed fuzzy set in $(X,\mathcal{T})$, then its lower fuzzy set A_2 is closed.

If "open" and "closed" are interchanged in (i), (ii), and (iii), the resulting statements are equivalent.

Theorem 4.8. If $(X,\mathcal{T})$ is C_{II} and P is F-open, then the quotient fts $(X/R,\mathcal{U})$ is C_{II}.

Theorem 4.9. If $(X,\mathcal{T})$ is compact (countably compact), then the quotient fts $(X/R,\mathcal{U})$ is compact (countably compact).

For reference see [4,6].

V. LOCAL PROPERTIES

In this section we introduce the concept of fuzzy points and state some results on local properties of fuzzy topological spaces.

Definition 5.1. A fuzzy point p in X is a fuzzy set with membership function

$$\mu_p(x) = y, \quad \text{for } x = x_0.$$
$$= 0, \quad \text{otherwise,}$$

where $0 < y < 1$. p is said to have support x_0 and value y.

Definition 5.2. Let p be a fuzzy point and A a fuzzy set in X. Then p is said to be in A or A contains p, denoted $p \in A$, iff $\mu_p(x) < \mu_A(x)$ for all $x \in X$.

Theorem 5.1. If $A = \cup_{i \in I} A_i$, where I is any index set, then $p \in A$ iff $p \in A_i$ for some $i \in I$.

In ordinary set theory, Theorem 5.1 is trivial. But in

the case of fuzzy set theory, this is not as trivial as one may imagine. In fact, if one replaces the inequality in definition 5.2 by $\mu_p(x) \leq \mu_A(x)$, then Theorem 5.1 is no longer true. On the other hand, should we restrict all fuzzy sets to take values {0,1} and hope that Definitions 5.1 and 5.2 would reduce to the corresponding definitions in ordinary set theory, we should have used $0 < y \leq 1$ and $\mu_p(x) \leq \mu_A(x)$ instead of $0 < y < 1$ and $\mu_p(x) < \mu_A(x)$. In other words, our current definitions will not reduce to the ordinary case even if we impose the restriction that all fuzzy sets will take values {0,1} only.

Theorem 5.2. Let $(X,\mathcal{T})$ be a fts. Then a subfamily $\mathcal{B}$ of $\mathcal{T}$ forms a base of $\mathcal{T}$ iff for every member A of $\mathcal{T}$ and for every fuzzy point $p \in A$, there exists a member B of $\mathcal{B}$ such that $p \in B \subset A$.

Definition 5.3. Let $(X,\mathcal{T})$ be a fts and p a fuzzy point. A subfamily $\mathcal{B}_p$ of $\mathcal{T}$ is called a local base of p iff $p \in B$ for every member B of $\mathcal{B}_p$, and for every member A of $\mathcal{T}$ such that $p \in A$ there exists a member B of $\mathcal{B}_p$ such that $p \in B \subset A$.

Definition 5.4. A fts $(X,\mathcal{T})$ is said to be C_I iff every fuzzy point in X has a countable local base.

Theorem 5.3. If $(X,\mathcal{T})$ is C_{II}, then it is C_I.

Definition 5.5. A fts $(X,\mathcal{T})$ is said to be separable iff there exists a countable sequence of fuzzy points $\{p_i\}$, $i = 1,2,\ldots$, such that for every member A of $\mathcal{T}$ and $A \neq \Phi$, there exists a p_i such that $p_i \in A$.

Theorem 5.4. If a fts $(X,\mathcal{T})$ is C_{II}, then it is separable.

In Section III, we note that a C_{II} fts is also Lindelöf. Together with Theorem 5.4, one sees that among the four types of countability properties, namely, C_{II}, C_I, Lindelöf and separable, C_{II} is the strongest. We also note that the

F-continuous image of a Lindelöf fts is also Lindelöf. Here, we have a similar result.

Theorem 5.5. Let f be an F-continuous function from a separable fts (X,T) onto a fts (Y,U). Then (Y,U) is also separable.

If f is F-open as well as F-continuous, we then have:

Theorem 5.6. Let f be an F-continuous function from a $C_{II}(C_I)$ fts (X,T) onto a fts (Y,U). If f is also F-open, then (Y,U) is $C_{II}(C_I)$.

The introduction of fuzzy points enables us to discuss convergence in a meaningful way.

Definition 5.6. Let p_n, $n = 1,2,\ldots$, be a sequence of fuzzy points in a fts (X,T) with supports x_n, $n = 1,2,\ldots$. Let p be a fuzzy point with support $x \neq x_n$, for all $n \geq n_0$, where n_0 is some number. Then p_n is said to coverge to p, written $p_n \to p$, iff for every member A of T such that $p \in A$, there exists a number m, such that $p_n \in A$ for all $n \geq m$.

Note that the restriction on the supports is necessary to make the definition meaningful. Note also that if p_n has value y_n and p has value y, in general $p_n \to p$ does not imply $y_n \to y$. In fact, we have the following observations. If $p_n \to p$ and p has support x_0 and value y, then $p_n \to q$ for all fuzzy points q with support x_0 and value $z \geq y$. In the theory of general topology, we have a similar sitaution. As a matter of fact, the uniqueness of limits of convergent nets is a characterization of a special type of topological space, namely, Hausdorff space.

Definition 5.7. Let p be a fuzzy point in (X,T) with support x_0. Let A be a fuzzy set in X. Then p is an accumulation point of A iff for every member B of T such that $p \in B$, $B \cap A_p \neq \Phi$, where A_p is the fuzzy set with membership function

$$\mu_{A_p}(x) = 0, \quad \text{for } x = x_0,$$
$$= \mu_A(x), \quad \text{otherwise.}$$

Similar to our previous remarks on convergence, we note that if p is an accumulation point of A and P has support x_0 and value y, then all fuzzy points q with the same support x_0 and value $z \geq y$ are accumulation points of A.

Theorem 5.7. Let (X,T) be a C_I fts. Let A be a fuzzy set and p a fuzzy point in X. Then p is an accumulation point of A iff there exists a sequence of fuzzy points p_n, $n = 1,2,\ldots$, such that $p_n \in A$ and $p_n \to p$.

Theorem 5.8. Let (X,T) be a fts. If there exists a countable sequence of fuzzy points $\{p_i\}$, $i = 1,2,\ldots$, in X such that every fuzzy point p in X is an accumulation point of the fuzzy set $A = \bigcup_i p_i$. Then (X,T) is separable.

However, the converse of the above theorem is in general not true, demonstrating yet another departure from general topology. We have the following counterexample.

Let X be a point set. Let $x_0 \in X$. Let A_α, $0 \leq \alpha \leq 1$, be fuzzy sets in X defined by

$$\mu_{A_\alpha}(x) = \alpha, \quad \text{for } x = x_0$$
$$= 0, \quad \text{otherwise.}$$

Let $T = \{\Phi, X, A_\alpha, 0 \leq \alpha \leq 1\}$. Then (X,T) is a fts. Consider the countable sequence of fuzzy points $\{p_\beta\}$ such that the support of each p_β is x_0 and β ranges over the set of rational numbers between 0 and 1. Any member B of T such that $B \neq \Phi$ will contain a member of $\{p_\beta\}$. Thus (X,T) is separable. Let p be a fuzzy point with support x_0 and value α_0, $0 < \alpha_0 < 1$. The p is not an accumulation point of the union A of any countable fuzzy points since $B \cap A_p = \Phi$ for all $B \in T$ containing p and $B \neq X$.

The localization of compactness is naturally local compactness.

Definition 5.8. A fts (X,T) is said to be locally compact iff for every fuzzy point p in X there exists a member $A \in T$ such that (i) $p \in A$ and (ii) A is compact, i.e. each open cover of A has a finite subcover.

Clearly, each compact fts is locally compact.

The next result demonstrates once more the ramification of fuzzy topology from general topology.

Theorem 5.9. A discrete fts (X,T) is not locally compact.

Like C_I fts, we have the following theorem.

Theorem 5.10. Let f be an F-continuous function from a locally compact fts (X,T) onto a fts (Y,U). If f is also F-open, then (Y,U) is locally compact.

Next we discuss the product and quotient spaces generated by C_I, separable and locally compact spaces.

Theorem 5.11. Let $\{(X_\alpha,T_\alpha)\}$, $\alpha \in I$, be a countable family of C_I fts's. Then the product fts (X,T) is C_I.

However, there exists an uncountable family of C_I spaces, whose product is not C_I.

Theorem 5.12. Let $\{(X_\alpha,T_\alpha)\}$, $\alpha \in I$, be an uncountable family of C_I spaces such that:

(i) none is indiscrete, i.e., for each $\alpha \in I$, there exists $U_\alpha \in T_\alpha$ such that $U_\alpha \neq \Phi, X_\alpha$;

(ii) for each $\alpha \in I$, there exists a fuzzy point $p_\alpha \in U_\alpha$ such that

$$p = \bigcap_{\alpha \in I} P_\alpha^{-1}[p_\alpha]$$

is a fuzzy point in X; and

(iii) in each fts (X_α,T_α), for any $A \in T_\alpha$ and $A \neq \Phi$, there exists a point $x \in X_\alpha$ such that $\mu_A(x) = 1$, where μ_A is the membership function of A.

Then the product fts (X,T) is not C_I.

Unlike general topology, given fuzzy points p_α in X_α, $\alpha \in I$,

$$p = \bigcap_{\alpha \in I} P_\alpha^{-1}[p_\alpha]$$

is not always a fuzzy point in X; it is either a fuzzy point or the empty fuzzy set Φ. For example, let $I = (0,1)$. In each X_α, let p_α be a fuzzy point with support x_α and value α. Then

$$p = \bigcap_{\alpha \in I} P_\alpha^{-1}[p_\alpha] = \Phi.$$

Theorem 5.13. Let $\{(X_\alpha, T_\alpha)\}$, $\alpha \in I$, be a countable family of separable spaces. Then the product fts (X,T) is also separable.

Theorem 5.14. Let $\{(X_\alpha, T_\alpha)\}$, $\alpha = 1,2,\ldots,n$, be a finite family of locally compact fts's. Then the product fts (X,T) is also locally compact.

Theorem 5.15. (i) If (X,T) is separable, then the quotient fts $(X/R,U)$ is separable.

(ii) If (X,T) is C_I and P is F-open, then the quotient fts $(X/R,U)$ is C_I.

(iii) If (X,T) is locally compact, and P is F-open, then the quotient fts $(X/R,U)$ is locally compact.

For reference see [5].

VI. NORMALITY AND UNIFORMITY

In this section we present results on separation properties of fuzzy topological spaces. For this purpose we need a fuzzy version of the unit interval.

Definition 6.1. The fuzzy unit interval I_F is the set

of all monotonic decreasing functions λ from the real line R to the unit interval I, satisfying:

$$\lambda(t) = 1 \quad \text{for } t < 0,\ t \in R$$

$$\lambda(t) = 0 \quad \text{for } t > 1,\ t \in R$$

after the identification of λ,μ: $R \to I$ if for every $t \in R$, $\lambda(t-) = \mu(t-)$ and $\lambda(t+) = \mu(t+)$, where $\lambda(t-) = \inf_{s<t} \lambda(s) = \lim_{s \uparrow t} \lambda(s)$ and $\lambda(t+) = \sup_{s>t} \lambda(s) = \lim_{s \downarrow t} \lambda(s)$.

We may define a partial ordering on I_F by $\lambda \leq \mu$ if for every $t \in R$, $\lambda(t-) \leq \mu(t-)$ and $\lambda(t+) \leq \mu(t+)$. We may embed the unit interval in the fuzzy unit interval by identifying $r \in I$ with the function $\alpha: R \to I$ where $\alpha(t) = 1$ for $t < r$ and $\alpha = 0$ for $t > r$.

We define a fuzzy topology on I_F by taking as subbase $\{L_t, R_t\}_{t \in R}$, where L_t, R_t are fuzzy sets on I_F defined by

$$L_t(\lambda) = 1 - \lambda(t-)$$

$$R_t(\lambda) = \lambda(t+),$$

for all $\lambda \in I_F$.

Note that if we replace I by $\{0,1\}$, then the fuzzy unit interval and its topology reduce to the unit interval and its usual topology.

<u>Definition 6.2</u>. A fts (X,T) is said to be normal if for every closed fuzzy set K and open fuzzy set U such that $K \subset U$, there exists a fuzzy set V such that $K \subset V^\circ \subset V^- \subset U$.

We can now state a characterization theorem for normal spaces, which is similar to Uryshon lemma in the theory of general topology.

<u>Theorem 6.1</u>. A fts (X,T) is normal iff for every closed fuzzy set K and open fuzzy set U such that $K \subset U$, there exists a F-continuous function $f: X \to I_F$ such that for every $x \in X$

Should $\mu_K(x) \leq f(x)(1-) \leq f(x)(0+) \leq \mu_U(x)$.

Definition 6.3. A fts $(X,\mathcal{T})$ is perfectly normal if for every closed fuzzy set K and open fuzzy set U such that $K \subset U$, there exists a F-continuous function $f: X \to I_F$ such that for every $x \in X$

$$\mu_K(x) = f(x)(1-) \leq f(x)(0+) = \mu_U(x).$$

Theorem 6.2. A fts is perfectly normal iff it is normal and every closed fuzzy set is a countable intersection of open fuzzy sets. Next we present some results on fuzzy quasi-uniformities and fuzzy uniformities. First let us consider a quasi-uniformity on X in general topology. An element D is a subset of $X \times X$. We may define a mapping $d: 2^X \to 2^X$ by $d(V) = \{y | x \in V \text{ and } (x,y) \in D\}$. It is obvious that $V \subset d(V)$ and $d(\underset{I}{\cup} V_i) = \underset{I}{\cup} d(V_i)$ for V and V_i in 2^X. Conversely, given $d: 2^X \to 2^X$ satisfying $V \subset d(V)$ and $d(\underset{I}{\cup} V_i) = \underset{I}{\cup} d(V_i)$ for V and V_i in 2^X, we may define $D \subset X \times X$ such that D contains the diagonal by $D = \{(x,y) | y \in d(\{x\})\}$. Thus in defining a quasi-uniformity for a fuzzy topology, we take our basic elements of the quasi-uniformity to be elements of the set Q of mappings $d: I^X \to I^X$ which satisfy:

$V \subset d(V)$ for $V \in I^X$.

$d(\underset{T}{\cup} V_i) = \underset{I}{\cup} d(V_i)$ for $V_i \in I^X$.

Let $d, e \in Q$. We say $d \subset e$ if $d(V) \subset e(V)$ for every $V \in I^X$. We define $d \circ e$ by composition of mappings.

Definition 6.4. A fuzzy quasi-uniformity of X is a subset $\mathcal{D}$ of Q such that

(a) $\mathcal{D} \neq \Phi$.

(b) If $d \in \mathcal{D}$, $e \in Q$ and $d \subset e$, then $e \in \mathcal{D}$.

(c) If $d \in \mathcal{D}$, $e \in \mathcal{D}$, then there exists $f \in \mathcal{D}$ such that

$f \subset d$ and $f \subset e$.

(d) If $d \in \mathcal{D}$, then there exists $e \in \mathcal{D}$ such that $e \circ e \subset d$.

The pair $(X,\mathcal{D})$ is called a fuzzy quasi-uniform space.

Note that any subset $\mathcal{B}$ of $\mathcal{Q}$ which satisfies condition (d) generates a fuzzy quasi-uniformity in the sense that the collection of all $d \in \mathcal{Q}$ which contain a finite intersection of elements of $\mathcal{B}$ is a quasi-uniformity. Such a set $\mathcal{B}$ is called a subbase for the quasi-uniformity generated. If $\mathcal{B}$ also satisfies (c), then $\mathcal{B}$ is called a base.

Theorem 6.3. Suppose a mapping $i:I^X \to I^X$ satisfies the interior axioms:

(a) $i(X) = X$.

(b) $i(V) \subset V$ for $V \in I^X$.

(c) $i(i(V)) = i(V)$ for $V \in I^X$.

(d) $i(V \cap W) = i(V) \cap i(W)$ for $V,W \in I^X$.

Then $T = \{V \in I^X | i(V) = V\}$ is a fuzzy topology and $i(V) = V°$.

Let $(X,\mathcal{D})$ be a fuzzy quasi-uniform space. Define $i:I^X \to I^X$ by

$$i(V) = \bigcup\{U \in I^X | d(U) \subset V \text{ for some } d \in \mathcal{D}\}.$$

Then i satisfies the interior axioms.

Definition 6.5. The fuzzy topology generated by $\mathcal{D}$ is the fuzzy topology generated by i. Let (X,T) be a fts. Let A be any open fuzzy set in T. Define

$$d_A(V) = X \quad \text{for } V \not\subset A$$
$$= A \quad \text{for } V \subset A.$$

So $d_A \circ d_A = d_A$. Then $\{d_A | A \in T\}$ forms a subbase for a fuzzy quasi-uniformity which generates the original fuzzy topology T. Therefore we have:

Theorem 6.4. Every fuzzy topology is fuzzy quasi-uniformizable.

Definition 6.6. Let $(X,\mathcal{D})$ and $(Y,\mathcal{E})$ be fuzzy quasi-uniform spaces. A function $f:X \to Y$ is said to be F-quasi-uniformly continuous if for every $e \in \mathcal{E}$, there exists a $d \in \mathcal{D}$ such that $d \subset f^{-1}(e)$.

Theorem 6.5. Every F-quasi-uniformly continuous function is F-continuous in the induced fuzzy topologies.

Next we state a theorem corresponding to the characterization of quasi-pseudo metrizability in terms of quasi-uniformities.

Theorem 6.6. Let $(X,\mathcal{D})$ be a fuzzy quasi-uniform space. Then $\mathcal{D}$ has a base $\{d_r | r \in R, r > 0\}$ such that $d_r \circ d_s \subset d_{r+s}$ for $r,s \in R, r,s > 0$ iff $\mathcal{D}$ has a countable base.

Finally, we turn to fuzzy uniformities.

Definition 6.7. Let $d:I^X \to I^X$. Define $d^{-1}:I^X \to I^X$ by $d^{-1}(V) = \bigcap\{U | d(U') \subset V'\}$. Then a fuzzy quasi-uniformity $\mathcal{D}$ is called a fuzzy uniformity if it also satisfies:

(e) If $d \in \mathcal{D}$, then $d^{-1} \in \mathcal{D}$.

If $\mathcal{D}$ is a fuzzy uniformity, then $(X,\mathcal{D})$ is called a fuzzy uniform space.

An F-uniformly continuous functions from a fuzzy uniform space to another fuzzy uniform space can be similarly defined as in the fuzzy quasi-uniform case.

Before stating the main result on fuzzy uniform spaces we have to construct a fuzzy uniform structure on I_F.

Define $b_\varepsilon:(I_F)^X \to (I_F)^X$ by $b_\varepsilon(V) = \bigcap\{R_{s-\varepsilon} | V \subset L'_s\}$. Then the set $\{b_\varepsilon, b_\varepsilon^{-1}\}_{\varepsilon>0}$ forms a subbase for a fuzzy uniformity on I_F. The topology generated by the fuzzy uniformity is the one introduced previously.

We are now in a position to characterize fuzzy uniformizability.

Theorem 6.7. Let $(X,\mathcal{D})$ be a fuzzy uniform space and let

$d \in \mathcal{D}$. Suppose $d(U) \subset V$. Then there exists a F-uniformly continuous function $f: X \to I_F$ such that

$$\mu_U(x) \leq f(x)(1-) \leq f(x)(0+) \leq \mu_V(x)$$

for all $x \in X$.

For reference see [7,8].

REFERENCES

1. L. A. Zadeh, "Fuzzy Sets," Information and Control, 8, 1965, pp. 338-353.
2. C. L. Chang, "Fuzzy Topological Spaces," J. Math. Anal. Appl.,24, 1968, pp. 182-190.
3. C. K. Wong,"Covering Properties of Fuzzy Topological Spaces," J. Math. Anal. Appl., 43, 1973, pp. 697-704.
4. C. K. Wong, "Fuzzy Topology: Product and Quotient Theorems," J. Math. Anal. Appl., 45, 1974, pp. 512-521.
5. C. K. Wong, "Fuzzy Points and Local Properties of Fuzzy Topology," J. Math. Anal. Appl., 46, 1974, pp. 316-328.
6. J. A. Goguen,"The Fuzzy Tychonoff Theorem," J. Math. Anal. Appl., 43, 1973, pp. 734-742.
7. B. W. Hutton, "Normality in Fuzzy Topological Spaces," Department of Math., U. of Auckland, Auckland, New Zealand, Report Series No. 49, Feb. 1974.
8. B. W. Hutton, "Uniformities on Fuzzy Topological Spaces," Dept. of Math., U. of Auckland, Auckland, New Zealand, Report Series No. 50, Feb. 1974.
9. A. B. Engel and V. Buonomano, "Towards a General Theory of Fuzzy Sets I, II," Instituto de Matematica, Universidade Estadual de Campinas, Campinas, Brazil, Technical Report, April 1974.
10. J. L. Kelly, General Topology, Van Nostrand, Princeton, New Jersey, 1955.

INTERPRETATION AND EXECUTION OF FUZZY PROGRAMS

C. L. Chang
IBM Research Laboratory
San Jose, California 95193 U.S.A.

ABSTRACT

In this paper, a fuzzy program is defined through a flowchart where each arc is associated with a fuzzy relation (called a fuzzy branching condition) and a fuzzy assignment. Input, program and output variables occurring in a fuzzy program represent fuzzy subsets. A fuzzy program is interpreted as implicitly defining a tree; and the execution of the fuzzy program is equivalent to searching a solution path in the tree, i.e., tree searching. Examples of fuzzy programs and their executions are given.

1. INTRODUCTION

Fuzzy algorithms and fuzzy programs were introduced in [2,4,5,6,10]. The purpose is to use them to describe complex or ill-defined problems [11]. It is often the case that while real-world problems can be solved easily by human, they are often too complex or too ill-defined to be handled by machines. Our aim is to try to use fuzzy algorithms to capture ill-defined procedures given by human for solving ill-defined and complex problems. For a more detailed account of motivation of fuzzy algorithms, the reader is encouraged to read [2,4]. In this paper, we shall first briefly introduce fuzzy sets and related concepts [1,4,7,8]. Then, we shall define fuzzy programs through flowcharts. We shall interpret that a fuzzy program implicitly defines a tree, and that the execution of the fuzzy program is equivalent to tree-searching [12,13], i.e., searching a solution path in the tree. Most definitions given in Sections 2 and 3 may be found in [1,4,8].

2. FUZZY SETS AND FUZZY RELATIONS

A fuzzy subset A of a univers of discourse U is characterized by a membership function $u_A: U \to [0,1]$ which associates with each element x of U a number $u_A(x)$ in the interval [0,1] which represents the grade of membership of x in A. The support of A, denoted by support(A), is the subset of U defined by

$$\underline{\text{support}}(A) \triangleq \{x \mid x \in U \text{ and } u_A(x) > 0\}.$$

Through the support of the fuzzy subset A, A can be conveniently written as

$$A \triangleq \{u_A(x)/x \mid x \in \underline{\text{support}}\ (A)\}.$$

That is, only those points whose grades of membership are greater than 0 are listed in A. For example, suppose that

$$U = \{1,2,3,\ldots,10\}.$$

Then, A = {0.5/3, 0.7/4, 1.0/5, 0.5/8} is a fuzzy subset of U.

Another example is

$$\underline{\text{young}} = \{u(x)/x \mid u(x) = 1 \text{ if } x \in [0,25], \text{ or } u(x) = (1 + (\tfrac{x-25}{5})^2)^{-1} \text{ if } x \in [25,100]\},$$

where young is a fuzzy subset of [0,100], with x representing age.

In a fuzzy subset A, if $u_1(x)/x$ and $u_2(x)/x$ are both in A, and if $u_1(x) \neq u_2(x)$, then the element with smaller grade of membership should be deleted. For example, if

A = {0.2/a, 0.5/a, 0.1/b}, then A should be reduced to {0.5/a, 0.1/b}.

A fuzzy relation R between a set X and a set Y is a fuzzy subset of the Cartesian product X x Y. R is characterized by a membership function $u_R(x,y)$ and is expressed

$$R \triangleq \{u_R(x,y)/(x,y) \mid x \in X \text{ and } y \in Y\}$$

More generally, an nary fuzzy relation R is a fuzzy subset of $X_1 \times X_2 \times \ldots \times X_n$, and is expressed

$$R \triangleq \{u_R(x_1,\ldots,x_n)/(x_1,\ldots,x_n) \mid x_i \in X_i,\ i = 1,\ldots,n\}.$$

If R is a relation from X to Y and S is a relation from Y to Z, then the composition of R and S is a fuzzy relation denoted by R ∘ S and defined by

$$R \circ S \triangleq \{u(x,z)/(x,z) \mid x \in X,\ z \in Z, \text{ and}$$
$$u(x,z) = \bigvee_y (u_R(x,y) \wedge u_S(y,z))\}$$

where v and ∧ denote, respectively, max and min.

Suppose A and B are fuzzy subsets of U and V, respectively. Suppose R is a fuzzy relation between U and V. Then, ARB, or more precisely R(A,B), is a fuzzy relation between U and V defined as

$$ARB \triangleq \{u(x,y)/(x,y) \mid u(x,y) = u_A(x) \wedge u_R(x,y) \wedge u_B(y),$$
$$u_A(x)/x \in A,\ u_B(y)/y \in B \text{ and } u_R(x,y)/(x,y) \in R\}.$$

The degree of the truthness of ARB, denoted by T(ARB), is the highest grade of membership in ARB. We note that T(ARB) can be conveniently obtained through the use of the max-min product*. For example, let

---

*In the max-min matrix product, the operations of addition and multiplication are replaced by max and min, respectively.

$A = \{0.2/a,\ 0.6/b\}$,
$B = \{0.8/c,\ 0.4/d,\ 0.1/e\}$, and
$R = \{0.3/(a,c),\ 0.9/(a,d),\ 0.5/(a,e),\ 0.5/(b,c),$
$0.3/(b,c),\ 0.8/(b,e)\}$.

A, B and R can be written in matrix forms as follows:

$$A:\ \begin{array}{cc} a & b \\ [0.2 & 0.6] \end{array}$$

$$B:\ \begin{array}{ccc} c & d & e \\ [0.8 & 0.4 & 0.1] \end{array}$$

$$R:\ \begin{array}{c} \\ a \\ b \end{array} \begin{array}{c} \begin{array}{ccc} c & d & e \end{array} \\ \begin{bmatrix} 0.3 & 0.9 & 0.5 \\ 0.5 & 0.3 & 0.8 \end{bmatrix} \end{array}$$

Then, T(ARB) is calculated as follows:

$$T(ARB) = [0.2\ \ 0.6] \begin{bmatrix} 0.3 & 0.9 & 0.5 \\ 0.5 & 0.3 & 0.8 \end{bmatrix} \begin{bmatrix} 0.8 \\ 0.4 \\ 0.1 \end{bmatrix}$$

$$= [0.2\ \ 0.6] \begin{bmatrix} 0.4 \\ 0.5 \end{bmatrix}$$

$$= 0.5.$$

Another example is shown in Fig. 1. We are given fuzzy subsets A and B, and a fuzzy relation R shown in Fig. 1(a). Thus, we obtain a fuzzy relation ARB shown in Fig. 1(b). Since 0.5 is the highest grade of membership in ARB, T(ARB) = 0.5.

More generally, if $A_1,\ldots,A_k$ are fuzzy subsets of $U_1,\ldots,U_k$, respectively, if R is a fuzzy relation for U_1 x.. .x U_k, then $R(A_1,\ldots,A_k)$ is a fuzzy relation for U_1 x...x U_k given by

$$R(A_1,\ldots,A_k) \triangleq \{u(x_1,\ldots,x_k)/(x_1,\ldots,x_k)\,|\,u(x_1,\ldots,x_k)$$
$$= u_{A_1}(x_1) \wedge \ldots \wedge u_{A_k}(x_k) \wedge u_R(x_1,\ldots,x_k),$$

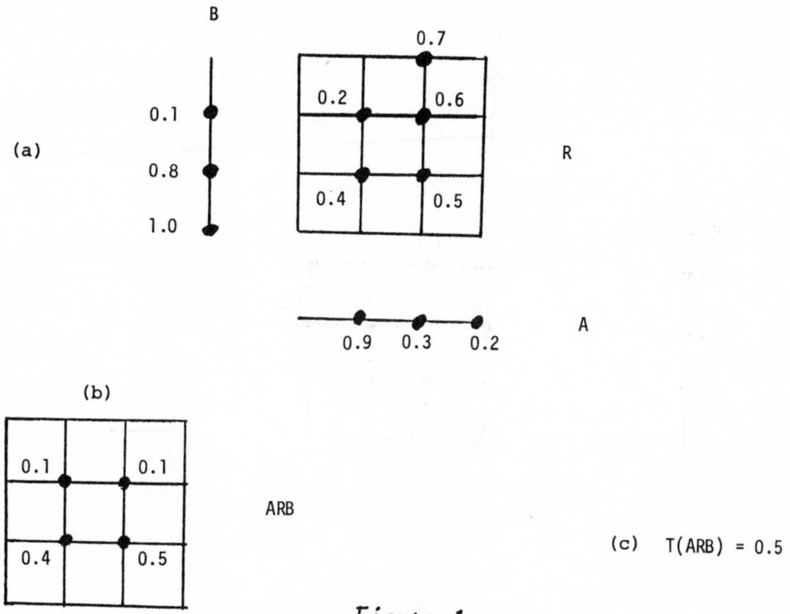

Figure 1

$u_{A_i}(x_i)/x_i \in A_i$ for $i=1,\ldots,k$ and

$u_R(x_1,\ldots,x_k)/(x_1,\ldots,x_k) \in R\}$.

The degree of the truthness of $R(A_1,\ldots,A_k)$ is the highest grade of membership in $R(A_1,\ldots,A_k)$.

If R is a fuzzy relation for $U_1 \times \ldots \times U_k$, then the <u>projection</u> of R on U_i is a fuzzy subset of U_i, denoted by <u>proj</u>$(R;U_i)$, defined by

$$\underline{\text{proj}}(R;U_i) \triangleq \{u(x_i)/x_i \mid x_i \in U_i,$$

$$u(x_i) = \bigvee_{\substack{y_j \in U_j \\ j=1,\ldots,k \\ j \neq i}} u_R(y_1,\ldots,y_{i-1},x_i,y_{i+1},\ldots,y_k)\}$$

For example, Fig. 2 shows a fuzzy relation R and its projections <u>proj</u>(R;X) and <u>proj</u>(R;Y).

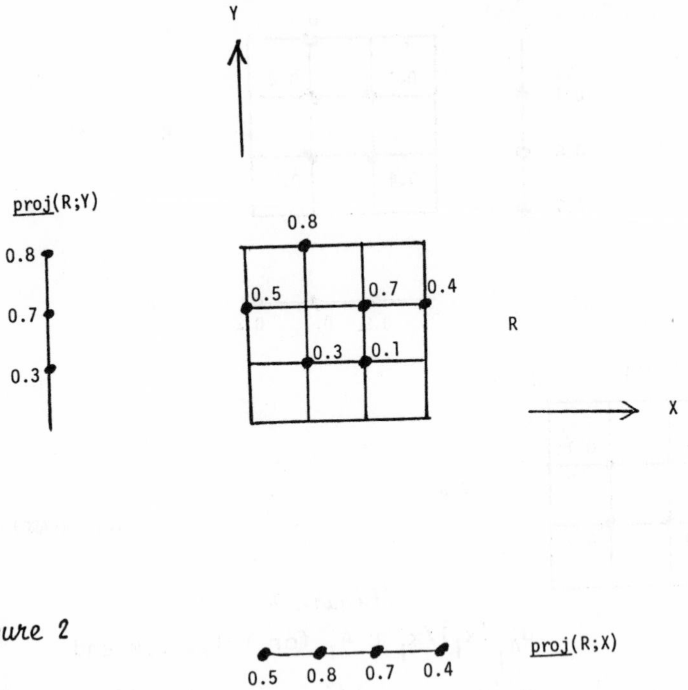

Figure 2

If R is a fuzzy relation from U to V, and A is a fuzzy subset of U, then the fuzzy subset B of V which is induced by A is given by the composition of R and A; that is

$$B \triangleq A \circ R$$

in which A plays the role of a unary relation.

For example, in Fig. 3, B is a fuzzy subset induced by a fuzzy subset A and a fuzzy relation R.

More generally, if R is a fuzzy relation for $U_1 \times \ldots \times U_k$, and $A_1, \ldots, A_{i-1}, A_{i+1}, \ldots, A_k$ are fuzzy subsets of $U_1, \ldots, U_{i-1}, U_{i+1}, \ldots, U_k$, respectively, then the fuzzy subset A_i^* of U_i which is induced by A_j, $j = 1, \ldots, k$, $j \neq i$, is given by

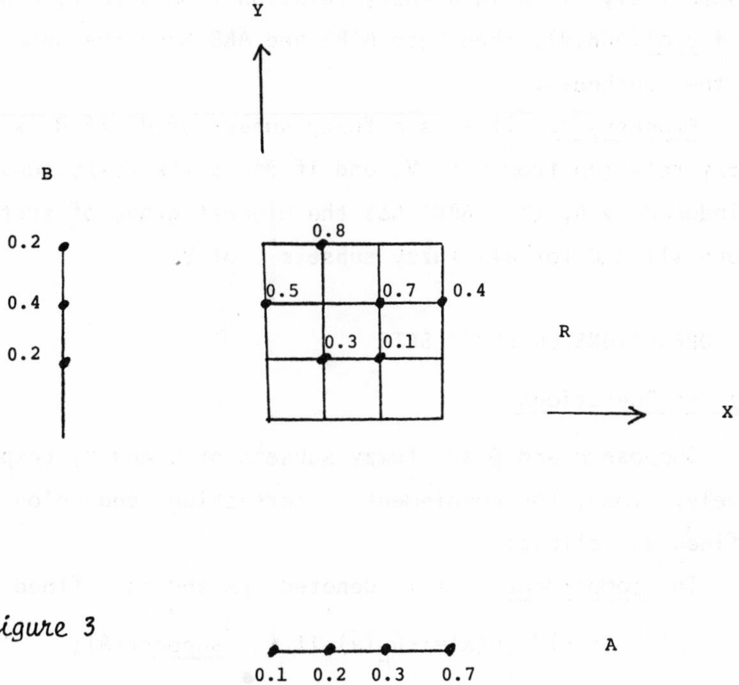

Figure 3

$$A_i^* \triangleq \{u(x_i)/x_i \mid x_i \in U_i \text{ and } u(x_i)$$

$$= \bigvee_{\substack{y_j \in U_j \\ j=1,\ldots,k \\ j \neq i}} (u_R(y_1,\ldots,y_{i-1},x_i,y_{i+1},\ldots,y_k) \wedge u_{A_1}(y_1) \wedge \ldots \wedge u_{A_k}(y_k))\}$$

<u>Property 1</u>. If A and B are fuzzy subsets of U and V, respectively, if R is a fuzzy relation from U to V, then

<u>proj</u>(ARB;U) = $A \cap A^*$, and

<u>proj</u>(ARB;V) = $B \cap B^*$,

where A^* is the fuzzy subset of U induced by B, and B^* is the fuzzy subset of V induced by A, that is $A^*=R\circ B$, and $B^*=A\circ R$.

<u>Property 2</u>. If A and B are fuzzy subsets of U and V,

respectively, if R is a fuzzy relation from U to V, and if A' = proj(ARB;U), then both A'RB and ARB have the same degree of the truthness.

Property 3. If A is a fuzzy subset of U, if R is a fuzzy relation from U to V, and if B* is the fuzzy subset of V induced by A, then ARB* has the highest grade of truthness among all ARC for all fuzzy subsets C of V.

3. OPERATIONS ON FUZZY SETS

(A) Set Operations

Suppose A and B are fuzzy subsets of U and V, respectively. Then, the complement, intersection, and union are defined as follows:

The complement of A is denoted $\neg A$ and is defined by

$$\neg A \triangleq \{u(x)/x \mid u(x)=1-u_A(x) \text{ if } x \in \underline{\text{support}}(A); \quad u(x)=1 \text{ otherwise.}\}$$

For example, let U = {1,2,3,4,5}, and let a fuzzy subset A of U be given as A = {0.6/2, 0.3/5}. Then, $\neg A$ = {1.0/1, 0.2/2, 1.0/3, 1.0/4, 0.7/5}.

The intersection of A and B is denoted by $A \cap B$ and is defined by

$$A \cap B \triangleq \{u(x)/x \mid x \in (\underline{\text{support}}(A) \cap \underline{\text{support}}(B)) \text{ and } u(x) = u_A(x) \wedge u_B(x)\}.$$

The union of A and B is denoted by A U B and is defined by

$$A \cup B \triangleq \{u(x)/x \mid u(x) = u_A(x) \text{ if } x \in \text{support}(A) \text{ and } x \notin \text{support}(B);\ u(x) = u_A(x) \vee u_B(x) \text{ if } x \in \text{support}(A) \text{ and } x \in \text{support}(B);\ u(x) = u_B(x) \text{ if } x \notin \text{support}(A) \text{ and } x \in \text{support}(B)\}.$$

For example, let U = {a,b,c,d,e}, and let the fuzzy subsets A and B of U be given as

$$A = \{0.1/a,\ 0.9/b\}$$
$$B = \{0.5/b,\ 0.3/c\}.$$

Then,

$$A \cap B = \{0.5/b\}$$
$$A \cup B = \{0.1/a,\ 0.9/b,\ 0.3/c\}.$$

(B) Logical Operations

The negation not and the connective and and or are logical operations. They are defined as follows:

Let A and B be fuzzy subsets of U and V, respectively. Then,

$$\underline{\text{not}}\ A \triangleq \{u(x)/x \mid u(x) = 1-u_A(x) \text{ and } u_A(x)/x \in A\}.$$
$$A\ \underline{\text{and}}\ B \triangleq A \cap B$$
$$A\ \underline{\text{or}}\ B \triangleq A \cup B.$$

(C) Fuzzification Operations

The operation of fuzzification has the effect of transforming a nonfuzzy set into a fuzzy set or increasing the fuzziness of a fuzzy set. Let A be a fuzzy subset. Fuzzification of A is obtained through the use of a kernel K(x), which is a fuzzy subset. The result of application of a fuzzification to A by using the kernel K(x) is denoted by F(A;K) and is defined by

$$F(A;K) = \bigcup_{x \in \text{support}(A)} u_A(x)K(x)$$

where $u_A(x)K(x)$ is a fuzzy set defined as

$$u_A(x)K(x)=\{u(y)/y \mid u(y)=u_A(x)u_{K(x)}(y) \text{ and } u_{K(x)}(y)/y \in K(x)\}.$$

We note that if $A = \{1.0/x\}$, then $F(A;K) = K(x)$. Therefore, the kernel $K(x)$ is the fuzzy set resulting from the fuzzification of a singleton $\{1.0/x\}$. Note that $K(x)$ plays a role similar to the role of impulse response in linear systems.

As an illustration of fuzzification, assume that U, A, and K(x) are defined by

$$U = \{1,2,3,4\}$$
$$A = \{0.8/1,\ 0.6/2\}$$
$$K(1) = \{1.0/1,\ 0.4/2\}$$
$$K(2) = \{1.0/2,\ 0.4/1,\ 0.4/3\}.$$

Then, the fuzzification of A is given by

$$\begin{aligned} F(A;K) &= 0.8\{1.0/1,\ 0.4/2\} \cup 0.6\{1.0/2,\ 0.4/1,\ 0.4/3\} \\ &= \{0.8/1,\ 0.32/2\} \cup \{0.6/2,\ 0.24/1,\ 0.24/3\} \\ &= \{0.8/1,\ 0.6/2,\ 0.24/3\}. \end{aligned}$$

(D) Hedge Operations

Examples of hedge operations are very, slightly, approximately, slightly, more or less, etc. A detail discussion of hedges may be found in [7,8].

First, we define very. If A is a fuzzy subset, then very A, (or more precisely, very(A),) may be defined

$$\text{very } A \triangleq \{(u_A(x))^2/x \mid u_A(x)/x \in A\}.$$

Next, we consider hedges approximately, slightly, more or less, etc. These hedges are defined through fuzzifications, which are in turn defined by kernels. Different kernels may give different interpretations of these hedges. We shall use an example to illustrate the method of defining the hedge approximately. Other hedges may be treated similarly.

Let the universe of discourse U be given as

$$U = \{1,\ 1.5,\ 4\}.$$

Let equal be a fuzzy relation for U x U given

equal = {1.0/(1,1), 1.0/(1.5,1.5), 1.0/(4,4)}.

Assume we choose a kernel K defined by

K(1,1) = {1.0/(1,1), 0.8/(1,1.5), 0.8/(1.5,1)}
K(1.5,1.5) = {1.0/(1.5,1.5), 0.8/(1.5,1), 0.8/(1,1.5)}
K(4,4) = {1.0/(4,4), 0.2/(4,1.5), 0.2/(1.5,4)}.

Then, using the kernel K, we may define a fuzzy relation approximately equal by

approximately equal = F(equal;K)
= 1.0{1.0/(1,1),0.8/(1,1.5),0.8/(1.5,1)}
U 1.0{1.0/(1.5,1.5),0.8/(1.5,1),0.8/(1,1.5)}
U 1.0{1.0/(4,4),0.2/(4,1.5),0.2/(1.5,4)}
= {1.0/(1,1),1.0/(1.5,1.5),1.0/(4,4),
0.8/(1,1.5),0.8/(1.5,1),0.2/(1.5,4),
0.2/(4,1.5)}.

We note that while the method of defining hedges is given as above, the same hedge may have different meanings in different problem domains. A proper interpretation of a hedge for a problem domain may be obtained by using a proper kernel for fuzzification.

(E) Function Operations

A function operation is a mapping that operates on elements of supports of fuzzy subsets. For example, let A and B be fuzzy subsets given by

A = {0.2/1, 0.8/5}
B = {0.4/2, 0.6/3}.

Then, A + B is defined by

$$A + B = \{0.2/1,\ 0.8/5\} + \{0.4/2,\ 0.6/3\}$$
$$= \{\frac{0.2 \wedge 0.4}{1+2},\ \frac{0.2 \wedge 0.6}{1+3},\ \frac{0.8 \wedge 0.4}{5+2},\ \frac{0.8 \wedge 0.6}{5+3}\}$$
$$= \{0.2/3,\ 0.2/4,\ 0.4/7,\ 0.6/8\}.$$

In general, let $A_1,\ldots,A_n$ be fuzzy subsets of $U_1,\ldots,U_n$, respectively. Assume g is a function that mapps $U_1 \times\ldots\times U_n$ into U. Then, $g(A_1,\ldots,A_n)$ is a fuzzy subset of U given by

$$g(A_1,\ldots,A_n) = \{u(x)/x | x=g(x_1,\ldots,x_n),\ u(x)=u_{A_1}(x_1)\wedge\ldots\wedge u_{A_n}(x_n), \text{ and } u_{A_i}(x_i)/x_i \in A_i,\ i=1,\ldots,n\}$$

4. FUZZY PROGRAMS

A fuzzy program can be defined as either a sequence of fuzzy statements, or as a flowchart. For clarity of presentation, we shall use a flowchart to define a fuzzy program. This is done by first defining a directed graph as follows:

Definition. A directed graph consists of a nonempty set V, a set A disjoint from V, and a mapping D from A into V x V. The elements of V and A are called vertices and arcs, respectively, and D is called the directed incidence mapping associated with the directed graph. If $a \in A$ and $D(a) = (v,v')$, then v and v' are called the initial and terminal vertices of arc a, respectively. A finite directed graph is a graph whose number of vertices and arcs is finite. A sequence of vertices $v_1,\ldots,v_q$ denotes a path if v_i and v_{i+1} are the initial and terminal vertices of an arc, respectively, for $i=1,\ldots,q-1$.

Now, we formally define a fuzzy program.

Definition. A fuzzy program consists of an input vector $\bar{x}=(x_1,\ldots,x_L)$, a program vector $\bar{y}=(y_1,\ldots,y_M)$, an output vector $\overline{z}=(z_1,\ldots,z_N)$, and a finite directed graph G such that the following conditions are satisfied:

(1) $x_1,\ldots,x_L,y_1,\ldots,y_m,z_1,\ldots,z_N$ are variables representing fuzzy subsets;

(2) In the graph G, there is exactly one vertex called the start vertex that is not a terminal vertex of any arc; and there is exactly one vertex called the halt vertex that is not an initial vertex of any arc; and every vertex is on some path from the start vertex S to the halt vertex H;

(3) In G, each arc a not entering H is associated with a fuzzy relation $R_{\underline{a}}(\overline{x},\overline{y})$ called a fuzzy branching condition, and a fuzzy assignment $\overline{y} \leftarrow f_{\underline{a}}(\overline{x},\overline{y})$; and each arc a entering H is associated with a fuzzy relation $R_{\underline{a}}(\overline{x},\overline{y})$ and a fuzzy assignment $\overline{z} \leftarrow f_{\underline{a}}(\overline{x},\overline{y})$, where $R_{\underline{a}}$ and $f_{\underline{a}}$ are a fuzzy relation and an operation for fuzzy sets as discussed in Section 2 and Section 3, respectively.

An example of a fuzzy program P is shown in Fig. 4, where several is an input variable, y_1 and y_2 are program variables, and z is an output variable. All these variables represent fuzzy subsets of real numbers. In the fuzzy program P, approximately equal is a fuzzy relation, and + and - are functions operations "addition" and "subtraction" for fuzzy subsets, respectively.

5. EXECUTIONS (INTERPRETATIONS) OF FUZZY PROGRAMS

To interpret the fuzzy program P shown in Fig. 1, we have to define the hedge approximately. We know

$$\text{equal} = \{1.0/(x,x) \mid x \text{ is a real number}\}.$$

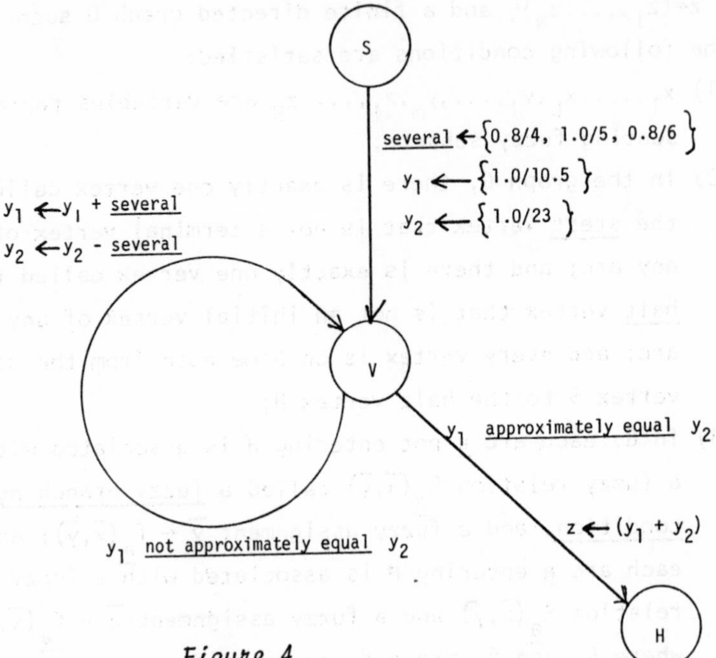

Figure 4

Suppose we choose a kernel given by

$K(x,x) = \{u(y,z)/(y,z)\,|\,u(y,z)=1-d \text{ if } y+z=2x, |y-z| \leq d$ and $0 \leq d \leq 1\}$.

If we define approximately equal by fuzzifying equal by K, we obtain

$$\underline{\text{approximately equal}} = \{u(y,z)/(y,z)\,|\,u(y,z)=1-d \text{ if } |y-z| \leq d \text{ and } 0 \leq d \leq 1\}.$$

(1) Control is assumed initially at vertex S and the values of y_1, y_2 and z are arbitrary. This is summarized in Table 1, where a dash sign "-" indicates an aribtrary value.

Table 1

Path having taken by control	y_1	y_2	z
S	-	-	-

(2) From vertex S, control can go to vertex V. The path is SV, and the values of y_1 and y_2 are {1.0/10.5} and {1.0/23}, respectively. The value of z is still arbitrary. When path SV is taken by control, Table 1 is no longer true. Therefore, we obtain Table 2, where a check mark "✓" is put in front of row S to indicate that this row is no longer available.

Table 2

	path having taken by control	y_1	y_2	z
✓	S	-	-	-
	SV	{1.0/10.5}	{1.0/23}	-

(3) In Table 2, row SV is unchecked. Therefore, from SV, control can take the paths SVH and SVV in parallel. Thus, we generate Talbe 3 as follows.

Table 3

	path having taken by control	y_1	y_2	z
✓	S	-	-	-
✓	SV	{1.0/10.5}	{1.0/23}	-
	SVH	Φ	Φ	Φ
	SVV	{0.8/14.5,1.0/15.5, 0.8/16.5}	{0.8/19,1.0/18, 0.8/17}	-

In row SVH of Table 3, the values of y_1, y_2 and z are determined as follows: Let AE stand for <u>approximately equal</u>. Note that y_1AEy_2 is a fuzzy relation. We shall evaluate (y_1AEy_2) as changing the values of y_1 and y_2 to the projections of (y_1AEy_2) on y_1 and y_2, respectively. However, according to Property 1 of Section 2, it is equivalent to changing the values of y_1 and y_2 by the following assignments

$$y_1 \leftarrow y_1 \cap y_1^*$$
$$y_2 \leftarrow y_2 \cap y_2^*$$

where y_1^* and y_2^* are induced fuzzy subsets given by

$$y_1^* = AE \circ y_2$$
$$y_2^* = y_1 \circ AE.$$

The matrix representations of y_1, y_2, AE and not AE are given by

$$y_1:\quad \begin{matrix} 10.5 \\ [1] \end{matrix}$$

$$y_2:\quad \begin{matrix} 23 \\ [1] \end{matrix}$$

$$AE:\quad 10.5\overset{23}{[0]}$$

$$\underline{\text{not}}\ AE:\quad 10.5\overset{23}{[1]}$$

Now, y_1^* and y_2^* are calculated by max-min matrix products as follows:

$$y_1^* = AE \circ y_2 = [0]\ [1] = [0]$$
$$y_2^* = y_1 \circ AE = [1]\ [0] = [0]$$

Therefore, $y_1^* = \Phi$ and $y_2^* = \Phi$. (Note that Φ denotes the empty fuzzy subset.) Thus, we have

$$y_1 \leftarrow y_1 \cap y_1^* \leftarrow \{1.0/10.5\} \cap \Phi \leftarrow \Phi,$$
$$y_2 \leftarrow y_2 \cap y_2^* \leftarrow \{1.0/23\} \cap \Phi \leftarrow \Phi.$$

After $y_1 AE y_2$ is executed, we then execute the following assignment

$$z \leftarrow y_1 + y_2 \leftarrow \Phi + \Phi \leftarrow \Phi.$$

Therefore, when the path SVH is taken by control, the values of y_1, y_2 and z are Φ, as indicated in row SVH of Table 3.

Similarly, the values of y_1, y_2 and z in row SVVV of Table 3 are computed as follows:

Now, the test is y_1(not AE)y_2. Again, we interpret y_1(not AE)y_2 as executing the following assignments:

$$y_1 \leftarrow y_1 \cap y_1^*$$
$$y_2 \leftarrow y_2 \cap y_2^*$$

where $y_1^* = (\underline{\text{not}}\ \text{AE}) \circ y_2$ and $y_2^* = y_1 \circ (\underline{\text{not}}\ \text{AE})$.

Since

$$y_1^* = (\underline{\text{not}}\ \text{AE}) \circ y_2 = [1]\ [1] = [1]$$
$$y_2^* = y_1 \circ (\underline{\text{not}}\ \text{AE}) = [1]\ [1] = [1],$$

we have $y_1^* = \{1.0/10.5\}$ and $y_2^* = \{1.0/23\}$.

Therefore,

$$y_1 \leftarrow y_1 \cap y_1^* \leftarrow \{1.0/10.5\} \cap \{1.0/10.5\} \leftarrow \{1.0/10.5\}$$
$$y_2 \leftarrow y_2 \cap y_2^* \leftarrow \{1.0/23\} \cap \{1.0/23\} \leftarrow \{1.0/23\}.$$

After y_1(not AE)y_2 is interpreted, we proceed to interpret the following assignments:

$$y_1 \leftarrow y_1 + \underline{\text{several}}$$
$$\leftarrow \{1.0/10.5\} + \{0.8/4,\ 1.0/5,\ 0.8/6\}$$
$$\leftarrow \{\frac{1.0 \wedge 0.8}{10.5 + 4}, \frac{1.0 \wedge 1.0}{10.5 + 5}, \frac{1.0 \wedge 0.8}{10.5 + 6}\}$$
$$\leftarrow \{0.8/14.5,\ 1.0/15.5,\ 0.8/16.5\}, \text{ and}$$

$$y_2 \leftarrow y_2 - \underline{\text{several}}$$
$$\leftarrow \{1.0/23\} - \{0.8/4,\ 1.0/5,\ 0.8/6\}$$
$$\leftarrow \{\frac{1.0 \wedge 0.8}{23-4}, \frac{1.0 \wedge 1.0}{23-5}, \frac{1.0 \wedge 0.8}{23-6}\}$$
$$\leftarrow \{0.8/19,\ 1.0/18,\ 0.8/17\}.$$

The values of y_1 and y_2 are thus recorded in row SVV of Table 3. In Table 3, generations of rows SVH and SVV from row SV is called the <u>expansion</u> of SV. SVH and SVV are called the <u>successors</u> of SV in the terminology of tree searching [12,13]. When a row is expanded, it is check-marked.

(4) In Table 3, SVH can not be further expanded. However, SVV can be expanded. SVVH and SVVV are successors of SVV. Expanding SVV, we obtain the following table:

Table 4

	path having taken by control	y_1	y_2	z
√	S	-	-	-
√	SV	{1.0/10.5}	{1.0/23}	-
	SVH	Φ	Φ	Φ
√	SVV	{0.8/14.5,1.0/15.5, 0.8/16.5}	{0.8/19,1.0/18, 0.8/17}	-
	SVVH	{0.5/16.5}	{0.5/17}	{0.5/33.5}
	SVVV	{0.8/18.5,0.8/19.5, 1.0/20.5,0.8/21.5, 0.8/22.5}	{0.8/15,0.8/14, 1.0/13,0.8/12, 0.8/11}	-

The values of y_1, y_2 and z in row SVVH of Table 4 are obtained as follows:

The matrix representations of y_1, y_2 and AE are:

$$y_1: \begin{matrix} 14.5 & 15.5 & 16.5 \\ [0.8 & 1.0 & 0.8] \end{matrix}$$

$$y_2: \begin{matrix} 19 & 18 & 17 \\ [0.8 & 1.0 & 0.8] \end{matrix}$$

$$AE: \begin{matrix} & 19 & 18 & 17 \\ 14.5 & 0 & 0 & 0 \\ 15.5 & 0 & 0 & 0 \\ 16.5 & 0 & 0 & 0.5 \end{matrix}$$

Therefore,

$$y_1^* = AE \circ y_2 = \begin{bmatrix} 0 & 0 & 0 \\ 0 & 0 & 0 \\ 0 & 0 & 0.5 \end{bmatrix} \begin{bmatrix} 0.8 \\ 1.0 \\ 0.8 \end{bmatrix} = \begin{bmatrix} 0 \\ 0 \\ 0.5 \end{bmatrix}$$

$$y_2^* = y_1 \circ AE = [0.8\ 1.0\ 0.8] \begin{bmatrix} 0 & 0 & 0 \\ 0 & 0 & 0 \\ 0 & 0 & 0.5 \end{bmatrix} = [0\ 0\ 0.5]$$

Thus, $y_1^* = \{0.5/16.5\}$ and $y_2^* = \{0.5/17\}$.

First, the interpretation of $y_1 AE y_2$ is

$$\begin{aligned} y_1 &\leftarrow y_1 \cap y_1^* \\ &\leftarrow \{0.8/14.5, 1.0/15.5, 0.8/16.5\} \cap \{0.5/16.5\} \\ &\leftarrow \{0.5/16.5\}\ , \text{ and} \end{aligned}$$

$$\begin{aligned} y_2 &\leftarrow y_2 \cap y_2^* \\ &\leftarrow \{0.8/19, 1.0/18, 0.8/17\} \cap \{0.5/17\} \\ &\leftarrow \{0.5/17\} \end{aligned}$$

Then, the interpretation of $z \leftarrow y_1 + y_2$ is given by

$$\begin{aligned} z &\leftarrow y_1 + y_2 \\ &\leftarrow \{0.5/16.5\} + \{0.5/17\} \\ &\leftarrow \left\{\frac{0.5 \wedge 0.5}{16.5 + 17}\right\} \\ &\leftarrow \{0.5/33.5\}. \end{aligned}$$

Therefore, we have the above values of y_1, y_2 and z shown in row SVVH of Table 4.

Similarly, the values of y_1, y_2 and z in row SVVV are computed as follows: The matrix representations of y_1, y_2 and <u>not</u> AE are

	14.5	15.5	16.5
y_1:	[0.8	1.0	0.8]
	19	18	17
y_2:	[0.8	1.0	0.8]

$$\underline{\text{not}}\ AE: \begin{array}{c} \\ 14.5 \\ 15.5 \\ 16.5 \end{array} \begin{array}{c} \begin{array}{ccc} 19 & 18 & 17 \end{array} \\ \begin{bmatrix} 1.0 & 1.0 & 1.0 \\ 1.0 & 1.0 & 1.0 \\ 1.0 & 1.0 & 0.5 \end{bmatrix} \end{array}$$

Therefore,

$$y_1^* = (\underline{\text{not}}\ AE) \circ y_2 = \begin{bmatrix} 1.0 & 1.0 & 1.0 \\ 1.0 & 1.0 & 1.0 \\ 1.0 & 1.0 & 0.5 \end{bmatrix} \begin{bmatrix} 0.8 \\ 1.0 \\ 0.8 \end{bmatrix} = \begin{bmatrix} 1.0 \\ 1.0 \\ 1.0 \end{bmatrix}$$

$$y_2^* = y_1 \circ (\underline{\text{not}}AE) = [0.8 \quad 1.0 \quad 0.8] \begin{bmatrix} 1.0 & 1.0 & 1.0 \\ 1.0 & 1.0 & 1.0 \\ 1.0 & 1.0 & 0.5 \end{bmatrix}$$

$$= [1.0 \quad 1.0 \quad 1.0].$$

Thus, $y_1^* = \{1.0/14.5,\ 1.0/15.5,\ 1.0/16.5\}$, and $y_2^* = \{1.0/19,\ 1.0/18,\ 1.0/17\}$. First, the interpretation of $y_1\,(\underline{\text{not}}\ AE)\,y_2$ is

$$\begin{aligned} y_1 &\leftarrow y_1 \cap y_1^* \\ &\leftarrow \{0.8/14.5, 1.0/15.5, 0.8/16.5\} \cap \\ &\qquad \{1.0/14.5, 1.0/15.5, 1.0/16.5\} \\ &\leftarrow \{0.8/14.5, 1.0/15.5, 0.8/16.5\}, \text{ and} \\ y_2 &\leftarrow y_2 \cap y_2^* \\ &\leftarrow \{0.8/19, 1.0/18, 0.8/17\} \cap \{1.0/19, 1.0/18, 1.0/17\} \\ &\leftarrow \{0.8/19, 1.0/18, 0.8/17\}. \end{aligned}$$

Then, the interpretation of $y_1 \leftarrow y_1 + \underline{\text{several}}$ and $y_2 \leftarrow y_2 - \underline{\text{several}}$ are given by

$$\begin{aligned} y_1 &\leftarrow y_1 + \underline{\text{several}} \\ &\leftarrow \{0.8/14.5, 1.0/15.5, 0.8/16.5\} + \{0.8/4, 1.0/5, 0.8/6\} \\ &\leftarrow \{0.8/18.5, 0.8/19.5, 1.0/20.5, 0.8/21.5, 0.8/22.5\}, \\ y_2 &\leftarrow y_2 - \underline{\text{several}} \\ &\leftarrow \{0.8/19, 1.0/18, 0.8/17\} - \{0.8/4, 1.0/5, 0.8/6\} \\ &\leftarrow \{0.8/15, 0.8/14, 1.0/13, 0.8/12, 0.8/11\}. \end{aligned}$$

The above computed values of y_1, y_2 are shown in row SVVV of Table 4.

We can keep extending Table 4 by expanding any unexpanded row of Table 4. The question now is when do we stop? The answer is that we should inspect the value of the output variable z. If z has an element which has a high grade of membership, we may stop the execution of the fuzzy program and output the element of z which has the highest grade of membership in z. If such an element is not unique, then a random or arbitrary choice can be made among the elements having the highest grade of membership. Of course, other stopping rules may be used. For example, if there are elements of z whose grades of membership are greater than a prespecified threshold value, output all these elements. Or we may want to use a criterion which minimizes the number of execution steps, etc. Alternately, we may use an interactive system. In this case, we can print out at a terminal the values of z whenever they are generated. A user sitting in front of the terminal may examine the printout and use whatever criterion to choose elements of z and stop the execution of the program.

For the example considered here, the first time z is not empty is when the path SVVH is taken by control. We may stop the execution and output the value 33.5 with its grade of membership being 0.5. However, if we are not satisfied with this value, we may continue expanding row SVVV of Table 4, hoping that better values may be generated for z.

There is a warning that we should mention here. That is, no matter what criterion we use, we should not accept the values of output variables blindly. For example, consider the fuzzy program shown in Fig. 5.

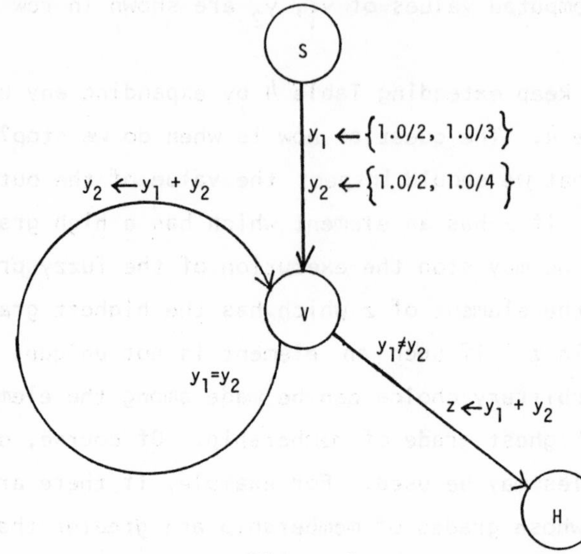

Figure 5

The execution of the above program can be summarized in Table 5.

Table 5

	path having taken by control	y_1	y_2	z
(1)	S	-	-	-
(2)	SV	{1.0/2,1.0/3}	{1.0/2, 1.0/4}	-
(3)	SVH	{1.0/2,1.0/3}	{1.0/2, 1.0/4}	{1.0/4,1.0/5, 1.0/6,1.0/7}
(3)	SVV	{1.0/2}	{1.0/4}	-
(4)	SVVH	{1.0/2}	{1.0/4}	{1.0/6}
(4)	SVVV	Φ	Φ	-

The leftmost numbers in Table 5 indicate the order that the rows are expanded. For this example, when row SVH is generated, we may just output the value of z and terminate the execution. However, we note that 1.0/4 is not a correct element for z. We note that 1.0/4 is obtained from 1.0/2 of y_1 and 1.0/2 of y_2. However, for $y_1 = \{1.0/2\}$ and $y_2 = \{1.0/2\}$, the grade of the truthness of the branching condition $y_1 \neq y_2$ is 0. Therefore 1.0/4 should not be an element of z. In other words, in order that an element, say u/x, to be considered an element of z, all the fuzzy branching conditions for deriving u/x must have the grades of the truthness greater than or equal to u. This means that we may have to trace how u/x is computed and make sure that the fuzzy branching conditions have proper grades of truthness.

Finally, we give some justifications of the way we interpret a fuzzy program. In general, there are two types of arcs in a fuzzy program. They are shown in Fig. 6. The interpretation of Fig. 6(a) is very simple. We just update

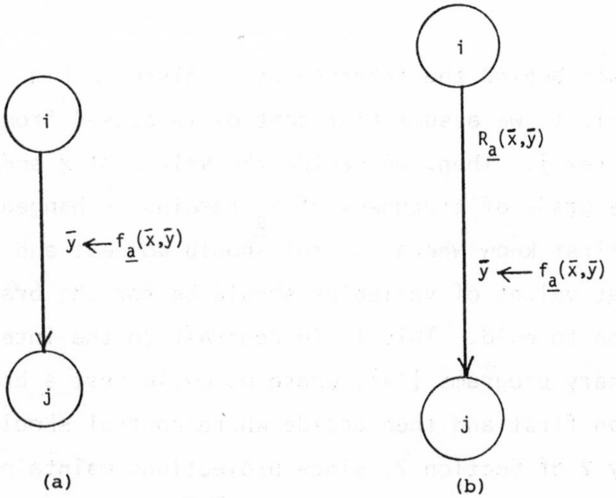

Figure 6

the value of the program vector $\bar{y}$ by performing the operation $f_{\underline{a}}$ on the existing values of the program vector $\bar{y}$ and the input vector $\bar{x}$. However, the interpretation of Fig. 6(b) is tricky. We interpret Fig. 6(b) as the one shown in Fig. 7.

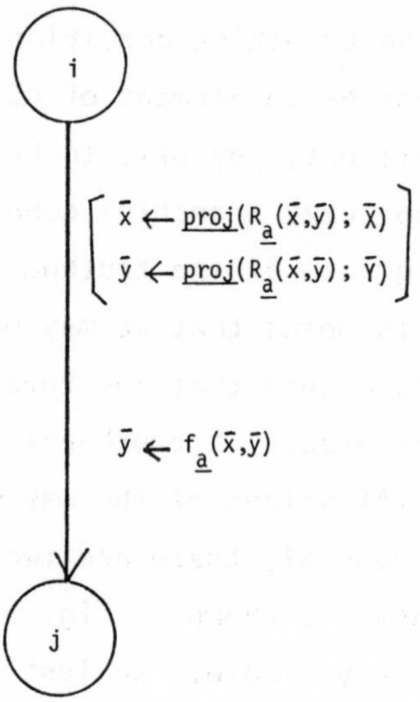

Figure 7

The reason behind the interpretation given in Fig. 7 is this: First, we assume that control is passed from vertex i to vertex j. Then, we decide the values of $\bar{x}$ and $\bar{y}$ such that the grade of truthness of $R_{\underline{a}}$ remains unchanged. That is, we first know where control should go next and then decide what values of variables should be for the branching condition to hold. This is in contrast to the interpretation of ordinary programs [14], where we would test a branching condition first and then decide where control should go. By Property 2 of Section 2, since projections maintain the same degree of the truthness of $R_{\underline{a}}(\bar{x},\bar{y})$ as the original fuzzy subsets, the interpretation of Fig. 7 is a reasonable interpretation. (We note that $\underline{\text{proj}}(R_a(\overline{x},\overline{y});\overline{X})$ and $\underline{\text{proj}}(R_a(\overline{x},\overline{y});\overline{Y})$ can be computed as $\underline{\text{proj}}(R_a(\overline{x},\overline{y});\overline{X}) = x \cap x^*$ and $\underline{\text{proj}}(R_a(\overline{x},\overline{y});\overline{Y}) = \overline{y} \cap \overline{y}^*$, respectively, where $\overline{x}^* = (x_1^*,\ldots,x_L^*)$ and $\overline{y}^* = (y_1^*,\ldots,y_M^*)$ are vectors of induced fuzzy subsets with respect to $R_a(\overline{x},\overline{y})$. See Property 1 of Section 2.) As an illustration, let us consider a simple example in a non-fuzzy case. Suppose we have an arc shown in Fig. 8. Assume control is at vertex i. If y_1=1 and y_2=2, or more precisely in our notation, $y_1 = \{1.0/1\}$ and $y_2 = \{1.0/2\}$, and if control is passed to vertex j, then y_1=y_2 is interpreted as

$$y_1 \leftarrow \underline{\text{proj}}(y_1{=}y_2;Y_1) \leftarrow y_1 \cap y_1^* \leftarrow \{1.0/1\} \cap \Phi \leftarrow \Phi,$$
$$y_2 \leftarrow \underline{\text{proj}}(y_1{=}y_2;Y_2) \leftarrow y_2 \cap y_2^* \leftarrow \{1.0/2\} \cap \Phi \leftarrow \Phi.$$

Therefore, $y_3 \leftarrow y_1 + y_2 \leftarrow \Phi + \Phi \leftarrow \Phi$. This means that since y_3 is empty, it will not be fruitful to go to vertex j. On the other hand, if $y_1 = 2$ and $y_2 = 2$, or more precisely $y_1 = \{1.0/2\}$ and $y_2 = \{1.0/2\}$, and if control is passed to vertex j, then $y_1 = y_2$ is interpreted as

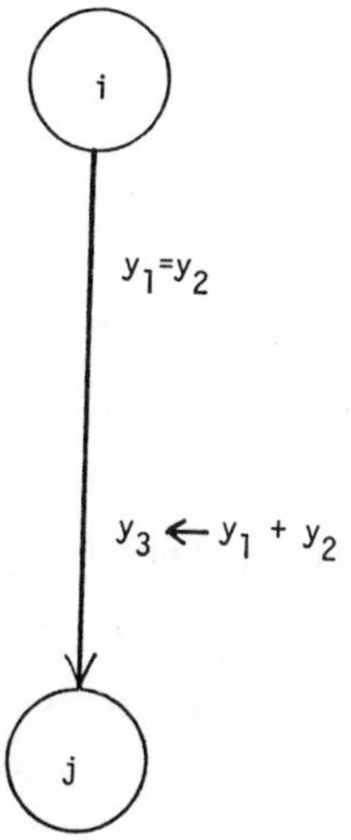

Figure 8

$y_1 \leftarrow \underline{proj}(y_1{=}y_2;Y_1) \leftarrow y_1 \cap y_1^* \leftarrow \{1.0/2\} \cap \{1.0/2\} \leftarrow \{1.0/2\}$,
$y_2 \leftarrow \underline{proj}(y_1{=}y_2;Y_2) \leftarrow y_2 \cap y_2^* \leftarrow \{1.0/2\} \cap \{1.0/2\} \leftarrow \{1.0/2\}$.
Thus, $y_3 \leftarrow y_1 + y_2 \leftarrow \{1.0/2\} + \{1.0/2\} \leftarrow \{1.0/4\}$. Therefore, it will be fruitful for control to go to vertex j, since y_3 is not empty.

6. MODELING ILL-DEFINED PROCEDURES BY FUZZY PROGRAMS

As discussed in [2,4], there are many ill-defined procedures used by human in the real world. For example, the procedures we use for driving a car, searching for an object, tieing a knot, parking a car, cooking a meal, composing music, building a house, etc., are all ill-defined. Even though these procedures are imprecise, we seem to have no trouble of executing them. However, when we want to implement these procedures on machines, we encounter tremendous difficulties. It is our hope that some of the ill-defined procedures encountered in real life may be implemented by fuzzy programs. In the sequel, we shall use an example to illustrate how this could be done.

Suppose we want to adjust a TV set for clear pictures. The following is a procedure P we usually follow: Turn an adjusting knob until a clear picture is obtained. If the knob is overturned, turn it back a bit.

In procedure P, <u>bit</u> and <u>clear</u> are fuzzy. To model procedure P, we first identify the universes of discourse. There are two universes of discourse, namely, the set U_1 of all knob angles and the set U_2 of all TV pictures. Then, <u>bit</u> is a fuzzy subset of U_1, and <u>clear picture</u> is a fuzzy subset of U_2. Let variables y_1 and y_2 denote fuzzy subsets of U_1 and U_2, respectively. That is, y_1 denotes knob angles and y_2 denotes pictures. Let <u>clearer than</u> be a fuzzy relation for $U_2 \times U_2$. Let PICTURE(θ) be a function operation that gets a picture for a knob angle θ. Then, Procedure P may be described by the fuzzy program shown in Fig. 9.

7. CONCLUDING REMARKS

There are many ways to interpret and execute a fuzzy program [2,4,5,6,10]. In this paper, we execute a fuzzy program by tree-searching methods [12,13]. In addition, we have shown that we can use fuzzy programs to describe ill-defined or complex problems (procedures).

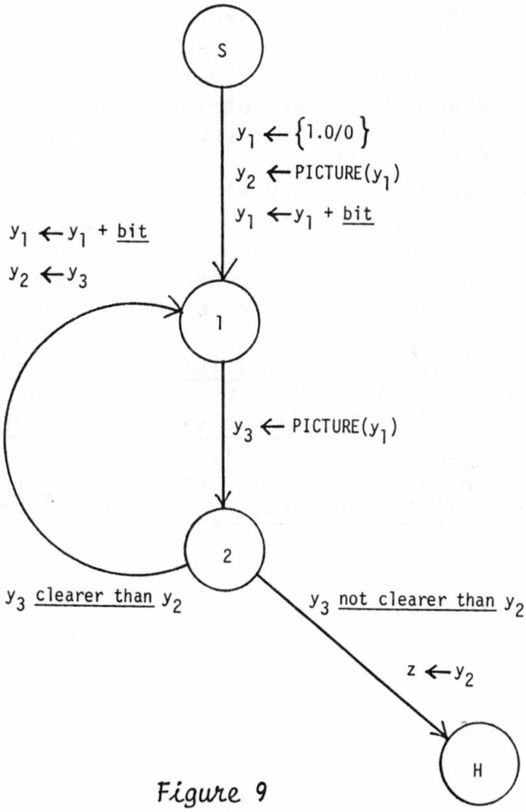

Figure 9

REFERENCES

1. L. A. Zadeh, "Fuzzy Sets," Information and Control, Vol. 8, pp. 338-353, 1965.
2. L. A. Zadeh, "Fuzzy Algorithms," Information and Control, Vol. 12, pp. 94-102, 1968.
3. L. A. Zadeh, "Quantitative Fuzzy Semantics," Information Science, Vol. 3, pp. 159-176, 1971.
4. L. A. Zadeh, "Outline of a New Approach to the Analysis of Complex Systems and Decision Processes," IEEE Trans. on Systems, Man and Cybernetics, Vol. SMC-3, No. 1, pp. 28-44, Jan. 1973.
5. E. Santos, "Fuzzy Algorithms," Information and Control, Vol. 17, pp. 326-339, 1970.
6. S. K. Chang, "On the Execution of Fuzzy Programs Using Finite-State Machines," IEEE Trans. on Computers, Vol. C-21, pp. 241-253, March 1972.
7. G. Lakoff, "Hedges: A Study in Meaning Criteria and the Logic of Fuzzy Concepts," in Proc. 8th Reg. Meeting Chicago Linguist. Soc., 1972.
8. L. A. Zadeh, "A Fuzzy-Set-Theoretic Interpretation of Hedges," Electron. Res. Lab., Univ. California, Berkeley, Memo. M-335, 1972.
9. L. A. Zadeh, "On Fuzzy Algorithms," Electron. Res. Lab., Univ. Calif., Berkeley, Memo. M-325, 1971.
10. K. Tanaka and M. Mizumoto, "Fuzzy Programs and Their Executions," appear elsewhere in this proceedings.
11. H. A. Simon, "The Structure of Ill Structured Problems," Artificial Intelligence, Vol. 4, pp. 181-201, 1973.
12. J. R. Slagle, Artificial Intelligence: The Heuristic Programming Approach, McGraw-Hill, 1971.
13. N. J. Nilsson, Problem-Solving Methods in Artificial Intelligence, McGraw-Hill, 1971.
14. C. L. Chang and R. C. T. Lee, Symbolic Logic and Mechanical Theorem Proving, Academic Press, 1973.

ON RISK AND DECISION MAKING IN A FUZZY ENVIRONMENT

Sheldon S. L. Chang
Department of Electrical Sciences
State University of New York
Stony Brook, New York 11794

ABSTRACT

A multistage decision process is a fuzzy mapping from $X \times U^N \to X$ where X and U are the state and policy spaces. Given the initial fuzzy set S_i on X, the set of final fuzzy sets S_f on X has certain inclusive properties. Risk and optimum decisions are defined in terms of the inclusive properties.

INTRODUCTION

As control theory progresses from relatively simple systems to large scale systems then to ill-defined socio-economic systems, the treatment of uncertainty or fuzziness becomes a factor of increasing significance. One important concept which emerges with uncertainty is that of risk. There is no longer an optimum policy but a class of optimum policies with varying degrees of risk. This concept will be clarified mathematically with the help of fuzzy set theory.

Heuristically one may describe uncertainty in our knowledge about the state of a system as a fuzzy sphere (convex fuzzy set) in state space, and uncertainty in the system model as a mapping from a point to a fuzzy sphere (convex fuzzy mapping). Obviously fuzziness will be propagated from time to time. One question then is "Will convexity be propagated or conserved?" or alternatively "Will our fuzzy sphere be eventually transformed into some odd shape by the system dynamics?" We shall answer these questions in the subsequent sections.

CONVEXITY AND FUZZY MAPPING

Definitions

1. Fuzzy Mapping [1]

A fuzzy mapping f from X to Y is a fuzzy set on X x Y with membership function $\mu_f(x,y)$. A fuzzy function f(x) is a fuzzy set on Y with membership function

$$\mu_{f(x)}(y) = \mu_f(x,y). \tag{1}$$

Its inverse $f^{-1}(y)$ is a fuzzy set on X with

$$\mu_{f^{-1}(y)}(x) = \mu_f(x,y). \tag{2}$$

2. Fuzzy Mapping of a Fuzzy Set [1]

Let A be a fuzzy set on X. The fuzzy set f(A) is defined as

$$\mu_{f(A)}(y) \equiv \sup_{x\varepsilon X} (\mu_A(x) \wedge \mu_f(x,y)) \tag{3}$$

3. Merged Fuzzy Set on Product Space

Let A and B be fuzzy sets on X, and Y respectively. C is said to be the merged fuzzy set of A and B on product space X X Y if its membership function satisfies the following relation

$$\mu_C(x,y) = \mu_A(x) \wedge \mu_B(y) \tag{4}$$

4. Convex Fuzzy Mapping

A fuzzy mapping f from X to Y is said to be a convex fuzzy mapping if it is a convex fuzzy set on X x Y.

5. Function of More Than One Variables.

A fuzzy mapping f from X_1 X_2 ... X_n to Y is a fuzzy set on $X_1 \times X_2 \times \ldots X_n \times Y$. The mapping f is said to be convex if it is a convex fuzzy set on $X_1 \times X_2 \ldots \times Y$.

6. Mapping of Two or More Fuzzy Sets.

Let A_1, A_2 ... A_n be fuzzy sets on X_1, X_2, ... X_n respectively. The function $f(A, A_2 \ldots A_n)$ is a fuzzy set on Y with membership function

$$\mu_f(A_1,A_2,,A_n)(y) = \max_{x_1,x_2..x_n} \mu_{A_1}(x_1)..\Lambda\mu_{A_n}(x_n)\Lambda\mu_f(x_1,x_2...x_n,y) \quad (5)$$

7. Projection of Fuzzy Sets

Let A be a fuzzy set on X and X_1, X_2 are subspaces of X such that $X_1 \times X_2 = X$. The projection of A on X_1 is defined by

$$\mu_{P(A)}(x_1) = \max_{x_2} \mu_A(x_1, x_2) \quad (6)$$

THEOREMS ON CONVEXITY

Theorem 1. If A and B are convex fuzzy sets on X and Y respectively, then its merged fuzzy set C is convex on X x Y.

Proof. Given (x_1, y_1), (x_2, y_2) such that

$$\mu_c(x_1, y_1) \geq \mu_c(x_2, y_2) \geq C_1$$

Then by definition $\mu_A(x_1) \geq C_1$, $\mu_B(y_1) \geq C_1$, $\mu_A(x_2) \geq C_1$, and $\mu_B(y_2) \geq C_1$. Since A and B are convex, given any λ, $o < \lambda < 1$

$$\mu_A(\lambda x_1 + (1-\lambda)x_2) \geq C_1$$

$$\mu_B(\lambda y_1 + (1-\lambda)y_2) \geq C_1$$

From (4), the above inequalities imply

$$\mu_c(\lambda x_1 + (1-\lambda)x_2), (\lambda y_1 + (1-\lambda)y_2) \geq C_1$$

Theorem 2. If f is a convex fuzzy mapping from X to Y and A is a convex fuzzy set on X then f(A) is a convex fuzzy set on Y.

Proof. Let B denote f(A). Let $o < \lambda < 1$ and $\mu_B(y_1) \geq C_1$, $\mu_B(y_2) \geq C_1$. From (3), there exist x_1 and x_2 such that

$\mu_A(x_1) \geq C_1$, $\mu_A(x_2) \geq C_1$, $\mu_f(x_1, y_1) \geq C_1$, and $\mu_f(x_2, y_2) \geq C_1$.

Because f and A are convex

$$\mu_A(\lambda x_1 + (1-\lambda)x_2) \geq C_1$$

$$\mu_f(\lambda x_1 + (1-\lambda)x_2, \lambda y_1 + (1-\lambda)y_2) \geq C_1$$

The theorem follows from (3).

Theorem 3. The projection P(A) of a convex fuzzy set A is convex.

Proof. Let a_1 and b_1 be points on X_1 with membership function greater than C:

$$\mu_{P(A)}(a_1) \geq C$$

$$\mu_{P(A)}(b_1) \geq C$$

From (6), there exist points a_2, b_2 on X_2 such that

$$\mu_A(a_1, a_2) = \mu_{P(A)}(a_1) \geq C$$

$$\mu_A(b_1, b_2) = \mu_{P(A)}(b_1) \geq C$$

For any , $o < \lambda < 1$

$$\mu_{P(A)}(\lambda a_1 + (1-\lambda)b_1) \max_{x_2} \mu_A(\lambda a_1 + (1-\lambda)b_1, x_2)$$

$$\geq \mu_A(\lambda a_1 + (1-\lambda)b_1, \lambda a_2 + (1-\lambda)b_2) \geq C$$

Theorem 4. If $A_1, A_2 \ldots A_n$ are convex sets on $X_1, X_2 \ldots X_n$ respectively, and f: $X_1, X_2, \ldots X_n$ to Y is a convex fuzzy mapping, then $f(A_1, A_2 \ldots A_n)$ is a convex fuzzy set on Y.

Proof. Let A be the merged set of $A_1, A_2 \ldots A_n$. Then

$$\mu_A(x_1, x_2 \ldots x_n) = \mu_{A_1}(x_1) \wedge \mu_{A_2}(x_2) \ldots \wedge \mu_{A_n}(x_n)$$

From (5)

$$f(A) = f(A_1, A_2 \ldots A_n)$$

From Theorem 1, A is convex. Form Theorem 2, f(A) is convex, and Theorem 4 is proved.

In the following corollaries, $A_1, A_2 \ldots A_n$ are assumed to be convex fuzzy sets on $X_1, X_2 \ldots X_n$, and f is an ordinary or deterministic mapping.

Corollary 1: If f is a linear mapping from $X_1 \times X_2 \times \ldots x_n$ to Y, $f(A_1, A_2 \ldots A_n)$ is a convex fuzzy set on Y.

Proof. The set f is a linear subspace. It is a convex fuzzy set with membership function 1 on the subspace and 0 elsewhere.

Corollary 2: If Y is the real line, and f is a continuous mapping from X^n to Y, then $f(A_1, A_2, \ldots A_n)$ is a convex fuzzy set on Y.

Proof. Given $\mu_{f(A)}(y_a) \geq C_1$ and $\mu_{f(A)}(y_b) \geq C_1$, there exist x_a and x_b in X^n such that $\mu_A(x_a) \geq C_1$ and $\mu_A(x_b) \geq C_1$, $f(x_a) = y_a$ and $f(x_b) = y_b$. Let λ varies from 0 to 1, then $f(\lambda x_a + (1-\lambda)x_b)$ varies continuously from y_b to y_a. Since $\mu_A(\lambda x_a + (1-\lambda)x_b) \geq C_1$, $\mu_{f(A)}(y) \geq C_1$ for all y between y_a and y_b. The corollary is proved.

DEFINITION OF BINARY OPERATION

The binary operations of two fuzzy sets: $A_1 + A_2$, $A_1 - A_2$, $A_1 \times A_2$, A_1/A_2, $A_1 \circ A_2$ are defined as mappings of $X_1 \times X_2$ to Y with the mapping f defined by

$y = x_1 + x_2$, $y = x_1 - x_2$, $y = x_1 \times x_2$, $y = x_1/x_2$, $y = x_1 \circ x_2$.

Corollary 3: The binary operations +, - x, o preserves convexity. The binary operation / preserves convexity if the support of A_2 is a segment on the real line not including 0.

Proof. Corollary 3 follows from Corollaries 1 and 2.

MULTISTAGE DECISION PROCESS

Let $x = E^n$ represent the state space. Let $U \subset E^m$

represent the set of allowed controls. The state of the system is a fuzzy set p(t) on x. The dynamic system is represented by a fuzzy mapping from X x U into X

$$f: \quad X \times U \to X$$

$$\mu_{p(t+1)}(x(t+1)) = \mu_f(p(t), u; x(t+1))$$

An N-state decision process is represented by choice of u(t), t=0,1,2...N-1. The criterion of choice is to minimize a coordinate x_o of X.

Theorem 5. If in a multistage decision process,

(i) p(0) is a convex fuzzy set,

(ii) Given any $u \in U$, the mapping f: $X \to X$ is a convex fuzzy mapping, then the state fuzzy sets p(t), t = 1,2...N are convex. The projection of p(N) on x_o is convex.

Proof. The Theorem is proved by repeated applications of Theorem 2, and then Theorem 4.

Let u(•) denote the sequence u(1), u(2)...u(N-1). The projection of p(N) on x_o is the total cost, and will be denoted as C(u(•)). Since C(u(•)) is a convex fuzzy set, given any m < 1, the set of x_o with $\mu_{C(u(*))} \geq m$ is an interval I(m, u(•)) on the real line. Because C(u(•)) is convex the following inclusive property holds

$$I(m_1, u(\cdot)) \supset I(m_2, u(\cdot)) \quad \text{if } m_1 < m_2 \tag{7}$$

RISK AND OPTIMUM DECISION

Definition. Let $I_s(m,u(\cdot))$ and $I_i(m,u(\cdot))$ denote the upper and lower bounds (sup and inf) of I(m,u(•)) respectively. A control policy u*(•) is said to be m-optional if for any other u(•) the following conditions are valid:

(a) $I_s(m, u^*(\cdot)) > I_s(m, u(\cdot)) \to I_i(m, u^*(\cdot)) < I_i(m, u(\cdot))$

(8)

(b) $I_i(m, u^*(\cdot)) > I_i(m, u(\cdot)) \to I_s(m, u^*(\cdot)) < I_s(m, u(\cdot))$

The following theorems have been proved: [2]

Theorem 6. The set of m-optimal cost intervals is simply ordered by the strict contained-in relationship.

Theorem 7. If a control policy $u(\cdot)^*$ minimizes the quantity

$$\alpha\, I_i(m, u(\cdot)) + (1-\alpha)\, I_s(m, u(\cdot)) \tag{9}$$

then $u(\cdot)^*$ is a member of the m-optimal set, and α is said to be the risk parameter.

Theorem 8. If $\alpha_1 < \alpha_2$, then
$I(m, u_1(\cdot)) \subset I(m, u_2(\cdot))$

Theorem 9. Let $u_\alpha(\cdot)$ denote the m-optimal control which minimizes (9) with parameter α. If either $I_s(m, u_\alpha(\cdot))$ or $I_i(m, u_\alpha(\cdot))$ is continuous in α, then every m-optimal control minimizes (9) with some α.

APPLICATIONS

The set of m-optimal policies allow a decision maker to make two choices, m and α. If he is trying to win a race, he should choose a policy with $\alpha = 1$. If he is trying to accomplish some goal with safety first, he should choose a policy with $\alpha = o$. The choice of m is a subjective matter and is determined by the odds one is willing to accept.

The set of m-optimal or nearly optimal policies with $\alpha = 0+$, are called guaranteed cost policies. These policies can be determined by fuzzy dynamic programming as shown in references [4] and [5].

ACKNOWLEDGEMENT

This work was sponsored by the National Science Foundation under Grant No. GK-16017.

REFERENCES

[1] S. S. L. Chang and L. A. Zadeh, "On Fuzzy Mapping and Control", IEEE Transactions on Systems, Man and Cybernetics, Vol. SMC-2, No. 1, January, 1972.

[2] S. S. L. Chang and P. E. Barry, "Optimal Control of Systems with Uncertain Parameters", Fifth IFAC Congress, Paris, 1972.

[3] R. E. Bellman and L. A. Zadeh, "Decision Making on a Fuzzy Environment", Management Science, Vol. 17, pp. B-141 - B-164, 1970.

[4] S. S. L. Chang and T. K. C. Peng, "Adaptive Guaranteed Cost Control of Systems with Uncertain Parameters", IEEE Transactions AC-17, No. 4, 1972.

[5] S. S. L. Chang, "Control and Estimation of Fuzzy Systems", IEEE Conference on Decision and Control, November, 1974.

AN AXIOMATIC APPROACH TO RATIONAL DECISION MAKING IN A FUZZY ENVIRONMENT[†]

L. W. Fung and K. S. Fu
School of Electrical Engineering
Purdue University
West Lafayette, Indiana 47907

ABSTRACT

An axiomatic approach to rational decision making under uncertainty is presented. The notion of fuzzy sets is used to the advantage that it is a convenient tool for unifying the axioms in various situations, such as statistical decision, group decision and decision problems in which several criteria of optimality are involved. The first theorem asserts that under these axioms the decision strategy can be either pessimistic, optimistic, or one of mixed type. The introduction of a stronger postulate restricts our choice to only the pessimistic type, which in effect is the minimax principle. Meanwhile, the intersection and union operations on fuzzy sets are deduced from the assumptions.

I. INTRODUCTION

In the past many people were interested in building a logical foundation for decision making under uncertainty. Numerous results can be found in the pioneering works of Chernoff, Rubin, Milnor, Hurwitz, Savage and Arrow, and in recent works of Fine (1972) and Finetti (1972). A majority of these authors tended to construct utility-type decision functions which depend on the use of a priori information or subjective probabilities about the source of uncertainty.

[†]This work was supported by the National Science Foundation Grant GK-36721.

The most controversial of all the principles for making decision is undoubtedly the minimax principle, which clearly is a more pessimistic strategy than most people would like to employ (except possibly in the situation of playing games). Luce and Raiffa (1957) and Savage (1954) discussed at length views from both sides.

A very general class of decision-making problems is concerned with decisions made by a group, which has been used as a model of decision making for various reasons. Democratic theorists, economic as well as political, have long wrestled with the intriguing ethical question of how "best" to aggregate individual choices into social preferences and choices. Recently, Fung and Fu (1973b) considered the group decision model as a means of reducing excessive subjectiveness due to idiocyncracy of a single individual. On the other hand, it seems to be an appropriate model (Chapter 10, Savage 1954) for interpreting the minimax principle, in particular, since it provides the kind of terminology and intuition based on economical and social phenomena which are easy to argue. In this paper, the authors favor interpretations of assumptions and results in terms of group decision problems partly because of the reasons presented above, and partly because the notion of fuzzy sets can be incorporated into the problems as a very handy tool.

The purpose of this paper is to attempt to lay a logical foundation for the minimax principle, and at the same time to give an axiomatic formalism of the theory of fuzzy sets first studied by Bellman and Giertz (1973). In Section II we shall list a collection of basic assumptions which a rational decision-making scheme must satisfy. In Section III some lemmas and a theorem are proved, which asserts that the minimax principle together with two other strategies, are the only three candidates which comply with the axioms postulated in Section II. Two assumptions will be given in Section IV to replace a weaker one in the previous list, and it is shown that the minimax principle is the only choice.

II. BASIC ASSUMPTIONS

The simplest notion of fuzzy sets is a generalization of the ordinary concept of sets to the case where the memberships of elements are not clearcut, or in other words, the "boundary" of the sets under consideration is not sharply defined. Hence, it is essential to talk about a universal space X which contains all entities we are interested in. A fuzzy set, according to its innovator (Zadeh 1965), is a mapping from the universal space to the [0,1] real interval. Thus we can think of a particular fuzzy set in X as a particular assignment of memberships to the elements in X. In this sense an ordinary set is a special case of assignment in which only 0 or 1 is attached to every x in X, and so the range of the mappings on the universal space is the set {0,1} instead of the [0,1] interval. Therefore, it seems to be natural to treat fuzzy sets as a kind of continuously valued logic (Preparata and Yeh 1972).

In group decision-making processes, we have a set D of actions to choose and m individuals are involved in the process. We can identify the universal space X as D and the preference pattern of every individual is represented by a fuzzy set in D: let $f_i(a)$ denote the degree of preference of action a by individual i, where $f_i(a) \varepsilon [0,1]$, and the preference

assignment of memberships from the action set D to the [0,1] interval. One important assumption about the assignments is that they are commensurate, and furthermore, for every $a \in D$ if $f_i(a) > f_j(a)$, then we say individual i prefers action a more than individual j does. In some interesting decision-making problems involving multiple criteria of optimality (see e.g. Fung and Fu 1973a), we have a set of actions D which is the universal space, and m criteria each of which is associated with a fuzzy set where the membership of action a is the value (in [0,1] interval) of the performance criterion which we want to maximize over all the actions. In statistical decision theory, the universal set X is a set of decisions $\mathcal{a}$ and the membership function is a decreasing function of the risk function $R(\theta,d)$, where $d \in \mathcal{a}$ and θ is in a set of states of nature (Ferguson 1967). Thus every state θ of nature is associated with a fuzzy set whose membership function $f_\theta(\cdot)$ is defined, for instance, by $f_\theta(d) \equiv 1 - R(\theta,d)$ for all $d \in \mathcal{a}$, provided the range of the risk function is the [0,1] interval.

To sum up these cases we see that the membership of an element in space X represents the degree of acceptability of the element by the individuals in the group, with respect to the criteria, or based on the assumption that the states of nature are some particular ones, in each of these contexts. It is appropriate to use the word admissibility to denote various degrees of membership. For example, we can think of an individual i who constructs his admissible set of actions A_i by assigning preferences to every action in D, with $f_i(a)$ representing his preference of action a, $f_i(a) \in [0,1]$; then the admissible set A_i of individual i is a fuzzy set which can be expressed by writing

$$A_i = \{(a, f_i(a)): \ a \in D\}$$

where $f_i(a)$ is the membership or admissibility of action a in the set A_i. We shall interchangeably use the words membership, preference and admissibility to denote a degree of belonging, and the words membership function, assignment, admissible set, and preference pattern to denote a mapping from the universal space to a range of membership values (e.g., the [0,1] interval).

Two important questions about fuzzy sets can be raised (Fung and Fu 1973a):

1. How can we assign memberships to elements in a fuzzy set? and
2. How can the notion of fuzzy sets be applied to practical problems?

Specifically, the first question concerns the construction of a numerical scale for membership values in such a way that the scale satisfies some conditions imposed on a rational measurement system. This is a very difficult problem and is still unsolved at this stage. The main difficulty is in constructing a homomorphism from a qualitative preference system into a quantitative preference system.[1] In some particular situations, such as the statistical decision-making problem described above, the membership functions are sometimes easily available. But the general theory of decision-making should be able to include the interesting problems in which only subjective membership assignments are possible. It was emphasized by Bellman and Zadeh (1970) that the membership of an element is not a statistical quantity assigned by some individual. In fact, the assignment of membership functions very much depends on the idiosyncracy, or the state of mind of the individual involved

[1]A detailed exposition of the theory of measurement is given in the book by Krantz, Luce, Suppes and Tversky (1971).

in the problem. By a rational assignment we mean a membership function which preserves some basic properties of the individual's qualitative preference structure on a numerical scale. A reason to study this problem is to eliminate excessive arbitrariness in the qualitative preference structure due to idiosyncracy, because we think that no meaningful decision can be made in a very chaotic situation.[1]

We can suggest four ways to study this membership assignment problem. First, a numerical measurement scale can be established as we described above. Secondly, we can reduce excessive subjectiveness of an individual by incorporating the expertise of a number of individuals; that is to say, we consider the problem of eliminating idiosyncracy as a collective or group decision problem. Some algorithms to solve this problem were given by Fung and Fu (1973b). Thirdly, Zadeh (1973) introduced fuzzy linguistic variables to evaluate the memberships of complicated actions or commands in terms of the memberships of simpler actions. His approach assumes that the memberships of simple entities are available. The fourth approach to the assignment problem is to consider a topological space which is much simpler than a real interval. We shall adopt this approach in the present paper as a basis for the rational aggregation problem which is the main theme of our dissertation.

The idea of using a topological structure instead of a numerical scale to describe psychological and social phenomena is not new (see e.g. Lewin 1936). A measurement structure which depends on the "closeness" of psychological and

[1] Two related problems about rationalizing estimation and pattern recognition when very little is known about the data were studied by Fine (1970) and Fine (1972).

social states was proposed by Shelly (1962). In his paper, measurement is simply the attachment of any symbol (numerical, geometric, linguistic, and so on) to a state in the set of states of interest, and the relative closeness of any two states can always be determined relative to some third state. We shall show that the fundamental operations in fuzzy set theory and the minimax principle in decision-theory can be deduced on a simple topological structure in which the range of membership functions is not the [0,1] interval, but can be any abstract set of symbols which is linearly ordered and connected. The topolotical structure S will be assumed to be a semilattice (Petrich 1973) and is a topology induced by a linear order on S (Kelley 1955). A feature of our approach is a generalization of fuzzy sets and decision theory to include the situations where the scale of memberships or risk functions is not necessarily established nor is a metric defined.[1]

Let the universal space X be a set D of actions. A fuzzy set A in D is defined as a set of ordered pairs, $A \equiv \{a, f_A(a)) : a \in D\}$ where $f_A: D \to S$, S being a topological space having the following properties:

AXIOM I. S is an order topology induced by a linear order $\leq$, and is a connected topological space.

Remark. It follows from this definition of fuzzy sets that two fuzzy sets A_1 and A_2 are identically equal, $A_1 = A_2$, if for all $a \in D$, $f_{A_1}(a) = f_{A_2}(a)$; that is, if the membership functions corresponding to these sets are identical.

[1]Unlike the characterization in Bellman and Giertz (1973) and Goguen (1967), our approach does not characterize two binary operations (viz. intersection and union for fuzzy sets) at the same time and so the range S does not have the sturcture of a lattice (Goguen 1967) which requires the two operations to be distributive on S.

Some brief explanations of the above axiom are in order. We say that the range set S is a linearly ordered set if there is a linear order $\leq$ defined on S, where $\leq$ satisfies the following four conditions:

1. $x \leq x$ for each x,
2. if $x \leq y$ and $y \leq x$, then $x = y$
3. if $x \leq y$ and $y \leq z$, then $x \leq z$,
4. for each pair s, y, either $x \leq y$ or $y \leq x$.

We use the symbol < to denote "$\leq$ but not =". Intervals in the ordered set S are defined as usual, e.g.

$$(a,b] = \{x:x \in S, a < x \leq b\}.$$

By neighborhood of an element a of S we mean an open interval containing a. (By convention, the sets below are considered to be open intervals:

$$\{x:x \in S, x < a\} \text{ and } \{x:x \in S, x < a\}$$

for any a in S.) Neighborhoods so defined form a basis for the open sets in the order topology of S, i.e., every non-empty member of the topological space of S is the union of open intervals, and each open interval is also a member of the space. By definition, a subset of S is closed if its complement in S is a member of the topological space of S. For example, the set $(a_1,a_2) \cup (a_3,a_4)$ with $a_1 < a_2 < a_3 < a_4$ is open, but the set $[a_1,a_2] \cup [a_3,a_4]$ is closed since its complement in S, $\{x \in S:x < a_1\} \cup (a_2,a_3) \cup \{x:x > a_4\}$, is open in S. The order topological space S is said to be connected if S is not the union of two open disjoint nonempty sets in the space. Note that any topological space on a discrete set is not connected, whereas the [0,1] real interval is. A more precise notation for a topolgical space of S is an ordered pair (S,T_S) where T_S is a collection of subsets of S such that: S and $\phi \in T_S$; if U_1 and U_2 belong to T_S, then $U_1 \cap U_2$ belong to T_S; and if $\{U_\alpha : \alpha \in \Delta\}$ is an indexed family of sets, each of which belongs to T_S, then $\bigcup_{\alpha \in \Delta} U_\alpha$ belongs to T_S. Another concept about a mapping from one topological space to another is the following: Let f be a mapping from a space (S,T_S) to (X,T_X), i.e., $f:S \to X$; then we say f is continuous if at every $a \in X$ for any open set $V \subset X$ containing f(a), there exists an open set $U \subset X$ containing a such that $f(U) \subset V$. This topological concept of continuity is weaker than the notion of continuity when both spaces are subsets of the real numbers such as the [0,1] intervals.[1]

Having assumed a topological structure for the range of the membership functions, we can proceed to study rational decision-making and rational aggregates of fuzzy sets. Consider a group decision problem in which every individual's preference pattern is represented by a fuzzy set in the action space D. We shall postulate conditions which an aggregation of fuzzy sets must satisfy so that the meaning of aggregation is consistent with our concept of a process of compromising or amalgamating the preference patterns of individuals in the group. We shall take the aggregate of the fuzzy sets each representing an individual's preference pattern to represent the preference pattern of the whole group, so that the group can choose an action which has the highest membership in the aggregate set. In other words, we have discussed above that each individual in the group has an admissible set of actions, which is represented by a fuzzy set with memberships being the admissibilities of the

---

[1]For other concepts in topology and basic results we use in our proofs, refer to Kelley 1955.

actions in the action set; now the aggregate of the fuzzy sets for all individuals in the group is regarded as an admissible set for the whole group, so that the optimal choice of action is made based on the admissibilities of actions in the aggregated admissible set. Such a process of making decisions by the group is said to be rational, and if the aggregation operation on fuzzy sets satisfies the following Axioms I-VII, it will be called a rational aggregation. The Axioms presented below will be the minimal set of conditions which a rational aggregation operation must comply with in order to best reflect the ideal process of amalgamating individual opinions in a group. In statistical decision and game theory, it should reflect the ideal process of taking into consideration the preference patterns of the decisions conditioned by all possible states of nature.

By an aggregate A of fuzzy sets $A_1,\ldots,A_m$ in the action space D, we mean a fuzzy set which in some sense (expounded below) represents a confluence of the sets $A_1,\ldots,A_m$ (Bellman and Zadeh 1970). More specifically, suppose A is written as $\{(a, f_A(a)):a \in D\}$ and for each $i = 1,\ldots,m$, A_i is written as $\{(a,f_{A_i}(a)):a \in D\}$; then action a in the aggregate set A has a membership $f_A(a)$ which is the result of amalgamating the memberships $f_{A_i}(a)$ of action a in each fuzzy set A_i. In ordinary set theory, the union operation is an aggregation of sets, say $K_1,\ldots,K_m$, such that the amalgamation process is to assign membership 1.0 to every action which belongs to at least one of the sets $K_1,\ldots,K_m$, i.e., which has membership 1.0 in at least one of the sets. The intersection operation on ordinary sets is another type of aggregation. Thus, our goal is to generalize the aggregations on ordinary sets to fuzzy sets in which every element has memberships taking on values between 0 and 1. In brief, we have replaced the real interval [0,1] which is the range of membership functions by a weaker topological space S induced by a linear order and assumed to be connected. Other assumptions on the aggregation are simplified as a consequence, and surprisingly we do not even have to assume that the range S is bounded. (Theorem 1)

In the following we shall list a collection of basic assumptions on the properties of a rational aggregation, and in the next two sections we shall show how to deduce the particular forms of a rational aggregate on the basis of these postulates.

Definition. Let $\mathcal{F}$ be the class of fuzzy sets in a set D of actions and taking values in a range S. An aggregation $\oplus$ is a binary operation on $\mathcal{F}$, i.e., $\oplus : \mathcal{F} \times \mathcal{F} \to \mathcal{F}$, and an aggregate of two fuzzy sets $A_1, A_2 \in \mathcal{F}$ is represented by $A_1 \oplus A_2$.

AXIOM II. (Law of independent components) There exists a function $\cdot : S \times S \to S$ such that

$$A_1 \oplus A_2 = \{(a, f_{A_1}(a) \cdot f_{A_2}(a)) : a \in D\}$$

for all fuzzy sets A_1, A_2 in $\mathcal{F}$, and $\cdot$ is continuous in the order topology S.

AXIOM III. (Idempotent law) For all $A \in \mathcal{F}$, $A \oplus A = A$.

AXIOM IV. (Commutative law) For all A, B $\in \mathcal{F}$, we have $A \oplus B = B \oplus A$.

Axiom II states that the membership of any action in an aggregate depends only on its memberships in every set in the aggregation. Axiom III states simply that aggregating two identical fuzzy sets should give the original set, and Axiom III asserts that the aggregation operation $\oplus$ is symmetrically defined.

AXIOM V. For $m \geq 3$ define $A_1 \oplus A_2 \oplus \cdots \oplus A_m$ inductively by $A_1 \oplus A_2 \oplus \cdots \oplus A_m \equiv (A_1 \oplus \cdots \oplus A_{m-1}) \oplus A_m$.

AXIOM VI. (Associative law) For every A, B, C ε $\mathcal{F}$, we have $(A \oplus B) \oplus C = A \oplus (B \oplus C)$.

These are two crucial postulates on the way the concept of aggregating two fuzzy sets is extended to the case of more than two sets. Although they are obviously acceptable in a set-theoretic approach, their role in group decision theory can be disputed. It is interesting to note that Arrow (1951) and other research workers in mathematical economics did not postulate these properties for an aggregation operation. They presented their axioms on a rational aggregation in such a way that these properties must be possessed by aggregations involving any number of individuals (i.e., fuzzy sets in our context).

The main problem concerning these two Axioms is whether we should define an aggregation involving more than two sets in an inductive manner. Once we can accept this definition, Axiom VI should become a natural condition to impose on the aggregation involving more than two sets. Axiom V in effect states that, in group decision-making involving more than two individuals, we can first aggregate the preference patterns of any two individuals, and then use this aggregated preference pattern to form another aggregation with a third individual; repeat this process of amalgamating two individuals at a time until all individuals involved are considered. We must point out that although this manner of aggregating more than two individuals looks reasonable enough to be accepted, it is indeed a rather restrictive property of the aggregation process. An interesting implication of Axioms IV, V and VI is that for $m \geq 2$, $A_1 \oplus A_2 \oplus \cdots \oplus A_m$

is invariant under any permutation of the subscripts {1,.. .,m}; conversely, this property and Axioms IV and V imply Axiom VI.

Before we proceed to deduce the forms of a rational aggregate $\oplus$, we need to postulate the following condition on $\oplus$ in order to make the aggregation process behave rationally:

AXIOM VII. (Non-decreasingness of $\oplus$) For any action $a \in D$ and any fuzzy sets B, C_1, $C_2 \in F$, with $A_1 = B \oplus C_1$ and $A_2 = B \oplus C_2$, if $f_{C_1}(a) > f_{C_2}(a)$ then $f_{A_1}(a) \geq f_{A_2}(a)$.

In other words, consider two individuals in a group decision-making problem; suppose one individual maintains his preference of action a from a lower degree to a higher degree, it must follow that the preference of action a in the aggregate does not decrease as a consequence.

We can give a set-theoretic interpretation of this Axiom. For ordinary sets, we say that set $A \supset B$ if some actions in D have membership 1 in A but membership 0 in B, and conversely actions which have memberships 1 in B must have memberships 1 in A. That is to say, the membership of every action in B is less than or equal to its membership in A. We can carry over this concept of inclusion to the case of fuzzy sets (Zadeh 1965). For fuzzy sets A and B in F, we say that $A \supseteq B$ if and only if $f_A(a) \geq f_B(a)$ for all $a \in D$. (A has the intuitive notion of a "larger" set than B in the space F.) Axiom VII therefore is equivalent to

AXIOM VII. For any fuzzy sets B, C_1, $C_2 \in F$, if $C_1 \supseteq C_2$ then $B \oplus C_1 \supseteq B \oplus C_2$.

Moreover, in view of Axiom II, Axiom VII is equivalent to saying that the function • is non-decreasing in the second component. The other Axioms can be rephrased as:

Axiom III. For all $a \in S$, $a \cdot a = a$.

Axiom IV. For every $a, b, \in S$, $a \cdot b = b \cdot a$.

Axiom VI. For every $a, b, c, \in S$, we have

$$(a \cdot b) \cdot c = a \cdot (b \cdot c).$$

We shall use these equivalent statement in the rest of this paper, and because of the equivalence the binary operation $\cdot$ on S will also be called an aggregate.

III. RATIONAL AGGREGATES

LEMMA 1. Let S be a connected order topological space (induced by a linear order <) and $F : S \to S$ be a continuous mapping. For every a and b in S with $f(a) < f(b)$, if $c \in S$ and $f(a) < c < f(b)$, then there is at least one point s in S such that $f(s) = c$.

Proof. Assume that there does not exist any s in S such that $f(s) = c$. Let $I_u = \{x \in S : x > c\}$ and $I_\ell = \{x \in S : x < c\}$, which are open sets in the order topology, with $f(a) \in I_\ell$ and $f(b) \in I_u$. Since $S = I_\ell \cup \{c\} \cup I_u$, and $f^{-1}(\{c\}) = \phi$, $f^{-1}(S) = f^{-1}(I_\ell) \cap f^{-1}(I_u)$. On the other hand, since I_ℓ and I_u are disjoint nonempty open sets in S, $f^{-1}(I_\ell) \quad f^{-1}(I_u) = \phi$. Hence, S is not connected, which is a contradiction.

Remark. Lemma 1 is essentially the intermediate-value theorem.

LEMMA 2. Let S be a connected order topological space in which every point is an idempotent point. If S is a commutative semigroup, i.e., closed under a continuous and associative binary operation $\cdot$, then for every a, b in S,

$$\min\{a,b\} \leq a \cdot b \leq \max\{a,b\}.$$

Proof. Let a, b be points in S and, without loss of generality assume $a \leq b$. Then the inequality $\min\{a,b\} \leq a \cdot b$

is equivalent to $b \leq a \cdot b$. We shall show that $a \cdot b > b$ leads to a contradiction.

Define $f_a : S \to S$ by $f_a(x) = a \cdot x = x \cdot a$ for all x in S. Since a is an idempotent point, $a \cdot a = a$ and so $f_a(a) = a$. Let $c = a \cdot b$. Then by the associativity of $\cdot$ and idempotency of a,

$$a \cdot c = a \cdot (a \cdot b) = (a \cdot a) \cdot b = a \cdot b = c$$

which implies $f_a(c) = c$. If we assume $c > b$, then

$$a \cdot b = f_a(b) = c < b \leq a = f_a(a),$$

and Lemma 1 implies that there exists a point x in S such that $f_a(x) = b$ or $a \cdot x = b$. Consequently we have

$$a \cdot b = a \cdot (a \cdot x) = (a \cdot a) \cdot x = a \cdot x = b.$$

which is a contradiction to the assumption that $a \cdot b = c > b$. Hence $c \not> b$, or equivalently, $a \cdot b \geq \min\{a,b\}$.

Second part of the theorem can be proved in a similar manner by considering the function $f_b : S \to S$ defined by $f_b(x) = b \cdot x = x \cdot b$ for all x in S.

Lemma 2 asserts that in forming an aggregate $a \cdot b$ we would not obtain a value which is less than or greater than both a and b. There is still the possibility that $a \cdot b$ is some value strictly between these two points (when $a \neq b$). To rule out this possibility for every distinct a and b in S, we shall show below that only a mild condition, viz. that $\cdot$ is non-decreasing in each variable, will be required.

Suppose $a, b \in S$ and $a > b$. If $x = a \cdot b$ and $a > x > b$, then we say that x is an unbiased point in S. More specifically, an unbiased point x in S is a point which can be expressed as $x = a \cdot b$ for some $a, b \in S$ and $x \neq a$ or b. Note that unbiased point (s) may not exist in S. In fact, our objective is to show that at most one unbiased point can exist in S under the same conditions as in Lemma 2. In the sequel we shall leave out the qualification "if it exists" when we talk about any unbiased point in S.

Let x be an unbiased point in S. Then by definition there exist a_α and b_α in S such that $x = a_\alpha \cdot b_\alpha$ and $x \neq a_\alpha$ or b_α. Suppose $a_\alpha > b_\alpha$; we shall define $I_\alpha(x)$ to be the closed interval $[b_\alpha, a_\alpha]$ and

$$J(x) = \bigcup_\alpha I_\alpha(x)$$

where the union is over all intervals $I_\alpha(x)$ about x. The following Lemma gives some useful properties of the intervals $I_\alpha(x)$ and $J(x)$.

LEMMA 3. In any $I_\alpha(x)$ about an arbitrary unbiased point x, for every $z_1 < x$ and $x < z_2$ we have $z_1 \cdot z_2 = x$ if the operation $\cdot$ is non-decreasing in each variable. Furthermore, the interval $J(x)$ also has the above property.

Proof. Let $I_\alpha(x) = [b,a]$ with $a > b$. Since $a \cdot b = x$ we have $a \cdot x = a \cdot (a \cdot b) = (a \cdot a) \cdot b = a \cdot b = x$, and similarly $b \cdot x = x$. Idempotency of x implies $x \cdot x = x$. Thus, we have three points in S such that $b < x < a$, and

$$x \cdot b = x \cdot x = x \cdot a = x.$$

Since $\cdot$ is non-decreasing we have $x \cdot y = x$ for all $y \in I_\alpha(x)$. Now suppose $z_1, z_2 \in I_\alpha(x)$ with $z_1 < x$ and $z_2 > x$. Since $z_2 < a$, $z_1 < x$, by non-decreasingness,

$$z_1 \cdot z_2 \leq z_1 \cdot a \leq x \cdot a = x$$

and also

$$z_1 \cdot z_2 \geq z_1 \cdot x \geq b \cdot x = x$$

since $z_2 > x$ and $z_1 > b$. Hence $z_1 \cdot z_2 = x$.

To prove the second part, consider two intervals

$I_{\alpha_1}(x) = [b_1,a_1]$ and $I_{\alpha_2}(x) = [b_2,a_2]$ about x. (Note that by definition x has at least one interval $I_\alpha(x)$ about it; if this is the only interval, then $J(x) = I_\alpha(x)$ and the second part of the Lemma follows from the first part.) In the case where $I_{\alpha_1}(x) \subseteq I_{\alpha_2}(x)$, $I_{\alpha_1}(x) \cup I_{\alpha_2}(x) = I_{\alpha_2}(x)$ which is again an interval about x. The case where $b_1 < b_2 < x < a_1 < a_2$ will be studied below. For every t_1 and t_2 such that $b_2 \leq t_1 \leq x$ and $x \leq t_2 \leq a_1$, clearly $t_1 \cdot t_2 = x$ since both t_1 and t_2 are in $I_{\alpha_1}(x)$. Now suppose $b_2 \leq t_1 \leq x$ and $a_1 \leq t_2 \leq a_2$. It still follows that $t_1 \cdot t_2 = x$ since both t_1 and t_2 are in $I_{\alpha_2}(x)$. The other case where $b_1 \leq t_1 \leq b_2$ and $a_1 \leq t_2 \leq a_2$ follows from the first part of this Lemma and from the non-decreasingness of $\cdot$. More specifically, $t_2 \geq a_1$ and $t_1 \cdot a_1 = x$ implies $t_1 \cdot t_2 \geq x$, and similarly $t_1 < b_2$ and $t_2 \cdot b_2 = x$ implies $t_2 \cdot t_1 \leq x$. Hence we have $t_1 \cdot t_2 = x$.

By induction on the number of $I_\alpha(x)$ intervals about x we have shown that $J(x) = \bigcup_\alpha I_\alpha(x)$ has the property that for every $z_1 < x$ and $z_2 > x$ where $z_1, z_2 \in J(x)$, $z_1 \cdot z_2 = x$.

Remark. In view of Lemma 3, it makes sense to talk about $J(x) = \bigcup_\alpha I_\alpha(x)$ as the largest interval $I_\alpha(x)$ containing x. Note that for every unbiased point x, there is always associated such an interval $\overline{J(x)} = [a,b]$ with $a < x < b$. Here we write $\overline{J(x)}$ in place of $J(x)$ for the sake of convenience. Note that even though every interval $I_\alpha(x)$ is closed, $J(x)$ may not be closed since there can be in infinite number of intervals $I_\alpha(x)$ about x. Since the only difference between $J(x)$ and $\overline{J(x)}$ is in its boundary points a and b, $\overline{J(x)}$ has the property that for all $z_1,, z_2 \in \overline{J(x)}$, if $a < z_1 < x$ and $b > z_2 > x$ then $z_1 \cdot z_2 = x$ where $\overline{J(x)} = [a,b]$. This property is inherited from $J(x)$ as a result of Lemma 3.

LEMMA 4. Let t_1 and t_2 be two distinct unbiased points, $t_1 < t_2$ and let $\overline{J(t_1)} = [a_1,b_1]$, $\overline{J(t_2)} = [a_2,b_2]$. It is not possible that $a_2 < t_1$ and $b_1 > t_2$.

Proof. Suppose $a_2 < t_1$ and $b_1 > t_2$. Then $t_1 \in \overline{J(t_2)}$ and $t_2 \in \overline{J(t_1)}$. Pick any points x and y so that

$$\max\{a_1,a_2\} < x < t_1$$

$$t_2 < y < \min\{b_1,b_2\}.$$

This is possible since S is a connected space. Clearly, $x \in \overline{J(t_2)} = [a_2,b_2]$ since $a_2 \leq \max\{a_1,a_2\} < x < t_1 < t_2 < b_2$, and similarly $y \in \overline{J(t_1)} = [a_1,b_1]$. On the other hand, $x \in \overline{J(t_1)}$ since $a_1 \leq \max\{a_1,a_2\} < x < t_1 < b_1$, and $y \in \overline{J(t_2)}$. These imply that we have both $x \cdot y = t_1$ and $x \cdot y = t_2$ which is impossible since $t_1 \neq t_2$.

Our next Lemma asserts that the set of unbiased points in S is dense if S contains more than one point.

LEMMA 5. Let t_1 and t_2 be two distinct unbiased points, $t_1 < t_2$. Then there exists another unbiased point $u \in (t_1,t_2)$.

Proof. We shall prove that in each of the following cases the assumption that no unbiased point exists in (t_1,t_2) leads to a contradiction. Let $\overline{J(t_1)} = [a_1,b_1]$ and $\overline{J(t_2)} = [a_2,b_2]$.

Case (1): $A_1 < t_1 < b_1$, $a_2 < t_2 < b$. Pick any points x and y such that $a_1 < x < t_1$ and $t_2 < y < b_2$, and denote $z = x \cdot y$. (Figure 1a). If we assume that no unbiased point exists in (t_1,t_2), then $z \notin (t_1,t_2)$ since if $z \in (t_1,t_2)$ it will be an unbiased point. There are several other possibilities:

(a) $z = t_1$. By Lemma 3 this implies that

$$\overline{J(t_1)} \supseteq [a_1,y] \supsetneq [a_1,b_1]$$

which is a contradiction to the definition $\overline{J(t_1)} = [a_1,b_1]$.

(b) $z = t_2$. This implies a contradiction as in part (a).

(c) $z < t_1$. It follows that $x \cdot y = z < t_1$ which is a contradiction to the hypothesis that the operation $\cdot$ is nondecreasing since we have $x \cdot t_1 = t_1$ (Lemma 3) and recall that $y > t_2 > t_1$.

(d) $z > t_2$. This implies a contradiction as in part (c). Hence we conclude that in Case (1) there always exists a distinct unbiased point between any two unbiased points.

Case (2): $a_2 < t_1$ and $b_1 > t_2$ (see Figure 1b). This case is ruled out by Lemma 4.

Case (3): $a_1 < t_1 < a_2 < t_2 < b_2$ and $b_1 > t_2$. Pick any point y, $t_2 < y < \min\{b_1, b_2\}$. Since $a_1 < y < b_1$, we have $y \in \overline{J(t_1)}$. Furthermore, let x be a point such that $t_1 < x < t_2$; clearly $x \in \overline{J(t_1)}$. Let $g\colon\ S \to S$ be a function defined by $g(x) = y \cdot x$ for $x \in (t_1, t_2)$.

First assume $a_2 < x < t_2$, which by Lemma 3 implies that $x \cdot y = t_2$. Next assume $t_1 < x < a_2$. We have three possibilities as a consequence of the assumption that (t_1, t_2) does not contain any unbiased point:

(a) $x \cdot y = y$, which is not possible since $\cdot$ is nondecreasing and $x \cdot y = t_2$ for $a_2 < x < t_2$. For the same reason it is not possible that $x \cdot y < t_2$.

(b) $x \cdot y = x$, which implies that there is a jump point of the function g at $x = a_2$, a contradiction to the continuity of g.

(c) $x \cdot y = t_2$, which implies that $g(x) = y \cdot x = t_2$ for $t_1 < x \leq t_2$. However, Lemma 3 implies that $g(t_1) = y \cdot t_1 = t_1$. Thus we have deduced that g has a jump point at $x = t_1$, a contradiction to the continuity of g.

Hence, we conclude that in Case (3) there always exists a distinct unbiased point between any two unbiased points. (See Figure 1c.)

Case (4): $a_1 < t_1 < b_1 < t_2 < b_2$ and $a_2 < t_1$ (see Figure 1d). This case follows by similar reasoning as in Case (3).

We shall distinguish two types of aggregates which are extreme cases of aggregating individual preferences. The aggregate which is defined by $a \cdot b = \min\{a,b\}$ for all $a,b \in X$ is said to be pessimistic, and that defined by $a \cdot b = \max\{a,b\}$ is said to be optimistic. The following Theorem states that Axioms I - VII imply that a rational aggregate can only be pessimistic, optimistic or of mixed type. These axioms are the mildest conditions on the space S and binary operation $\cdot$ which we know can lead to this conclusion.

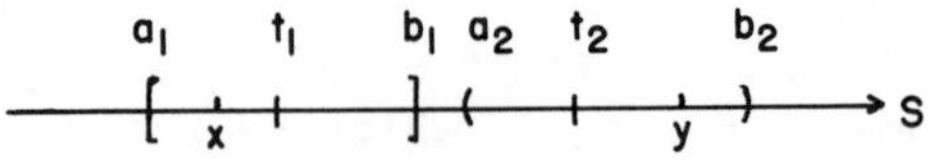

Ia (Case I)

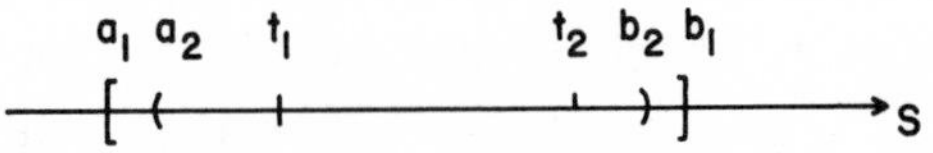

Ib (Case 2)

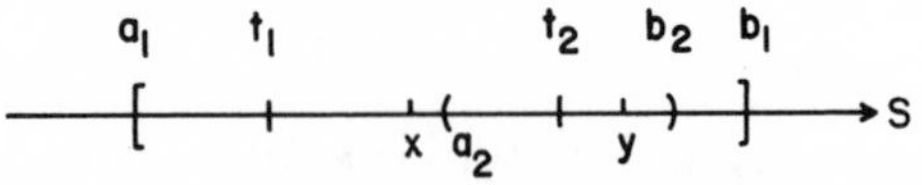

Ic (Case 3)

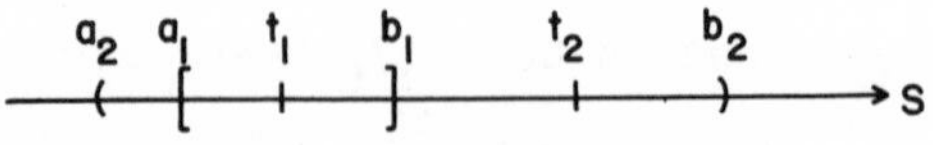

Id (Case 4)

Figure 1

The aggregate of mixed type is a very interesting type of aggregate: without the assumption of any quantitative measure and arithmetic operations, the aggregate represented by the operation • assigns a point in S for every a and b in such a way that a • b can take on a value different from a or b. More precisely, an aggregate of mixed type is a binary operation • on S such that $z_1 \cdot z_2 = x$ for $z_1 \leq x$ and $z_2 \geq x$, $z_1 \cdot z_2 = \max\{z_1, z_2\}$ for $z_1, z_2 \leq x$, and $z_1 \cdot z_2 = \min\{z_1 \cdot z_2\}$ for $z_1, z_2 \geq x$, where x is some arbitrary fixed threshold point. In other words, an aggregate of mixed type is a pessimistic aggregate if all the variables are above a threshold value, and it is an optimistic aggregate if all the variables are below the threshold value; in case some variables are below and some are above the threshold value, the aggregate takes on the threshold value. This type of aggregate seems to be a reasonable one since when the values of all variables are large enough, we would be willing to adopt a pessimistic aggregate for the sake of guarding against the possibility of adverse situations as informed by some unfavorite individual preferences,[5] and when the values of all variables are small enough, we would be willing to be a bit more optimistic or else the admissibility of the action under consideration would be too low to be of any competitive value when compared with other actions in the aggregate; in case some individuals strongly prefer an action and some are strongly against it, the group could be expected to make a compromise by choosing a fixed threshold value as the aggregate value for the action.

THEOREM 1. Let S be a connected order topological space induced by a linear order <. Suppose every point of S is an

[5] This view is shared by Savage (1954) (p.174) who argued in terms of utility, or personal income of an action, instead of the generalized notion of membership.

idempotent point, and S is a commutative semigroup, i.e. closed under a continuous and associative binary operation $\cdot$ which is non-decreasing in each variable. Then for every a and b in S, we can have either

$$a \cdot b = \min\{a,b\} \quad \text{(pessimistic)},$$

or

$$a \cdot b = \max\{a,b\} \quad \text{(optimistic)},$$

or that the aggregate is of mixed type.

Proof. Let x be an unbiased point with $\overline{J}(x) = [a,b]$. Lemma 4 implies that $x \cdot y = x$ for all $a < y < b$. Suppose x is the only unbiased point in S. Then $x \cdot z$ takes on values which are either x or z, for every $z > b$. Clearly $x \cdot z$ cannot be z because this implies a jump point of the function g defined by $g(z) = x \cdot z$ at z, a contradiction to the continuity of g. Hence $x \cdot z = x$ for all $z \geq x$; by similar reasoning we can show that $x \cdot y = x$, for all $y \leq x$. Furthermore, for every $y \leq x$ and $z \geq x$, we have $y \cdot z = x$ since x is assumed to be the only unbiased point. This implies that $I_\alpha(x) = J(x) = S$. By the non-decreasingness and continuity of the operation $\cdot$, it follows that $z_1 \cdot z_2 = \max\{z_1,z_2\}$ for $z_1,z_2 \leq x$, $z_1 \cdot z_2 = \min\{z_1,z_2\}$ for $z_1,z_2 \geq x$, and $z_1 \cdot z_2 = x$ for $z_1 \leq x$ and $z_2 \geq x$. Hence the aggregate represented by operation $\cdot$ is of mixed type when there is exactly one unbiased point in S. It is easy to show that Axioms I - VII are necessary conditions for the aggregate to be of mixed type.

Next we shall show that it is not possible to have more than one unbiased point in S. Suppose $x_1 < x$ is another unbiased point in S. By Lemma 5 we know that there exists another unbiased point $x_2 \in (x_1,x)$. Inductively we have a sequence of unbiased points $\{x_n\}$ which converges to x from below. Note that all x_n's in the sequence are distinct and are different from x. Since S is connected, the space is T_2 and so the sequence $\{x_n\}$ converges uniquely to x. Let $\{y_n\}$ be a subsequence of $\{x_n\}$ such that $y_n \in \overline{J(x)} = [a,b]$ for all n = 1,2,... Let each y_n be associated with the interval $\overline{J(y_n)} \equiv [a_n,b_n]$. Clearly $y_n < b_n < b$. Furthermore, $b_n \not\geq x$ because this implies

$$a \leq a_n < y_n < x < b_n \leq b$$

from which a contradiction follows as a result of Lemma 4. Thus for all n, we have $y_n < b_n < x$. Recall that $y_n \to x$ from below, so that $b_n \to x$. Consequently, in the limit $n \to \infty$ we have $\overline{J(x)} - [a_0,x]$ where $a_n \to a_0$ since $b_n \to x$ implies $\overline{J(y_n)} \to [a_0,x]$ and $y_n \to x$ implies $\overline{J(y_n)} \to \overline{J(x)}$. But we have assumed that $\overline{J(x)} = [a,b]$ with $b \neq x$ which is an unbiased point. Thus a contradiction is derived from the assumption that there exists more than one unbiased point in S.

The last alternative is that there is no unbiased point at all in S. In this case $a \cdot b = \min\{a,b\}$ or $a \cdot b = \max\{a,b\}$ in view of Lemma 2. Furthermore, the continuity of $\cdot$ and its symmetric property lead to the conclusion that either $a \cdot b = \min\{a,b\}$ or $a \cdot b = \max\{a,b\}$ for every a and b in S, i.e. it is not possible that the operation $a_1 \cdot b_1 = \min\{a_1,b_1\}$ for some pair a_1, b_1 in S but $a_2 \cdot b_2 = \max\{a_2,b_2\}$ for another pair a_2, b_2 in S.

Remark. If the space S is bounded from below and above, i.e. if there exist $u \equiv \ell.u.b.\{x \in S\}$ and $\ell \equiv g.\ell.b.\{x \in S\}$ such that $u,\ell \in S$, then the pessimistic and optimistic aggregates are two special cases of the aggregate of mixed type, with the threshold value x set to ℓ and u respectively.

In the following section we shall show that we can eliminate the possibility of a mixed-type aggregate by replacing the assumption of non-decreasingness of $\cdot$ with two other stronger axioms, thus establishing a more restrictive conclusion that a rational aggregate must be the pessimistic (or optimistic) aggregate.

IV. ALTERNATIVE ASSUMPTIONS

The axioms which will be assumed in place of Axiom VII (on the non-decreasingness of the mapping $\cdot$) are the following:

AXIOM VIII. There exists a point α in S and a lower limit of S, denoted by 0, such that for all x, $0 \leq x < \alpha$, we have $0 \cdot x = 0$.

This axiom states that every element x of S satisfies the boundedness condition, i.e. $x \geq 0$ for some special element 0 which we call zero, and that a point α exists in S, such that for all x smaller than α, the aggregate of x and the zero element will give the zero element. In other words, we say that the zero element dominates in the aggregate with x if x is smaller than α. This assumption is roughly consistent with our concept of a pessimistic aggregate. In analogy to this we have the following axiom for an optimistic aggregate:

AXIOM IX. There exists a point β in S and an upper limit of S, denoted by 1, such that for all x, $\beta < x \leq 1$, we have $1 \cdot x = 1$.

<u>Remark</u>. These axioms do not exclude the possibilities that, respectively, $0 \cdot x = 0$ for some (or all) $x \geq \alpha$, and $1 \cdot x = 1$ for some (or all) $x \leq \beta$.

THEOREM 2. If a rational aggregate $\oplus$ satisfies the conditions in Theorem 1 with the exception that Axiom VII is replaced by Axiom VIII, then $\oplus$ must be a pessimistic aggregate. Moreover, if $\oplus$ satisfies the conditions in Theorem 1 with IX in place of VII, then $\oplus$ must be an optimistic aggregate.

<u>Proof</u>. Let $Q = \{x \in S: \; 0 \cdot x \neq 0\}$. Then,

$$0 \cdot (0 \cdot x) = (0 \cdot 0) \cdot x = 0 \cdot x \neq 0,$$

which implies that if $x \in Q$, then $0 \cdot x \in Q$. Axiom VIII implies that Q is bounded from below by α, $\alpha \neq 0$. That is to say, for every $x \in Q$, we have $x \geq \alpha$. Since $\{0\}$ is a closed set, its inverse image defined by $I(0) = \{x \in S: \ 0 \cdot x = 0\}$ is also closed. Hence, $I(0)$ is the union of a family of closed intervals. Since $0 \cdot 0 = 0$ implies $0 \in I(0)$, one of these closed intervals must contain 0. Let it be denoted by $[0,\alpha'] \subseteq I(0)$. From our previous discussion we see that for every $x \in Q$, we have $x > \alpha$ (strict inequality) with $\alpha \equiv \alpha'$.

Consider the function $h: \ S \to S$ defined by $h(x) = 0 \cdot x$ for all $x \in S$. It follows from the above paragraph that $h(\alpha) = 0 \cdot \alpha = 0$ and $h(x) = 0 \cdot x > \alpha$ for all $x \in Q$. According to the formulation of Clifford (1958), S is connected if and only if S is complete and dense. Completeness of S implies that every bounded (from below) subset of S has a greatest lower bound. Hence, Q has a greatest lower bound, denoted by μ, so that if x_n is a sequence of points in Q which converges to μ from above, $h(x_n)$ converges to a point in Q greater than μ. Thus the function h has a jump point at μ, which is contradictory to the continuity assumption. Hence the set Q must be empty. In other words, we obtain the property that $0 \cdot x = 0$ for all $x \in S$.

Now for every $a, b \in S$, we assume without loss of generality that $a > b$. Define $f: \ S \to S$ by $f(x) = a \cdot x$ for all $x \in S$. By idempotency, we have $f(a) = a \cdot a = a$, and we have just shown that $f(0) = a \cdot 0 = 0$. Hence, by Lemma 1, there exists a point $c \in S$, $0 \leq c < b$ such that $f(c) = a \cdot c = b$. Consequently $a \cdot b = a \cdot (a \cdot c) = a \cdot c = b$. Furthermore, suppose $c > b$. Then since $b < c < a$, by Lemma 1 there exists a point $b < d < a$ such that $f(d) = a \cdot d = c$. It follows that $a \cdot c = a \cdot (a \cdot d) = (a \cdot a) \cdot d = a \cdot d = c$, which is a contradiction to our previous assertion that $a \cdot c = b$. Similarly, it is not possible that $c < b$. Hence, we conclude that $c = b$ uniquely, establishing the equality $a \cdot b = \min\{a,b\}$.

Second part of the theorem can be proved following similar arguments.

Remarks. In the above proof, after it is shown that $x \cdot 0 = 0$ for all $x \in S$, the rest can be proved in a rather different manner (Faucett 1955, Clifford 1958). It involved the notion of a special type of topological semigroup called thread,and was based on the argument that, given $S = [0,1]$, for every $x \in S$, we have $xS = Sx = [0,x]$ where xS is defined as $\{x \cdot s: \ s \in S\}$ and Sx as $\{s \cdot x: \ s \in S\}$. It follows from this result that, for any x and y in S, if $y > x$ then $x \cdot y \in xS = [0,x]$ or $x \cdot y \leq x$; in general, $x \cdot y \leq \min\{x,y\}$. Then using Lemma 1 it can be shown that $x \cdot y = \min\{x,y\}$.

We must note at this point that the rational decision scheme based on the pessimistic aggregation is a maximin-membership principle. As we pointed out earlier, the membership function of an action is always expresses as a decreasing function of the risk or loss, this in effect corresponds to the minimax-risk (or loss) principle.

A simple corollary to Theorems 1 and 2 immediately follows:

COROLLARY. Under the conditions of Theorem 1, when the rational aggregate is extended to the case which involves more than two fuzzy sets by virture of Axiom V, the aggregate can take either one of the following three forms: Let $A = A_1 \oplus A_2 \oplus \ldots \oplus A_m$, $m \geq 2$ and let $f_a(a)$ denote the membership of action a in the set A. Then for all $a \in X$,

(i) $f_A(a) = \min\{f_{A_1}(a), f_{A_2}(a), \ldots, f_{A_m}(a)\}$,

(ii) $f_A(a) = \max\{f_{A_1}(a), f_{A_2}(a), \ldots, f_{A_m}(a)\}$,

(iii) For any fixed x in S, $f_A(a) = x$, if there exists some i and j, $1 \leq i, j \leq m$, such that $f_{A_i}(a) > x$ and $f_{A_j}(a) < x$,

or $f_A(a) = \min\{f_{A_1}(a), \ldots, f_{A_m}(a)\}$ if $f_{A_i} > x$ for all $i = 1,2,\ldots,m$,

or $f_A(a) = \max\{f_{A_1}(a), \ldots, f_{A_m}(a)\}$ if $f_{A_i} < x$ for all $i = 1,2,\ldots,m$.

As a corollary to Theorem 2, the aggregate can take the form of either (i) or (ii).

V. DISCUSSIONS

In conclusion, we have three important aspects of the results obtained in this paper. First, Theorem 1 asserts that if we are content withtthe weaker implication that a rational aggregation can be either pessimistic, optimistic, <u>or</u> of mixed type, then there is no need to assume the range of membership functions bounded from above or below. A class of fuzzy optimal control problems has been recently studied based on this formulation (Fung and Fu, 1974). If we want to restrict the implication to only the pessimistic aggregation, such that the minimax principle is an immediate consequence, the range is necessarily bounded from below by some point which satisfies a further condition (Axiom VIII). Furthermore, if we are interested in deducing the intersection and union operations of fuzzy sets, represented respectively by the pessimistic and optimistic aggregates in Theorem 2, the range must then be bounded from above and from below.

Secondly, as pointed out in Luce and Raiffa (1957),

there are two distinct types of axiomatic approach to rational decision-making in the literature, one requiring the establishment of a complete ordering of the actions, the other isolating an "optimal" subset of actions but not attempting to rank non-optimal ones. In this paper we take the first approach since the latter cannot be conviently applied to fuzzy sets.

Thirdly, considering that basic assumptions postulated here (and thus the minimax principle) may not be appealing to some people, we must emphasize that it is difficult to give formal analysis of the concept of democratic decision for a group, a point discussed at length by Snow (1951), Hildreth (1953) and others. More specifically, the following criteria for a democratic group decision scheme, described verbally for brevity, are not logically consistent. They are: positive association of social and individual values, independence of irrelevant alternatives, citizen's sovereignty, and non-dictatorship. This assertion is the well-known Arrow's impossibility theorem (Arrow 1951).

REFERENCES

ARROW, K. J. (1951), "Social Choice and Individual Values," Cowles Commission Monograph No. 12; John Wiley & Sons, New York.

BELLMAN, R., and GIERTZ, M. (1973), On the Analytic Formalism of the Theory of Fuzzy Sets, Information Sci. 5, 149-156.

BELLMAN, R., and ZADEH, L.A. (1970), Decision Making in a Fuzzy Environment, Management Science, 17, No. 4, B141-164.

CHANG, C. L. (1968), Fuzzy Topological Spaces, J. Math. Anal. Appl., 24, 182-190.

CLIFFORD, A. H. (1958), Connected Ordered Topological Semigroups With Idempotent Endpoints, I, Trans. Amer. Math. Soc. 88, 80-98.

FAUCETT, W. M. (1955), Compact Semigroups Irreducibly Connected Between Two Endpoints, Proc. Amer. Math. Soc. 6, 741-747.

FINE, T. (1970), Extrapolation When Very Little is Known About the Source, Inform, & Control, 16, 331-359.

FINE, T. (1972), Rantional Classification When Very Little is Known, (submitted to Inform. and Control).

de FINETTI, B. (1972), "Probability, Induction, and Statistics: the Art of Guessing," John Wiley and Sons, New York.

FERGUSON, T. S. (1967), "Mathematical Statistics, a Decision Theoretic Approach,"Academic Press, New York.

FUNG, L. W. and FU, K. S. (1973a), The kth Optimal Policy Algorithm for Decision Making in Fuzzy Environments, Proc. Third IFAC Symposium, the Hugue, Netherlands.

FUNG, L. W. and FU, K. S. (1973b), Decision Making in a Fuzzy Environment, TR-EE 73-22, May 1973, School of Electrical Engineering, Purdue University, Lafayette, Indiana.

FUNG, L. W. and FU, K. S. (1974), "Characterization of a Class of Fuzzy Optimal Control Problems", Proc. Eighth Annual Princeton Conference on Information Science and Systems, March 28-29, 1974.

GOGNEN, J. A. (1967), L - Fuzzy Sets, J. Math. Anal. Appl., 18, 145-174.

HILDREDTH, C. (1953), Alternative Conditions for Social Orderings, Econometrica, 21, 81-94.

KELLEY, J. L. (1955), "General Topology," Van Nostrand Co., New Jersey.

DRANTZ, D. H., LUCE, R. D., SUPPES, P., AND TVERSKY, A. (1971), "Foundation of Measurement, Vol. I Additive and Polynomial Representations," Academic Press, New York.

LEWIN, K. (1936), "Principles of Topological Psychology," McGraw-Hill, New York.

LUCE, R. D., AND RAIFFA, H. (1957), "Games and Decisions," John Wiley and Sons, Inc., New York.

PETRICH, M. (1973), "Introduction to Semigroups," Charles E. Merrill Publishing Co., Ohio.

PREPARATA, F. P. AND YEH, R. T. (1972), Contrinuously Valued Logic, J. Comput. System Sci. 6, 397-418.

SAVAGE, L. J. (1954), "The foundation of Statistics," John Wiley and Sons, Inc., New York.

SHELLY, M. W. (1962), A Topological Approach to the Measurement of Social Phenomena, in "Mathematical Methods in Small Group Processes," (J. H. Criswell, H. Solomon and P. Suppes, Eds.), pp. 305-321, Stanford University Press, Stanford, Calif.

ZADEH, L. A. (1965), Fuzzy Sets, Inform. and Control, 8, 338-353.

ZADEH, L. A. (1973), Outline of a New Approach to the Analysis of Complex Systems and Decision Processes, IEEE Trans. on System, Man and Cybernetics, SMC-3, No. 1, 28-44.

DECISION-MAKING AND ITS GOAL IN A FUZZY ENVIRONMENT

Kiyoji Asai, Hideo Tanaka, and Tetsuji Okuda
Department of Industrial Engineering
University of Osaka Prefecture
Sakai, Osaka 591, Japan

1. INTRODUCTION

Recently the problems of decision-making in a fuzzy environment have attracted special interest in such a sense that it is becoming increasingly clear that in many real world problems we have more to do with fuzziness rather than randomness for the major sources of imprecision. The properties of fuzzy decision-making problems have been studied by Bellman and Zadeh [1]. In [1] fuzzy goals G and fuzzy constraints C are defined precisely as fuzzy sets, that is, as membership functions $\mu_G(x)$ and $\mu_C(x)$, and a fuzzy decision D may be viewed as an intersection of given goals and constraints, i.e. $\mu_D(x) = \mu_C(x) \wedge \mu_G(x)$. An optimal decision x^* is defined as any alternative in X which maximizes $\mu_D(x)$, i. e. $\sup_x \mu_C(x) \wedge \mu_G(x) = \mu_D(x^*)$. Roughly speaking, this definition of optimal decision means that the grade of membership of the optimal decision x^* in C is the same as the grade of membership of x^* in G. In the case of $\mu_D(x^*) < 0.5$, this optimal decision x^* belongs scarcely to C and also to G. Therefore, it seems that the optimal decision x^* loses its meaning.

To remove such a meaningless case that $\mu_D(x^*) < 0.5$, we will view this problem as follows. Even if we take a present decision x_C^* as action, it is not necessary that a goal is attained by the x_C^*. If we take an x_C^*, there will exist an optimal goal x_G^* associated with the x_C^*. We can obtain a formulation of fuzzy decision-making problems in which we can decide what goal must be chosen for given x_C^*. In this formulation, we do not regard the decision problems under consideration as the problems defined in [1], but we may regard the problems as the decision problems with a given term of planning. Precisely speaking, this problem is to decide an ordered pair (x_C^*, x_G^*) where x_C^* is a present optimal decision as action and x_G^* is an estimated optimal goal at present in such a sense that (x_C^*, x_G^*) maximizes $\mu_D(x_C, x_G : N) = \mu_C(x_C) \wedge \mu_G(x_G) \wedge \mu_R(x_C, x_G : N)$. Here $\mu_R(x, x' : N)$ denotes the grade of membership of an ordered pair (x, x') in a similarity relation R.

The decision problems with N periods are called N-decision problems for brevity. Let us define $\mu_R(x, x' : 0)$ as

$$\mu_R(x, x' : 0) = \begin{cases} 1 \ ; \ x = x' \\ 0 \ ; \ x \neq x'. \end{cases}$$

Hence, 0-decision problems are equivalent to the problems defined in [1]. Note that N-decision problems are different from the ordinary problems of N-stage decision processes in such a sense that in N-decision problems we can decide only a present optimal pair (x_C^*, x_G^*) at the first stage and at the second stage we may obtain a new optimal pair $(x_C^{*\prime}, x_G^{*\prime})$ such that Sup $\mu_C'(x_C) \wedge \mu_G'(x_G) \wedge \mu_{R'}(x_C, x_G : N\text{-}1)$ where $\mu_C(x)$ and $\mu_G(x)$ may be different from $\mu_C'(x)$ and $\mu_G'(x)$ because of elapsed time, and so on. In short, it is only question to decide a present optimal pair (x_{C1}^*, x_{G1}^*), and we are not concerned with future optimal pairs $(x_{C2}^*, x_{G3}^*), \ldots, (x_{Cn}^*, x_{Gn}^*)$, because we can not know how $\mu_C(x)$ and $\mu_G(x)$ vary in future.

Our main assertions are the following two points:

(1) Although many decisions in real world problems

arising from public, govermental and industrial systems have been done, it does not seem that the goal associated with a decision has been clearly shown. Nevertheless, it is perhaps natural that it is necessary to show a pair of decision and estimated goal associated with the decision. In our problems a decision x_C does not make sense without some estimated goal x_G and conversely an estimated goal x_G does not make sense without some decision x_C.

(II) There are many problems in a fuzzy environment where it is necessary to decide a present estimated goal such as the federal air quality act proposed by E. Muskie in 1970. Hence it is necessary to formulate decision problems in such a sense that we can decide an estimated goal.

In this paper, we will define N-decision problems from the above point of view and then discuss the properties of optimal decision and goal with a view to solve N-decision problems.

2. PSEUDO SIMILARITY RELATIONS

We will assume that the grade of membership of a decision and a goal in similarity relations depends on N periods. For example, this assumption means such a concept that the similarity relation between two persons x and x' in the sense of height depends on the point p where they are observed. Let us define the membership function of the above similarity relation as $\mu_{R_H}(x, x' : p)$. If $p \to \infty$, it can be assumed that $\lim_{p\to\infty} \mu_{R_H}(x, x' : p) = 1$, since two persons look like two points. Such a concept described above may be applied to N-decision problems.

In order to obtain a similarity relation R between x_C

and x_G from given $\mu_C(x)$ and $\mu_G(x)$, it is assumed that the relation R satisfies the following axioms:

(i) In the case of $N = \infty$, assume that an estimated goal x_G^0 is given. In such a sense that an arbitrary decision x_C is the first step to the estimated goal x_G^0, we can regard $\forall x_C$ and x_G^0 as the similar one because of $N = \infty$, that is, $\mu_R(x_C, x_G^0 : \infty) = 1$ for $\forall x_C$. Conversely, assume that a decision x_C^0 is given. Since an x_C^0 will be carried out in order to attain to some goal after N periods where N is finite, the goal associated with x_C^0 is not arbitrary. But in the case of $N = \infty$, it can be assumed that the goal associated with x_C^0 is arbitrary. Thus, we can regard x_C^0 and $\forall x_G$ as the similar one that is, $\mu_R(x_C^0, x_G : \infty) = 1$ for $\forall x_G$. Hence,

$$\lim_{N \to \infty} \mu_R(x_C, x_G : N) = 1 \qquad \text{for } \forall x_C \text{ and } \forall x_G.$$

(ii) Even if a fixed pair (x_C, x_G) is given, we can suppose that the longer the term of planning, the larger the grade of membership of (x_C, x_G) in the similarity relation. Thus, the following inequality is satisfied:

$$\mu_R(x_C, x_G : i) \leq \mu_R(x_C, x_G : j) \quad \text{for } i \leq j.$$

(iii) If the grade of membership of (x_C, x_G) in R with i periods is larger than one of $x_C', x_G')$, we can suppose that the same matter holds for any j periods. Thus,

$$\mu_R(x_C, x_G : i) \geq \mu_R(x_C', x_G' : i) \Longleftrightarrow \mu_R(x_C, x_G : j) \geq \mu_R(x_C', x_G' : j)$$

for $\forall i$ and $\forall j$.

(iv) Let us assume that the grade of difference between x_C and x_G in the sense of C may be expressed by $| \mu_C(x_C) - \mu_C(x_G) |$. If the grade of difference between x_C and x_G in the sense of C is larger than one of difference between x_C' and x_G', we can suppose that the grade of similarity between x_C and x_G

with one period is smaller than one of x_C' and x_G', and vice versa. Since the same relation in the sense of G holds, we can suppose that the following relation is satisfied:

$$\mu_R(x_C,x_G:1) \leq \mu_R(x_C',x_G':1) \iff |\mu_C(x_C)-\mu_C(x_G)| \vee |\mu_G(x_C)-$$

$$-\mu_G(x_G)| \geq |\mu_C(x_C')-\mu_C(x_G')| \vee |\mu_G(x_C')-\mu_G(x_G')|.$$

These axioms lead to the followings:

(a) Reflexivity: $\mu_R(x, x : i) = 1$ for $\forall i$.

It follows from (ii) that (a) is satisfied.

(b) Symmetry: $\mu_R(x, x' : i) = \mu_R(x', x : i)$ for $\forall i$.

It follows from (iii) and (iv) that (b) is satisfied.

<u>Definition 1</u>. Let us define recursively the membership function of the pseudo similarity relation which satisfies the above axioms as follows:

$$\mu_R(x_C, x_G:N) = \mu_{R_C}(x_C,x_G:N) \wedge \mu_{R_G}(x_C, x_G:N),$$

$$\mu_{R_i}(x_C,x_G:N) = \sup_x \mu_{R_i}(x_C, x:1) \wedge \mu_{R_i}(x, x_G:N-1),$$

$$\mu_{R_i}(x_C,x_G:1) = 1 - (|\mu_i(x_C) - \mu_i(x_G)|),$$

where $i \in \{C, G\}$.

In the following, let us assume that $\mu_C(x)$ and $\mu_G(x)$ are continuous and that there exists at least an x such that $\mu_C(x) = 1$ and $\mu_G(x) = 1$, respectively. It is easy to prove that the relation R in Definition 1 satisfies the axioms. We will call this relation R a pseudo similarity relation, since though the similarity relation defined by Zadeh in [2] satisfies reflexivity, symmetry and transitivity, this R does not satisfy transitivity.

If we will define $\mu_{R'}(x_C, x_G : N)$ as follows:

$$\mu_{R'}(x_C,x_G:N) = \mu_{R_C'}(x_C,x_G:N) \vee \mu_{R_G'}(x_C,x_G:N),$$

$$\mu_{R_i}'(x_C,x_G:N) = \operatorname*{Inf}_x \mu_{R_i}'(x_C,x:1) \vee \mu_{R_i}'(x,x_G:N-1),$$

$$\mu_{R_i}'(x_C,x_G:1) = 1 - \mu_{R_i}(x_C,x_G:1),$$

where $i \in \{C,G\}$, it is easy to show that the $\mu_R'(x, x' :j)$ satisfies the property of a pseudo metric in [3], that is, (i) $\mu_R'(x, x' :j) \geq 0$, (ii) $\mu_{R'}(x, x : j) = 0$, (iii) $\mu_{R'}(x, x' : j) = \mu_{R'}(x' , x : j)$ and (iv) $\mu_{R'}(x, x' :j) \leq \mu_{R'}(x, x'' : j) + \mu_{R'}(x'', x' : j)$ for $\forall j$.

3. 0-DECISION PROBLEMS

0-decision problems are equivalent to the problem defined in [1] since we define $\mu_R(x, x': 0)$ as $\mu_R(x, x' :0) = 1$ for $x = x'$ and $\mu_R(x , x' : 0) = 0$ for $x = x'$. From the point of view of level sets, this problem had been discussed in [4]. In this section, we will deal with this problems in the variational version.

Lemma 1.

$$(a + b) \vee (c + d) \geq (a \wedge c) + (b \vee d) \geq (a + b) \wedge (c + d).$$

Since it is enough to consider only four cases (i) $a > c$, $b > d$, (ii) $a > c$, $b < d$, (iii) $a \leq c$, $b > d$ and (iv) $a < c$, $b < d$, it is easy to prove this lemma 1. We may obtain the following propositions under the assumptions that functions $f_1,\ldots,f_n$ are differentiable and continuous at all $x \in R^n$.

Proposition 1.

$$|f_1(x+\delta x) \wedge f_2(x+\delta x) - (f_1(x) + <\nabla_x^1,\delta x>) \wedge (f_2(x)+<\nabla_x^2,\delta x>)| \leq 0(\delta x^2),$$

where we denote the inner product of two vectors by $<\cdot,\cdot>$ and

$$\nabla_x = \partial f/_{\partial x} \text{ and } \lim_{\|\partial x\| \to 0} 0(\delta x^2)/_{\|\delta x\|} \to 0.$$

$$|f_1(x+\delta x) \vee f_2(x+\delta x) - (f_1(x)+<\nabla_x^1,\delta x>) \vee (f_2(x)+<\nabla_x^2,\delta x\ )$$
$$\leq 0(\delta x^2).$$

The proofs of Proposition 1 and 2 are shown in Appendix.

Propostiton 3.

$$|f_1(x+\delta x) \wedge \cdots \wedge f_n(x+\delta x) - (f_1(x)+<\nabla_x^1, \delta x>) \wedge$$
$$\cdots \wedge (f_n(x)+<\nabla_x^n,\delta x>)\ | \leq 0(\delta x^2).$$

Proposition 4.

$$|f_1(x+\delta x) \vee \cdots \vee f_n(x+\delta x) - (f_1(x) + <\nabla_x^1,\delta x>) \vee$$
$$\cdots \vee (f_n(f_n(x)\ +<\nabla_x^n,\delta x>)| \leq 0(\delta x^2).$$

Propositions 3 and 4 can be proved by induction on n as follows: since it can be assumed that Proposition 3 of n - 1 functions holds, we can set $\tilde{f}(x+\delta x)$ and $\tilde{f}'(x+\delta x)$ as

$$\tilde{f}(x+\delta x) = f_1(x+\delta x) \wedge \cdots \wedge f_{n-1}(x+\delta x),$$
$$\tilde{f}'(x+\delta x) = (f_1(x) + <\nabla_x^1,\delta x>) \wedge \cdots \wedge (f_{n-1}(x)+<\nabla_x^{n-1},\delta x>),$$

and then we can apply the same procedure as in the proof of Proposition 1 to the proof of Proposition 3.

From these propositions, we may approximate a logical function ("and" and "or" function) $f_1(x+\delta x)*\cdots*f_n(x+\delta x)$ by $(f_1(x)+<\nabla_x^1,\delta x>)*\cdots*(f_n(x)+<\nabla_x^n,\delta x>)$ if $||\delta x||$ is sufficiently small, where the symbol * denotes $\wedge$ or $\vee$.

Let us consider the optimal x^* such that $\operatorname{Sup}_x \mu_C(x) \wedge \mu_G(x) = \mu_C(x^*) \wedge \mu_G(x^*)$. If $x^* \notin K = \{x|\nabla^C(x) \overset{x}{=} 0$ or $\nabla^G(x) = 0\}$, the following theorem holds.

Theorem 1. If $x^* \notin K$, $\mu_C(x^*) = \mu_G(x^*)$ and there exists an λ^* such that $(1-\lambda^*)\ \nabla^C(x^*) + \lambda^*\nabla^G(x^*) = 0$ and $0 < \lambda^* < 1$.

Proof. From Proposition 1 and the optimality of x^*, it is necessary to hold the following inequality for $^\forall\delta x$:

$$\mu_C(x^*) \wedge \mu_G(x^*) - (\mu_C(x^*)+<\nabla^C(x^*),\delta x>)\wedge(\mu_G(x^*)+<\nabla^G(x^*),\delta x>)\geq 0.$$

We can consider two following cases where the above inequality is satisfied:

(i) $\mu_C(x^*) \wedge \mu_G(x^*) - (\mu_C(x^*) + \langle\nabla^C(x^*),\delta x\rangle) \geq 0,$

$\mu_C(x^*) \wedge \mu_G(x^*) - (\mu_G(x^*) + \langle\nabla^G(x^*),\delta x\rangle) \geq 0.$

(ii) $[\mu_C(x^*) \wedge \mu_G(x^*) - (\mu_C(x^*) + \langle\nabla^C(x^*),\delta x\rangle)].$

$[\mu_C(x^*) \wedge \mu_G(x^*) - (\mu_G(x^*) + \langle\nabla^G(x^*),\delta x\rangle)] \leq 0.$

Since $\nabla^C(x^*) = 0$ or $\nabla^G(x^*) = 0$ in the case (i), this case contradicts $x^* \notin K$. Hence, it is necessary that the case (ii) is satisfied. It follows from the case (ii) that

$$\langle\nabla^C(x^*),\delta x\rangle\ (\mu_G(x^*)-\mu_C(x^*)\wedge\mu_G(x^*))+\langle\nabla^G(x^*),\delta x\rangle(\mu_C(x^*) - \mu_C(x^*) \wedge \mu_G(x^*))+\langle\nabla^C(x^*),\delta x\rangle\cdot\langle\nabla^G(x^*),\delta x\rangle \leq 0.$$

Since this inequality must be satisfied for $^\forall\delta x$, it implies that $\mu_C(x^*) = \mu_G(x^*)$ and $[\nabla^C(x^*)\nabla^{G'}(x^*)] \leq 0$, where $\nabla^{G'}(x^*)$ denotes the transpose of $\nabla_G(x^*)$. Since the condition of $\mu_C(x) = \mu_G(x)$ is necessary, $\operatorname{Sup}_x \mu_C(x) \wedge \mu_G(x) = \operatorname{Sup}_{x\in T} \mu_C(x)$ is satisfied, where $T = \{x|\mu_C(x)-\mu_G(x) = 0\}$. According to the theorem of Kuhn-Tucher, it is necessary that the optiaml x^* and λ^* satisfy the following equation: $(1-\lambda^*)\nabla^C(x^*)+\lambda^*\nabla^G(x^*) = 0$. This necessary condition implies that $\nabla^C(x^*),\nabla^G(x^*)$ are linearly dependent. Hence, it follows from $[\nabla^C(x^*)\nabla^{G'}(x^*)] \leq 0$ that there exists an λ^* such that $0 < \lambda^* < 1$.

<u>Corollary 1</u>. If $\mu_C(x)$ and $\mu_G(x)$ are strongly fuzzy convex, then the condition in Theorem 1 is necessary and sufficient.

In the following, we will consider the decision problems which are defined by r membership functions i.e. $\mu_D(x) = \mu_1(x) \wedge \cdots \wedge\mu_r(x)$. It is assumed that $x \in R_n$ and $r < n$.

<u>Theorem 2</u>. If $x^* \notin K' = \{x|\nabla^1(x) = 0$ or $\cdots$ $\nabla^r(x) = 0\}$. then $\nabla^1(x^*),\cdots,\nabla^r(x^*)$ are linearly dependent.

Proof. Let us assume that $\nabla^1(x^*),\cdots,\nabla^r(x^*)$ are linearly independent and let W_r denote the positive convex cone spanned by $\nabla^1(x^*),\cdots,\nabla^r(x^*)$. The cone W_r is pointed, because if some vector $x \in W_r$, then $-x \notin W_r$. We can apply the well-known separation theorem to the cone W_r and then there is an v such that $<v, w> > 0$ for $\forall w \in W_r$. If let v be δx, the following inequality holds:

$$\mu_1(x^*) \wedge \cdots \wedge \mu_r(x^*) - (\mu_1(x^*) + <\nabla^1(x^*), \delta x>) \wedge \cdots$$
$$\wedge (\mu_r(x^*) + <\nabla^r(x^*), \delta x>) < 0$$

which contradicts the optimality of x^*.

<u>Definition 2</u>. x^0 is called a confluent point of $\mu_1(x)$ and $\mu_2(x)$ if and only if x^0 satisfies the equality $\mu_1(x^0) = \mu_2(x^0) = c$ and the inequality

$$(\mu_1(x^0+\delta x)-c)(\mu_2(x^0+\delta x)-c) \leq 0 \text{ for } \forall \delta x.$$

<u>Theorem 3</u>. If $x^* \notin K'$, then there exists a pair (i, j) such that

$$\sup_x \mu_D(x) = \mu_1(x^*) \wedge \cdots \wedge \mu_r(x^*) = \mu_i(x^*) \wedge \mu_j(x^*)$$

and x^* is confluent point of $\mu_i(x)$ and $\mu_j(x)$.

Proof. There exists an k at least such that $k \geq 2$ and

$$\mu_e(x^*) > \mu_D(x^*) = \mu_{i_1}(x^*) = \cdots = \mu_{i_k}(x^*),$$

since $k = 1$ contradicts $x^* \notin K'$. From the optimality of x^*,

$$\mu_{i_1}(x^*) \wedge \cdots \mu_{i_k}(x^*) - \mu_{i_1}(x^*+\delta x) \wedge \cdots \wedge \mu_{i_k}(x^*+\delta x) \geq 0$$

is satisfied for $\forall \delta x$. If $\mu_{i_j}(x^*) \geq \mu_{i_j}(x^*+\delta x)$ for all j, this assumption contradicts $x^* \notin K'$. Hence, there are μ_{i_e} and $\mu_{i_{e'}}$ at least such that

$$(\mu_{i_e}(x^*+\delta x)-\mu_D(x^*))(\mu_{i_{e'}}(x^*+\delta x)-\mu_D(x^*)) \leq 0.$$

If $x^* \notin K'$ and $\mu_1,\ldots,\mu_r$ are strongly fuzzy convex, it follows from the theorem 3 that x_1^* such that $\sup_x \mu_1(x) \wedge \cdots \wedge \mu_r(x)$ is equal to x_2^* such that $\sup_x \mu_i(x) \wedge \mu_j(x)$, where the pair (i, j) satisfies Theorem 3.

4. 1-DECISION PROBLEMS

In this section we will discuss some properties of 1-decision problems and then we will show that 1-decision problems can be reduced to simply 0-decision problems. In other words, this problem can be solved by the method for solving 0-decision problem.

Definition 3. A pair (x_C^*, x_G^*) is called the optimal solution of 1-decision problem if and only if

$$\sup_{x_C, x_G} \mu_C(x_C) \wedge \mu_G(x_G) \wedge \mu_R(x_C, x_G:1) = \mu_D(x_C^*, x_G^*:1).$$

An x_C^* is called the optimal decision and an x_G^* is called the estimated optimal goal at present. Let the set U be defined by

$$U = \{(x_C, x_G) \mid \mu_C(x_C) \geq \mu_C(x_G) \text{ and } \mu_G(x_G) \geq \mu_G(x_C)\}.$$

Proposition 5.

$$\sup_{(x_C, x_G) \in X \times X} \mu_D(x_C, x_G:1) = \sup_{(x_C, x_G) \in U} \mu_D(x_C, x_G:1).$$

Proof. Let us assume that the optimal (x_C^*, x_G^*) belongs to $X \times X - U$, i.e. $(x_C^*, x_G^*) \in X \times X - U$. Since $(x_C^*, x_G^*) \notin U$, (x_C^*, x_G^*) satisfies at least either one of the following two cases: (i) $\mu_C(x_C^*) < \mu_C(x_G^*)$ or (ii) $\mu_G(x_C^*) > \mu_G(x_G^*)$. In the case of (i),

$$\mu_C(x_C^*) \wedge \mu_G(x_G^*) \wedge (1-\mu_R(x_C^*, x_G^*:1)) \leq \mu_C(x_G^*) \wedge \mu_G(x_G^*).$$

In the case of (ii),

$$\mu_C(x_C^*) \wedge \mu_G(x_G^*) \wedge (1-\mu_R(x_C^*, x_G^*:1)) \leq \mu_C(x_G^*) \wedge \mu_G(x_G^*).$$

Since (x_G^*, x_G^*), $(x_C^*, x_C^*) \in U$, this proposition is proved.

As an illustration, let us consider the case of Figure 1. In this case, Proposition 5 makes the exception of the pair $(x_C^{*\prime}, x_G^*)$. The pair (x_C^*, x_G^*) is more desirable than $(x_C^{*\prime}, x_G^*)$, since $\mu_C(x_C^*) > \mu_C(x_C^{*\prime})$ nevertheless $\mu_D(x_C^*, x_G^*:1) = \mu_D(x_C^{*\prime}, x_G^*:1)$ and $\mu_R(x_C^{*\prime}, x_G^*:1) = \mu_R(x_C^*, x_G^*:1)$. Hence, it seems that this exception is reasonable.

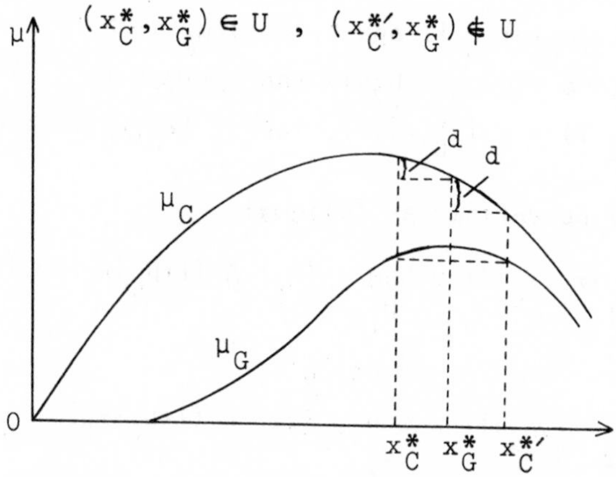

Figure 1. Explanation of set U.

<u>Proposition 6</u>.

$$\sup_{x_C, x_G} \mu_D(x_C\ ,\ x_G\ :\ 1) \geq 0.5.$$

Proof. From Proposition 5,

$$\sup_{X \times X} \mu_D(x_C, x_G:1) = \sup_U \mu_D(x_C, x_G:1)$$

$$\geq \sup_U \mu_C(x_C) \wedge (1-\mu_C(x_C)) \wedge (1-\mu_G(x_G)) \wedge \mu_G(x_G) = 0.5.$$

<u>Corollary 2</u>. If $\tilde{C}_0 \wedge \tilde{G}_0 = \phi$,

$$\sup_{X \times X} \mu_D(x_C, x_G:1) = 0.5,$$

where $\tilde{C}_0 = \{x|\mu_C(x) > 0\}$ and $\tilde{G}_0 = \{x|\mu_G(x) > 0\}$.

Since Proposition 6 implies that $\mu_D(x_C^*, x_G^*:1)$ for all 1-decision problems is more than 0.5, x_C^* and x_G^* are not unreasonable, even if $\mu_D(x^*) < 0.5$. In general, the inequality

$$\sup_{x_C, x_G} \mu_D(x_C, x_G:1) \geq \sup_x \mu_D(x)$$

is always satisfied. If $\mu_C(x)$ and $\mu_G(x)$ are pseudo convex$^+$, the following proposition holds.

<u>Proposition 7</u>.

If $C_1 \cap G_1 = \phi$,

$$\sup_{x_C, x_G} \mu_D(x_C, x_G : 1) > \sup_{x} \mu_D(x),$$

where $C_1 = \{x | \mu_C(x) \geq 1\}$ and $G_1 = \{x | \mu_G(x) \geq 1\}$.

Now with a view to obtain the optimal (x_C^*, x_G^*) such that $\sup_U \mu_D(x_C, x_G : 1) = \mu_D(x_C^*, x_G^* : 1)$, let $\mu_{D_1}(x_C, x_G : 1)$ and $\mu_{D_2}(x_C, x_G : 1)$ be defined as follows:

$$\sup_{U_1} \mu_{D_1}(x_C, x_G : 1) = \sup_{U_1} \mu_C(x_C) \wedge (1 + \mu_C(x_G) - \mu_C(x_C)) \wedge \mu_G(x_G),$$

$$\sup_{U_2} \mu_{D_2}(x_C, x_G : 1) = \sup_{U_2} \mu_C(x_C) \wedge (1 + \mu_G(x_C) - \mu_G(x_G)) \wedge \mu(x_G),$$

where $U_1 = \{(x_C, x_G) | \mu_C(x_C) \geq \mu_C(x_G)\}$ and
$U_2 = \{(x_C, x_G) | \mu_G(x_G) \geq \mu_G(x_C)\}$.

The decision problems of μ_D, μ_{D_1} and μ_{D_2} are different from each other only with respect to the metric induced by the function $\mu_C(x) \wedge \mu_G(x)$, $\mu_C(x)$ and $\mu_G(x)$ respectively.

<u>Proposition 8.</u>

$$\sup_{U_1} \mu_{D_1}(x_C, x_G : 1) = \sup_{x_G} \frac{1 + \mu_C(x_G)}{2} \wedge \mu_G(x_G).$$

$$\sup_{U_2} \mu_{D_2}(x_C, x_G : 1) = \sup_{x_C} \frac{1 + \mu_G(x_C)}{2} \wedge \mu_C(x_C).$$

Proof. $\sup_{U_1} \mu_C(x_C) \wedge (1 + \mu_C(x_G) - \mu_C(x_C)) \wedge \mu_G(x_G)$

$$= \sup_{x_G} \{ \sup_{x_C \in \{x | \mu_C(x) \geq \mu_C(x_G)\}} \mu_C(x_C) \wedge (1 + \mu_C(x_G) -$$

+ A function $\mu_A(x)$ is a pseudo convex if and only if $\mu_A(\underline{x}) < \mu_A(z)$ holds for $\forall z \in (\underline{x}, \overline{x})$, where $\mu_A(\underline{x}) < \mu_A(\overline{x})$.

$$\mu_C(x_C))\} \wedge \mu_G(x_G)$$

$$= \sup_{x_G} \frac{1 + \mu_C(x_G)}{2} \wedge \mu_G(x_G).$$

Furthermore, it is clear that $\mu_C(x_C^*) \leq \mu_C(x_G^*)$ is satisfied. The equality of μ_{D_2} can be proved by the same procedure.

Definition 4. (μ_C, μ_G) is called a pseudo complement if and only if $\mu_C(x) \leq \mu_C(x')$ implies $\mu_G(x) \geq \mu_G(x')$ and is implied by $\mu_G(x) \geq \mu_G(x')$.

If (μ_C, μ_G) is a pseudo complement, it is evident that $(x_{C_i}^*, x_{G_i}^*)$ such that $\sup_{U_i} \mu_{D_i}(x_C, x_G:1) = \mu_{D_i}(x_C^*, x_G^*:1)$ belongs to U, where i = 1, 2.

Let us define the sets S^*, S_1^* and S_2^* as follows:

$$S^* = \{(x_C^*, x_G^*) \mid \sup_U \mu_D(x_C, x_G:1) = \mu_D(x_C^*, x_G^*:1)\},$$

$$S_1^* = \{(x_{C_1}^*, x_{G_1}^*) \mid \sup_{U_1} \mu_{D_1}(x_C, x_G:1) = \mu_{D_1}(x_C^*, x_G^*:1)\},$$

$$S_2^* = \{(x_{C_2}^*, x_{G_2}^*) \mid \sup_{U_2} \mu_{D_2}(x_C, x_G:1) = \mu_{D_2}(x_C^*, x_G^*:1)\}.$$

Theorem 4. If (μ_C, μ_G) is a pseudo complement and $\mu_C(x_{C_1}^*) \geq \mu_C(x_{G_2}^*)$ and $\mu_C(x_{C_2}^*) \geq \mu_C(x_{G_1}^*)$, there exists a pair (i,j) such that $(x_{C_i}^*, x_{G_j}^*) \in S^*$, where i, j $\in \{1,2\}$.

Proof. Let us consider the following four cases:

Case 1: $\mu_C(x_{C_1}^*) \leq \mu_C(x_{C_2}^*)$, $\mu_G(x_{G_1}^*) \leq \mu_G(x_{G_2}^*)$

Let us take $(x_{C_1}^*, x_{G_1}^*)$. Hence,

$$\mu_D(x_{C_1}^*, x_{G_1}^*:1) \geq \mu_{D_1}(x_{C_1}^*, x_{G_1}^*:1) \wedge (1 + \mu_G(x_{C_2}^*) - \mu_G(x_{G_2}^*))$$

$$\geq \mu_{D_1}(x_{C_1}^*, x_{G_1}^*:1) \wedge \mu_{D_2}(x_{C_2}^*, x_{G_2}^*:1).$$

Case 2: $\mu_C(x^*_{C_1}) \geq \mu_C(x^*_{C_2})$, $\mu_G(x^*_{G_1}) \leq \mu_G(x^*_{G_2})$.

Let us take $(x^*_{C_2}, x^*_{G_1})$. Hence,

$$\mu_D(x^*_{C_2}, x^*_{G_1}:1) \geq \mu_C(x_{C_2}) \wedge \mu_G(x^*_{G_1}) \wedge (1 + \mu_C(x^*_{G_1}) - \mu_C(x^*_{C_1}))$$

$$\wedge (1 + \mu_G(x^*_{C_2}) - \mu_G(x^*_{G_2})) \geq \mu_{D_1}(x^*_{C_1}, x^*_{G_1}:1) \wedge \mu_{D_2}(x^*_{C_2}, x^*_{G_2}:1).$$

Case 3: $\mu_C(x^*_{C_1}) \leq \mu_C(x^*_{C_2})$, $\mu_G(x^*_{G_1}) \geq \mu_G(x^*_{G_2})$.

Let us take $(x^*_{C_1}, x^*_{G_2})$. Hence,

$$\mu_D(x^*_{C_1}, x^*_{G_2}:1) \geq \mu_C(x^*_{C_1}) \wedge \mu_G(x^*_{G_2}) \wedge (1 + \mu_C(x^*_{G_1}) - \mu_C(x^*_{C_1}))$$

$$\wedge (1+\mu_G(x^*_{C_2})-\mu_G(x^*_{G_2})) \geq \mu_{D_1}(x^*_{C_1}, x^*_{G_1}:1) \wedge \mu_{D_2}(x_{C_2}, x^*_{G_2}:1).$$

Case 4: $\mu_C(x^*_{C_1}) \geq \mu_C(x^*_{C_2})$, $\mu_G(x^*_{G_1}) \geq \mu_G(x^*_{G_2})$.

Let us take $(x^*_{C_2}, x^*_{G_2})$. Hence,

$$\mu_D(x^*_{C_2}, x^*_{G_2}:1) \geq \mu_{D_2}(x^*_{C_2}, x^*_{G_2}:1) \wedge (1 + \mu_C(x^*_{G_1}) - \mu_C(x^*_{C_1}))$$

$$\geq \mu_{D_2}(x^*_{C_2}, x^*_{G_2}:1) \wedge \mu_{D_1}(x^*_{C_1}, x^*_{G_1}:1).$$

Since the inequality

$$\sup_{X \times X} \mu_D(x_C, x_G:1) \leq \sup_{U_1} \mu_{D_1}(x_C, x_G:1) \wedge \sup_{U_2} \mu_{D_2}(x_C, x_G:1)$$

holds in any case, there exists a pair (i,j) such that $(x^*_{C_i}, x^*_{G_j})$ is optimal with regard to $\mu_D(x_C, x_G:1)$.

From Theorem 4 and Proposition 8, we can find the optimal pair (x^*_C, x^*_G) from four candidates of pairs $(x^*_{C_i}, x^*_{G_j})$, where $i, j \in \{1,2\}$.

5. N-DECISION PROBLEMS

We can consider N-decision problems as the extension of 1-decision problems.

Definition 5. A pair (x_C^*, x_G^*) is called the optimal solution of N-decision problems if and only if

$$\operatorname*{Sup}_{x_C, x_G} \mu_C(x_C) \wedge \mu_G(x_G) \wedge \mu_R(x_C, x_G : N) = \mu_D(x_C^*, x_G^* : N).$$

First, let us define U_1^n and U_2^n as follows:

$$U_1^n = \{(x_C, x_1, \ldots, x_{N-1}, x_G) \mid \mu_C(x_C) \geq \mu_C(x_1) \cdots \geq \mu_C(x_{N-1}) \geq \mu_C(x_G)\},$$

$$U_2^n = \{(x_C, x_1, \ldots, x_{N-1}, x_G) \mid \mu_G(x_C) \leq \mu_G(x_1) \cdots \leq \mu_G(x_{N-1}) \leq \mu_G(x_G)\}.$$

Proposition 9.

$$\operatorname*{Sup}_{X \cdots X} \mu_C(x_C) \wedge (1 + |\mu_C(x_C) - \mu_C(x_1)|) \wedge \cdots \wedge \mu_G(x_G)$$

$$= \operatorname*{Sup}_{U_1^n} \mu_C(x_C) \wedge (1 + |\mu_C(x_C) - \mu_C(x_1)|) \wedge \cdots \wedge \mu_G(x_G).$$

Proof. This proposition can be proved by the same procedure as in the proof of Proposition 5. For example, assume that the optimal solution $(x_C, x_1, \ldots, x_{N-1}, x_G)$ for the left side equation belongs to $X \cdots X - U_1^n$ and $\mu_C(x_C) \geq \cdots \geq \mu_C(x_i) < \mu_C(x_{i+1}) > \mu_C(x_{i+2}) \cdots \geq \mu_C(x_G)$, $\mu_C(x_{i-1}) \geq \mu_C(x_{i+1})$ and $\mu_C(x_i) \geq \mu_C(x_{i+2})$. The sequence $\{x_C, \cdots, x_{i+1}, x_i, x_{i+2}$ $x_{i+2}, \cdots, x_G\}$ which satisfies $\mu_C(x_C) \geq \cdots \geq \mu_C(x_{i+1}) \geq \mu_C(x_i) \geq \mu_C(x_{i+2}) \geq \cdots \geq \mu_C(x_G)$ is also optimal.

Next, we will consider the properties of the followings:

$$\mu_{D_1}(x_C^*, x_G^* : N) = \operatorname*{Sup}_{x_C, x_G} \mu_C(x_C) \wedge \mu_G(x_G) \wedge \mu_{R_1}(x_C, x_G : N),$$

$$\mu_{D_2}(x_C^*, x_G^* : N) = \operatorname*{Sup}_{x_C, x_G} \mu_C(x_C) \wedge \mu_G(x_G) \wedge \mu_{R_2}(x_C, x_G : N).$$

Proposition 10.

$$\operatorname*{Sup}_{X \cdot\cdot X} \mu_{D_1}(x_C, x_G : N) = \operatorname*{Sup}_{x_G} \frac{N + \mu_C(x_G)}{N + 1} \wedge \mu_G(x_G),$$

$$\sup_{X\cdots X} \mu_{D_2}(x_C, x_G : N) = \sup_{x_C} \mu_C(x_C) \wedge \frac{N + \mu_G(x_C)}{N+1} .$$

Proof.

$$\sup_{U_1^n} \mu_D(x_C, x_G : N) = \sup_{U_1^n} \frac{1 + \mu_C(x_1)}{2} \wedge (1 + \mu_C(x_2) - \mu_C(x_1)) \wedge \cdots \wedge \mu_G(x_G)$$

$$= \sup_{U_1^n} \frac{2 + \mu_C(x_2)}{3} \wedge (1 + \mu_C(x_3) - \mu_C(x_2)) \wedge \cdots \wedge \mu_G(x_G)$$

$$\vdots$$

$$= \sup_{x_G} \frac{N + \mu_C(x_G)}{N+1} \wedge \mu_G(x_G).$$

Proposition 11.

$$\lim_{N \to \infty} \sup_{U_i^n} \mu_{D_i}(x_C, x_G : N) = 1 \ ; \quad i = 1, 2.$$

This proposition is clear form Proposition 10 and the assumption of functions μ_C, μ_G. If N is sufficiently large, our formulation implies that our problem may be solved completely after N periods.

Let us define the sets S^{N^*}, $S_1^{N^*}$ and $S_2^{N^*}$ as follows:

$$S^{N^*} = \{(x_C^*, x_G^*) \mid \sup \mu_D(x_C, x_G : N) = \mu_D(x_C^*, x_G^* : N)\},$$

$$S_1^{N^*} = \{(x_{C_1}^*, x_{G_1}^*) \mid \sup \mu_{D_1}(x_C, x_G : N) = \mu_{D_1}(x_{C_1}^*, x_{G_1}^* : N)\},$$

$$S_2^{N^*} = \{(x_{C_2}^*, x_{G_2}^*) \mid \sup \mu_{D_2}(x_C, x_G : N) = \mu_{D_2}(x_{C_2}^*, x_{G_2}^* : N)\}.$$

The following theorem which corresponds to Theorem 4 holds in N-decision problems.

Theorem 5. If (μ_C, μ_G) is a pseudo complement and $\mu_C(x_{C_1}^*) \geq \mu_C(x_{G_2}^*)$ and $\mu_C(x_{G_2}^*) \geq \mu_C(x_{G_1}^*)$, then there exists a pair (i,j) such that $(x_{C_i}^*, x_{G_j}^*) \in S^{N^*}$, where $i, j \in \{1,2\}$.

The optimal $(x^*_{C_i}, x^*_{C_j})$ satisfies $\mu_C(x^*_{C_i}) = \min\{\mu_C(x^*_{C_1}), \mu_C(x^*_{C_2})\}$ and $\mu_G(x^*_{G_j}) = \min\{\mu_G(x^*_{G_1}), \mu_G(x^*_{G_2})\}$.

Proof. It is clear from the assumption of pseudo complement that $x_{C_1}, x_{C_2}, x_{G_1}, x_{G_2}$ is well-ordered set of R^1 in both senses of $\mu_C(x)$ and $\mu_G(x)$. The four cases are listed in Figure 2.

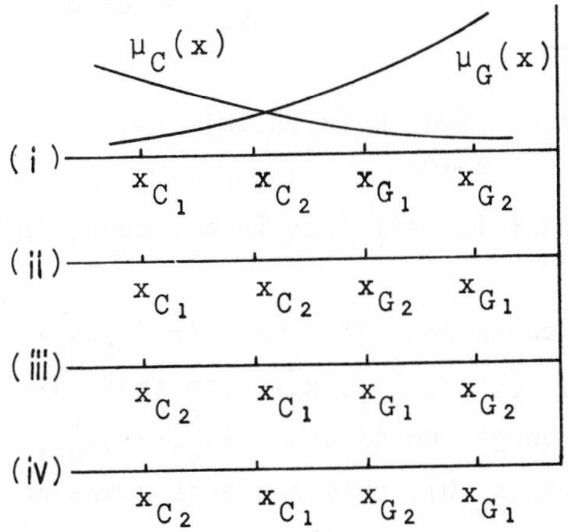

Figure 2. Four cases of well-ordered set $\{x_{C_1}, x_{C_2}, x_{G_1}, x_{G_2}\}$.

In the case of (i), $\mu_C(x^*_{C_1}) \geq \mu_C(x^*_{C_2})$ and $\mu_G(x^*_{G_2}) \geq \mu_G(x^*_{G_1})$. Now, we will take $(x^*_{C_2}, x^*_{G_1})$ as the optimal solution in the case of (i). It is evident that

$$\sup_{x_C, x_G} \mu_D(x_C, x_G : N) \geq \mu_C(x^*_{C_2}) \wedge \mu_G(x^*_{G_1}) \wedge \mu_{R_1}(x^*_{C_2}, x^*_{G_1} : N) \wedge \mu_{R_2}(x^*_{C_2}, x^*_{G_1} : N).$$

Furthermore, the following inequality is satisfied:

$$\mu_R(x^*_{C_2}, x^*_{G_1} : N) \geq \mu_{R_i}(x^*_{C_i}, x^*_{G_i} : N), \quad i = 1, 2,$$

since $|\mu_C(x^*_{C_1}) - \mu_C(x^*_{G_1})| \geq |\mu_C(x^*_{C_2}) - \mu_C(x^*_{G_1})|$ and

$| \mu_G(x^*_{G_2}) - \mu_G(x^*_{C_2})| \geq \mu_G(x^*_{G_1}) - \mu_G(x^*_{C_2})|$ are satisfied.

Thus,

$$\mu_D(x^*_{C_2}, x^*_{G_1} : N) \geq \mu_{D_1}(x^*_{C_1}, x^*_{G_1} : N) \wedge \mu_{D_2}(x^*_{C_2}, x^*_{G_2} : N)$$
$$\geq \sup_{x_C, x_G} \mu_D(x_C, x_G : N).$$

This leads to

$$\mu_D(x^*_{C_2}, x^*_{G_1} : N) = \sup_{x_C, x_G} \mu_D(x_C, x_G : N).$$

Since this fact is satisfied in any case, this theorem was proved.

We are only concerned with the pair (x^*_C, x^*_G) in the optimal sequence $\{x^*_C, x^*_1, \cdots, x^*_{N-1}, x^*_G\}$ such that maximizes $\mu_D(x_C, x_G : N)$. Although the sequence $\{x^*_1, \cdots, x^*_{N-1}\}$ is necessary for obtaining $\mu_R(x_C, x_G : N)$, this sequence makes no real sense. Thus, since it is necessary only to decide the optimal pair (x^*_C, x^*_G), we can get the optimal (x^*_C, x^*_G) for $\mu_D(x_C, x_G : N)$ in consequence of finding the optimal pairs for $\mu_{D_1}(x_C, x_G : N)$ and $\mu_{D_2}(x_C, x_G : N)$, which may be obtained more simply from Proposition 10.

6. CONCLUDING REMARKS

Since it seems that almost real world problems involving economic systems, public systems, etc, satisfy a pseudo complement in the domain under consideration, almost N-decision problems can be solved by the method for solving 0-decision problems. We may regard 0-decision problems as optimization problems of logical functions. Therefore we have discussed on the properties of optimization problems including logical functions, for it seems that they have not yet investigated.

Our approach has obvious limitations due to the necessity of satisfying the pseudo similarity relation defined in this paper. But if a similarity relation is given in any way, we may obtain the optimal pair of the decision and goal by using the similar procedure as in this paper.

REFERENCES

[1] R. E. Bellman and L. A. Zadeh: "Decision-Making in a Fuzzy Environment", Management Science, Vol. 17, No. 4, 1970.

[2] L. A. Zadeh: "Similarity Relations and Fuzzy Orderings", Information Sciences, Vol. 3, No. 2, 1971.

[3] A. DE Luca and S. Termini: "A Definition of a Nonprobabilistic Entropy in the Setting of Fuzzy Sets Theory", Information and Control, Vol. 20, No. 4, 1972.

[4] H. Tanaka and K. Asai: "On Fuzzy Mathematical Programming", Preprints of the 3rd IFAC Symposium, The Hague, 1973.

APPENDIX

Proof of Proposition 1: From the assumption of Proposition 1, there exists an δx such that

$$| f_1(x + \delta x) - f_1(x) - < \nabla_x^1, \delta x > | \leq 0_1(\delta x^2),$$

$$| f_2(x + {}^2x) - f_2(x) - < \nabla_x^2, \delta x > | \leq 0_2(\delta x^2).$$

From Lemma 1,

$$0(\delta x^2) = 0_1(\delta x^2) + 0_2(\delta x^2) \geq (f_1(x+\delta x) - f_1(x) - <\nabla_x^1, \delta x>) \vee$$
$$(f_2(x+\delta x) - f_2(x) - <\nabla_x^2, \delta x>)$$

$$\geq f_1(x+\delta x) \wedge f_2(x+\delta x) + (-f_1(x) - \langle\nabla^1_x, \delta x\rangle) \vee (-f_2(x) - \langle\nabla^1_x, \delta x\rangle)$$

$$= f_1(x+\delta x) \wedge f_2(x+\delta x) - (f_1(x) + \langle\nabla^1_x, \delta x\rangle) (f_2(x) + \langle\nabla^2_x, \delta x\rangle).$$

Conversely from Lemma 1,

$$-0(\delta x^2) \leq (f_1(x+\delta x) - f_1(x) - \langle\nabla^1_x, \delta x\rangle) \wedge (f_2(x+\delta x) - f_2(x) - \langle\nabla^2_x, \delta x\rangle),$$

$$0(\delta x^2) \geq (f_1(x) + \langle\nabla^1_x, \delta x\rangle - f_1(x+\delta x)) \vee (f_2(x) + \langle\nabla^2_x, \delta x\rangle - f_2(x+\delta x))$$

$$\geq (f_1(x) + \langle\nabla^1_x, \delta x\rangle) \wedge (f_2(x) + \langle\nabla^2_x, \delta x\rangle) + (-f_1(x+\delta x) \vee - f_2(x+\delta x)),$$

$$-0(\delta x^2) \leq f_1(x+\delta x) \wedge f_2(x+\delta x) - (f_1(x) + \langle\nabla^1_x, \delta x\rangle) \wedge (f_2(x) + \langle\nabla^2_x, \delta x\rangle).$$

From the above inequalities,

$$|f_1(x+\delta x) \wedge f_2(x+\delta x) - (f_1(x) + \langle\nabla^1_x, \delta x\rangle) \wedge (f_2(x) + \langle\nabla^2_x, \delta x\rangle)| \leq 0(\delta x^2).$$

<u>Proof of Proposition 2</u>: From Lemma 1,

$$-0(\delta x^2) \leq (f_1(x+\delta x) - f_1(x) - \langle\nabla^1, \delta x\rangle) \wedge (f_2(x+\delta x) - f_2(x) - \langle\nabla^2_x, \delta x\rangle) \leq f_1(x+\delta x) \vee f_2(x+\delta x) - (f_1(x) + \langle\nabla^1_x, \delta x\rangle) \vee (f_2(x) + \langle\nabla^2_x, \delta x\rangle).$$

Conversely from Lemma 1,

$$0(\delta x^2) \geq (f_1(x+\delta x) - f_1(x) - \langle\nabla^1_x, \delta x\rangle) \vee (f_2(x+\delta x) - f_2(x) - \langle\nabla^2_x, \delta x\rangle),$$

$$-0(\delta x^2) \leq (f_1(x) + \langle\nabla^1_x, \delta x\rangle - f_1(x+\delta x)) \wedge (f_2(x) + \langle\nabla^2_x, \delta x\rangle),$$

$$- f_2(x+\delta x)) \leq f_1(x+\delta x) \vee f_2(x+\delta x) - (f_1(x) + \langle\nabla_x^1, \delta x\rangle) \vee (f_2(x) + \langle\nabla_x^2, \delta x\rangle).$$

Thus,

$$|f_1(x+\delta x) \vee f_2(x+\delta x) - (f_1(x)+\langle\nabla_x^1, \delta x\rangle) \vee (f_2(x) + \langle\nabla_x^2, \delta x\rangle)| \leq 0(\delta x^2).$$

RECOGNITION OF FUZZY LANGUAGES

Namio Honda and Masakazu Nasu
Research Institute of Electrical Communication
Tohoku University
Katahira, Sendai 980
Japan

ABSTRACT

In this paper, we propose the concept of recognition of fuzzy languages by machines such as Turing machines, linear bounded automata, pushdown automata and finite automata. It is shown that it is a reasonable extension of the ordinary concept of recognition of languages by machines. Basic results are given about the recognition theory of fuzzy languages.

1. INTRODUCTION

In the formal language theory, languages are classified by the complexities of machines which recognize them. The most typical ones of such machines are finite automata, pushdown automata, linear bounded automata and Turing machines. As for fuzzy languages, it is also interesting to develop the theory of recognition of fuzzy languages by machines and their classification by the complexities of machines which recognize them. Of course, the theory should be a reasonable extension of the ordinary language recognition theory. In the ordinary language recognition theory, a machine is said to recognize a language L if and only if for every word in L he machine decides that it is a member of L, and for any ord not in L, the machine either decides that it is not a ember of L or loops forever. In other words, a machine may be said to recognize a language if and only if the machine computes the characteristic function of the language. So it is natural to define a machine to recognize a fuzzy language if and only if the machine computes its fuzzy membership function. But what does it mean that a machine computes a fuzzy membership function? In the ordinal language theory, it is defined that for a given input word a machine computes the characteristic function value at 1, if and only if it takes one of special memory-configurations, such as configurations with a final state and configurations with the empty stack for cases where the machine has pushdown stacks. Therefore it is a straightforward extension to define each fuzzy membership function value (which is an element of some lattice) to be represented by some memory configuration of the machine. That is to say, we require that the memory configuration which the machine moves into after a sequence of moves for a given input word should be uniquely associated with the membership function value of the word.

Furthermore it is considered to be essential that the value represented by a memory configuration of the machine is an element in a lattice. We also require that the machine can compare it with any cut-point which takes one of elements in the lattice. It is reasonable to assume that a cut-point will be represented by an infinite sequence of symbols in a finite alphabet, following the infinite expansions of decimals. We require that for any cut-point λ the machine having the memory configuration associated with the fuzzy membership function value $f(w)$ of a given input word w should be able to read, as new input, the **infinite** sequence corresponding to the cut-point λ sequentially, and after a finite step of moves, it can determine which of the following four cases is valid, (i) $f(w) > \lambda$, (ii) $f(w) < \lambda$, (iii) $f(w) = \lambda$, and (iv) $f(w)$

and λ are incomparable.

In this paper, a first step will be given toward the fuzzy language recognition theory on the line stated above.

2. FUZZY LANGUAGES

Let Σ be a finite set of symbols called an alphabet. The set of all strings of symbols in Σ including the null string ε will be denoted by Σ^*. An element of Σ^* will be called a word over Σ. A subset of Σ^* will be called a language over Σ. Let L be a lattice with minimum element 0. An L-fuzzy language over Σ is defined to be a mapping from Σ^* to L (following Goguen [2]). L will be often omitted if we are not confused. Let f be an L-fuzzy language over Σ, then $f(x)$ for x in Σ^* represents the grade for x to be a member of the fuzzy language.

3. CUT-POINTS AND THEIR REPRESENTATION

Let L be a lattice with minimum element 0. Usually a value in L is specified and used as a cut-point. It is interesting to consider language associated with an L-fuzzy language f and a cut-point λ such as;

$$L_G(f,\lambda) = \{x \in \Sigma^* | f(x) > \lambda\}$$

$$L_{GE}(f,\lambda) = \{x \in \Sigma^* | f(x) \geqq \lambda\}$$

Such languages will be called cut-point languages for f and λ.

Let Δ be a finite alphabet. Let Δ^∞ be the set of all infinite sequences of symbols in Δ extending infinitely to the right. We will define a representation of L over Δ. A one to one mapping r from L to Δ^∞ is a representation of L over Δ if and only if it suffices the following conditions.

(i) For ℓ, m in L, if $\ell \neq m$ there exists $w_1, w_2 \in \Delta^*$ with

$|w_1| = |w_2|$[†] and α, $\beta \in \Delta^\infty$ such that $r(\ell) = w_1\alpha$, $r(m) = w_2\beta$ and for all α', β' in Δ^∞ either following (a) or (b) holds;

(a) Either $w_1\alpha'$ or $w_2\beta'$ is not in $r(L)$.

(b) Both $w_1\alpha'$ and $w_2\beta'$ are in $r(L)$, and

$$r^{-1}(w_1\alpha') > r^{-1}(w_2\beta') \quad \text{if} \quad \ell > m,$$

$$r^{-1}(w_1\alpha') < r^{-1}(w_2\beta') \quad \text{if} \quad \ell < m$$

and $r^{-1}(w_1\alpha')$ and $r^{-1}(w_2\beta')$ are incomparable, if ℓ and m are incomparable.

Such $|w_1|$ ($= |w_2|$) will be called D-length of $r(\ell)$ and $r(m)$.

(ii) Let $d(r(\ell), r(m))$ be the minimum D-length of $r(\ell)$ and $r(m)$, then for any ℓ, m, and n in L with $\ell > m > n$,

$$d(r(\ell), r(n)) \leq \min \{d(r(\ell), r(m)), d(r(m), r(n))\}$$

holds.

We will say that for $\ell \in L$, $r(\ell)$ is the representation of ℓ with respect to r. A lattice cannot always have its representation. We will consider from now on only lattices which can have a representation over some finite alphabet. Condition (ii) means that representations of lattices are restricted to the type of one such as decimal expansions of real numbers. However there may be many representations for a lattice.

Example 1. Let L be a lattice with finite elements. If L has elements $\ell_1, \ldots, \ell_k$, let $\Delta = \{\ell_1, \ldots, \ell_k\}$ and $r(\ell_i) = \ell_i \ell_i \ell_i \ldots$ for $i \leq i \leq k$, then r is a representation of L over Δ.

Example 2. Let $L_{[0,1]}$ be the set of all real numbers in [0,1] with the ordinary ordering. A representation r_1 is given as follows: $\Delta = \{0, 1, \dot{0}, \dot{1}\}$. Let $e(\ell)$ for ℓ $L_{[0,1]}$ be

[†] $|w|$ represents the length of a word w.

the binary expansion of ℓ not of the form $w111\cdots$ with $w \in \{0,1\}^*0$. For any rational number ℓ in [0,1], we set $e(\ell) = w_0w_1w_1w_1\cdots$ such that if $e(\ell) = w_0'w_1'w_1'w_1'\cdots$, then $|w_0| \leq |w_0'|$ and $|w_1| \leq |w_1'|$. Let $\dot{e}(\ell) = w_0\dot{w}_1w_1w_1w_1\cdots$ where $\dot{w}_1 = \dot{a}_1\dot{a}_2\cdots\dot{a}_k$ with $w_1 = a_1\cdots a_k$, $a_i \in \{0,1\}$ $(1 \leq i \leq k)$.

$r_1(\ell) = e(\ell)$ if ℓ is irrational.

$r_1(\ell) = \dot{e}(\ell)$ if ℓ is rational.

Example 3. Let $L_{[0,1]_R}$ be the set of all rational numbers in [0,1] with the usual ordering relation. A representation r_2 of $L_{[0,1]_R}$ is such that $\Delta = \{0,1,\dot{0},\dot{1}\}$ and $r_2(\ell) = e(\ell)$ $(\ell \in L_{[0,1]_R})$ except for the following cases: $r_2(0) = \dot{0}\dot{0}\dot{0}\cdots$, $r_2(1) = \dot{1}\dot{1}\dot{1}\cdots$, $r_2(\ell) = w\dot{0}\dot{0}\dot{0}\cdots$ for ℓ such that $e(\ell) = w000\cdots$ with $w \in \{0,1\}^*1$.

4. F-RECOGNITIONS BY MACHINES.

Machines treated hereafter may be finite automata, push-down automata, linear bounded automata and Turing machines, which can generally be represented as follows:

A machine has an input terminal which read input symbols and ε sequentially, a memory storing and processing device and an output terminal. Formally a machine is given by 8-tuple $M = \langle \Phi,\Gamma,\Psi,\theta,\delta,\Omega,\kappa,\gamma_0 \rangle$ where;

Φ: a finite set of input symbols

Γ: a finite set of memory-configuration symbols

Ψ: a finite set of output symbols

Ω: a finite set of partial function $\{\omega_i\}$ from Γ^* to Γ^*

θ: a partial function from Γ^* to Γ^n, for some $n \geq 1$. (For a memory configuration $\gamma \in \Gamma^*$, $\theta(\gamma)$ indicates the accessible information of γ by M.)

δ: a partial function from $(\Phi \cup \{\varepsilon\}) \times \Gamma^n$ to 2^Ω

κ: a partial function from Γ^n to $\Psi \cup \{\varepsilon\}$

γ_0: an element in Γ^* (called the initial memory configuration).

A memory configuration ξ is said to be derived from a memory configuration γ by $\sigma \in \Phi \cup \{\varepsilon\}$ and is denoted by $\gamma \underset{\sigma}{\Rightarrow} \xi$, if and only if there exists $\rho = \theta(\gamma)$ and $\omega_i \in \delta(\sigma,\rho)$ such that $\omega_i(\gamma) = \xi$. For a word $w \in \Phi^*$, a memory configuration ξ is said to be derived from a memory configuration γ by w and is denoted by $\gamma \underset{w}{\Rightarrow} \xi$ if and only if there exists $\sigma_1,\sigma_2,\ldots,\sigma_\ell$ with σ_i in $\Phi \cup \{\varepsilon\}$ such that $w = \sigma_1\sigma_2 \cdots \sigma_\ell$, and $\gamma_0,\gamma_1,\cdots,\gamma_\ell$ with $\gamma_i \in \Gamma^*$ such that $\gamma_0 = \gamma$, $\gamma_\ell = \xi$ and $\gamma_i \underset{\sigma_i}{\Longrightarrow} \gamma_{i+1}$ for all $0 \leq i \leq \ell - 1$. ($\gamma \underset{\varepsilon}{\Rightarrow} \gamma$ is valid for all $\gamma' \in \Gamma^*$). Given an input word $w \in \Phi^*$ and having the initial memory configuration γ_0 first, M reads input symbols or ε sequentially along w, changes step by step memory configurations possibly in a nondeterministic way and reaches into γ such that $\gamma_0 \underset{w}{\Rightarrow} \gamma$, emitting output $\kappa(\theta(\gamma))$.

Obviously a machine $M = \langle \Phi, \Gamma, \Psi, \theta, \delta, \Omega, \kappa, \gamma_0 \rangle$ can be restricted to a specified family of automata such as Turing machines, pushdown automata, finite automata for appropriate choices of Γ, θ, δ, Ω, κ, and γ_0.

Furthermore we will define a deterministic machine. A machine $M = \langle \Phi, \Gamma, \Psi, \theta, \delta, \Omega, \kappa, \gamma_0 \rangle$ will be called a deterministic machine if for any memory configuration γ such that $\gamma_0 \underset{x}{\Rightarrow} \gamma$ for some x in Φ^*, if $\delta(\varepsilon,\theta(\gamma)) \neq \phi$, then $\delta(\varepsilon,\theta(\gamma))$ contains at most one element and for any $\sigma \in \Phi$, $\delta(\sigma,\theta(\gamma)) = \phi$, and if $\delta(\varepsilon,\theta(\gamma)) = \phi$, then for any $\sigma \in \Phi$, $\delta(\sigma,\theta(\gamma))$ contains at most one element.

Now we will define recognition of a fuzzy language by a machine. Let L be a lattice with a minimum element 0 and $f: \Sigma^* \to L$ be a fuzzy language over an alphabet Σ. Let r be a representation of L over an alphabet Δ. A machine $M = \langle \Phi, \Gamma, \Psi, \theta, \delta, \Omega, \kappa, \gamma_0 \rangle$ f-recognizes f with r if and only if the following conditions hold:

1. $\Phi = \Sigma \cup \Delta \cup \{c\}$, where c is an element not in $\Sigma \cup \Delta$.
2. $\Psi = \{>, <, =, !\}$
3. There exists a partial function ν from Γ^* to L which satisfies the following conditions (i) ∿ (iv).

(i) For any $x \in \Sigma^*$, $S_x \subset \text{Dom } \nu^\dagger$, where

$$S_x = \{\gamma \mid \gamma_0 \underset{xc}{\Longrightarrow} \gamma\}$$

If $S_x = \phi$, then max $\{\nu(\gamma) \mid \gamma \in S_x\}$ always exists. Let $\nu_x = \max \{\nu(\gamma) \mid \gamma \in S_x\}$ if $S_x \neq \phi$, and otherwise let ν_x not be defined. Then it holds that if $f(x) \neq 0$, then $S_x \neq \phi$ and $f(x) = \nu_x$.

(ii) Let γ be any memory configuration in S_x. Let a machine M_γ be $\langle \Delta, \Gamma, \Psi, \theta, \delta', \Omega, \kappa, \gamma \rangle$, where δ' is the restriction of δ over $(\Delta \cup \{\varepsilon\}) \times \Gamma^n$. Then M_γ is a deterministic machine.

(iii) For any memory configuration $\gamma \in \Gamma^*$, if $\kappa(\theta(\gamma))$ is in Ψ, that is, $\kappa(\theta(\gamma)) \neq \varepsilon$, then for any $\sigma \in \Phi \cup \{\varepsilon\}$ $\delta(\sigma, \theta(\gamma))$ is empty. And $\kappa(\theta(\gamma))$ is in Ψ only if $\gamma' \underset{y}{\Rightarrow} \gamma$ for some $\gamma' \in S_x$, x in Σ^* and y in PRE $r(L)^{\dagger\dagger}$.

(iv) Let γ be any element in S_x. For any $\ell \in L$, there exists a prefix v of $r(\ell)$ such that

$$\gamma \underset{v}{\Rightarrow} \gamma'$$

---

† Dom $\nu = \{\gamma \in \Gamma^* \mid \nu(\gamma) \text{ is defined}\}$.

†† w_1 will be called a prefix of a word or an infinite sequence α if $\alpha = w_1\beta$ for some β. Let Π be either a set of words or a set of infinite sequences. PRE Π is the set of all prefixes of elements in Π.

$\kappa(\theta(\gamma'))$ is >, if $\nu(\gamma) > \ell$,
$\kappa(\theta(\gamma'))$ is =, if $\nu(\gamma) = \ell$,
$\kappa(\theta(\gamma'))$ is <, if $\nu(\gamma) < \ell$,
and $\kappa(\theta(\gamma'))$ is !, if $\nu(\gamma)$ and ℓ are incomparable.

Given an input sequence xc in Σ^* where c indicates the end of the input sequence, a machine M moves possibly non-deterministically into some memory configuration γ such that $\nu(\gamma)$ is defined. S_x is the set of all such γ's. We consider the maximum value ν_x of $\{\nu(\gamma) | \gamma \in S_x\}$ as the value of x computed by M. If $S_x = \phi$, we consider that the value of x cannot be computed by M. We will call a sequence of moves from the initial memory configuration to a memory configuration in S_x a value computation for x.

Let γ be any memory configuration in S_x. Then we require that M_γ should be able to compare $\nu(\gamma)$ with any element ℓ in L. M_γ moves deterministically reading input symbols in $\{\varepsilon\} \cup \Delta$ along the infinite sequence $r(\ell)$ and emits one of >, <, = and ! following the order of $\nu(\gamma)$ and ℓ in L after reading a finite length of prefix of $r(\ell)$, and halts. (See (iii).) We will call a sequence of moves of M from γ in S_x to a halting configuration an order-comparing computation for γ. If a fuzzy language $f: \Sigma^* \to L$ is the function such that

$$f(x) = \nu_x \quad \text{if } \nu_x \text{ is defined}$$
$$f(x) = 0 \quad \text{otherwise},$$

then f is said to be f-recognized by the machine M with the representation r.

Let T_0, T_1, T_2 and T_3 be the classes of Turing machines, linear bounded automata, pushdown automata and finite automata respectively. And let DT_0, DT_1, DT_2 and DT_3 be the classes of deterministic Turing machines, deterministic linear bounded automata, deterministic pushdown automata and deterministic finite automata respectively. A fuzzy language f is said to be f-recognized by a machine in T_i (DT_i) if and only if f is f-recognized by a machine in T_i (DT_i) with some representation r, for i = 0, 2 and 3. And we will say that a fuzzy language f: $\Sigma^* \to L$ is f-recognized by a (deterministic) linear bounded automaton if and only if f is f-recognized by a (deterministic) Turing machine M with some representation r as follows: For any $x \in \Sigma^*$, if $\gamma_0 \underset{y_1}{\Longrightarrow} \gamma$ for some prefix y_1 of xc, or

$\gamma_0 \underset{xcy}{\Longrightarrow} \gamma$ for some y in PRE (L), then $|\gamma| \leq C\,|x|$ for some constant C, where γ_0 is the initial configuration of M. (This means that lengths of memory configurations in M for any $x \in \Sigma^*$ are always not greater than some constant time of $|x|$ throughout the value computation and the order-comparing computation of x with any cut-point in L.)

Example 4. Let $\Sigma = \{a,b\}$. For $w \in \Sigma^*$, let $n_a(w)$ and $n_b(w)$ be the numbers of occurences of a and b in w respectively. A fuzzy language f_1: $\Sigma^* \to L_{[0,1]_R}$ defined by

$$f_1(w) = \frac{1}{2} + \left(\frac{1}{2}\right)^{|n_a(w)-n_b(w)|+1} \qquad (w \in \Sigma^*)$$

is f-recognized by a deterministic pushdown automaton.

A pushdown automaton $M = \langle \Phi,\Gamma,\Psi,\theta,\delta,\Omega,\kappa,\sigma_0 \rangle$ with γ_2 in Example 3 f-recognizes f_1, where $\Phi = \Sigma \cup \Delta \cup \{c\}$ with $\Sigma = \{a,b\}$ and $\Delta = \{0,1,\dot{0},\dot{1}\}$, $\Gamma = Q \cup \{z_0,a,b,1\}$ where $Q = \{q_0,q_1,q_>, q_<, q_=\}$, $\Psi = \{>, <, =\}$, θ is a partial function from $\Gamma^* \to \Gamma^2$ such that $\theta(qx\sigma) = (q,\sigma)$ for all $q \in Q$, $x \in \{\varepsilon\} \cup z_0\{a,b\}^*$ and $\sigma \in \{z_0,a,b,1,\dot{1}\}$, $\Omega = \{\omega_{1a},\omega_{1b},\omega_a,\omega_b, \omega_-,\omega_1,\omega_{\dot{1}},\omega_>,\omega_<,\omega_=\}$ where

$\gamma_0 = q_0 z_0$
$\omega_{1\sigma}(x) = x1\sigma$ for $x \in \Gamma^*$ and $\sigma \in \{a,b\}$
$\omega_\sigma(x) = x\sigma$ for $x \in \Gamma^*$ and $\sigma \in \{a,b\}$
$\omega_-(x\sigma) = x$ for $x \in \Gamma^*$ and $\sigma \in \Gamma$
$\omega_1(q_0 x) = q_1 x 1$ for $x \in \Gamma^*$
$\omega_i(q_0 z_0) = q_1 z_0 i$
$\omega_>(q_1 x) = q_> x \quad \omega_<(q_1 x) = q_< x \quad \omega_=(q_1 x) = q_= x$
for $x \in \Gamma^*$

$\delta : \Phi \times \Gamma^2 \to 2^\Omega$ is defined as:

$\delta(c,(q_0,z_0)) = \{\omega_i\}$
$\delta(\sigma,(q_0,z_0)) = \{\omega_{1\sigma}\}$ for $\sigma \in \{a,b\}$
$\delta(\sigma,(q_0,\sigma)) = \{\omega_\sigma\}$ for $\sigma \in \{a,b\}$
$\delta(a,(q_0,b)) = \delta(b(q_0,a)) = \{\omega_-\}$
$\delta(c,(q_0,\sigma)) = \{\omega_1\}$ for $\sigma \in \{a,b\}$
$\delta(1,(q_1,1)) = \delta(0,(q_1,\sigma)) = \{\omega_-\}$ for $\sigma \in \{a,b\}$
$\delta(0,(q_1,1)) = \{\omega_>\}$, $\delta(1,(q_1,\sigma)) = \{\omega_<\}$ for $\sigma \in \{a,b\}$
$\delta(\dot{0},(q_1,z_0)) = \omega_=$, $\delta(\sigma,(q_1,\dot{1})) = \{\omega_>\}$ for $\sigma \in \{0,1,\dot{0}\}$
$\delta(\dot{1},(q_1,\dot{1})) = \omega_=$.

$\kappa(q_\eta,\sigma) = \eta$ for $\eta \in \Psi$ and $\sigma \in \{z_0,a,b,1,\dot{1}\}$
We set $\nu : \Gamma^* \to [0,1]_R$ defined by

$\nu(q_1 z_0 \dot{1}) = 1$
and $\nu(q_1 z_0 \sigma^n 1) = \frac{1}{2} + (\frac{1}{2})^{n+1}$ for $\sigma \in \{a,b\}$ and $n \geq 1$.

5. ISOLATED CUT-POINTS

Let f be an L-fuzzy language over Σ, where L be a lattice with a minimum element 0. $\ell \in L$ will be called an isolated cut-point of f if one of the following (i), (ii), and (iii) holds;

(i) There exists ℓ_1 and ℓ_2 in L such that $\ell_1 < \ell < \ell_2$ and for any $f(x)$ $(x \in \Sigma^*)$ with $f(x) = \ell$, either $f(x) \leq \ell_1$ or $f(x) \geq \ell_2$.

(ii) ℓ is a maximum element of L and there exists $\ell_1 \neq \ell$ in L such that for any $f(x)$ $(x \in \Sigma^*)$ with $f(x) \neq \ell$, $f(x) \leq \ell_1$ holds.

(iii) $\ell = 0$ and there exists $\ell_2 \neq 0$ in L such that for any $f(x)$ $(x \in \Sigma^*)$ with $f(x) \neq 0$, $f(x) \geq \ell_2$ holds.

Theorem 1. Let L be a lattice with minimum element 0. Let $f: \Sigma^* \to L$ be a fuzzy language and let ℓ be an isolated cut-point of f. Then if f is f-recognized by a machine in T_i, each of $L_{GE}(f,\ell)$ and $L_G(f,\ell)$ is recognized by a machine in T_i for $0 \leq i \leq 3$.

(Proof) Assume that f is f-recognized by a machine M $= \langle \Sigma \cup \Delta \cup \{c\}, \Gamma, \Psi, \theta, \delta, \Omega, \kappa, \gamma_0 \rangle$ in T_i with a representation r over Δ. Since ℓ is an isolated cut-point of f, either (i), (ii) or (iii) holds. We will only prove the case where (i) holds. (Proofs for other cases are similar.) Let ℓ_1 and ℓ_2 in L be such that $\ell_1 < \ell < \ell_2$ and for any $f(x)$ $(x \in \Sigma^*)$ with $f(x) = \ell$, either $f(x) \geq \ell_2$ or $f(x) \leq \ell_1$ holds. Let d_1 and d_2 be the minimum D-length of $r(\ell_1)$ and $r(\ell)$ and of $r(\ell)$ and $r(\ell_2)$ respectively. Let $d_3 = \max(d_1, d_2)$ and $w \in \Sigma^*$ be the prefix of $r(\ell)$ of length d_3. From the definition of f-recognition, the set $L[M,\geq]$ is recognized by a machine in T_i, where

$$L[M,\geq] = \{xcy \mid x \in \Sigma^*,\ y \in \Delta^*,\ \gamma_0 \underset{xcy}{\Longrightarrow} \gamma \text{ such that } \kappa(\theta(\gamma)) \text{ is } (=) \text{ or } (>)\}.$$

And it holds that

$$\{x \in \Sigma^* \mid f(x) \geq \ell\} = \{x \in \Sigma^* \mid xcy \in L[M,\geq] \text{ for some } y \text{ in } w\Delta^*\}.$$

This is proved as follows: If $f(x) \geq \ell$, there exists $\gamma \in \Gamma^*$ and $y \in \Delta^*$ such that y is a prefix of $r(\ell)$, $\gamma_0 \underset{xcy}{\Longrightarrow} \gamma$ and $\kappa(\theta(\gamma))$ is (>) or (=). Conversely assume that $\gamma_0 \underset{xcy}{\Longrightarrow} \gamma$ and $\kappa(\theta(\gamma))$ is (>) or (=) for some y in $w\Delta^*$ and γ in Γ^*. Then there exists $\ell' \in L$ such that $f(x) \geq \ell'$ and for some $\alpha \in \Delta^\infty$ and $w' \in \Delta^*$, $\gamma(\ell') = y\alpha = ww'\alpha$. From the definition of D-length, neither ℓ' and ℓ_1 nor ℓ' and ℓ_2 are incomparable. Also from the definition of D-length, neither $\ell' < \ell_1$ nor $\ell_2 < \ell'$ is valid. Thus $\ell_1 \leq \ell' \leq \ell_2$ and so $f(x) = \ell$ or $f(x) \geq \ell_2$. Hence $f(x) \geq \ell$.

It is obvious that there exists a gsm-mapping G such that

$$L_{GE}(f,\ell) = G(L[M,\geq] \cap \Sigma^* cw\Delta^*).$$

Since recursively enumerable sets, context free languages and regular sets are closed under a gsm-mapping operation respectively, Theorem has been proved for $i = 0$, 2 and 3.

For the case of $i = 1$, the machine M is a Turing machine such that for some constant C, $|\gamma| \leq C\,|x|$ for any $x \in \Sigma^*$ and for any memory configuration γ such that $\gamma_0 \underset{z}{\Rightarrow} \gamma$ with z in PRE $(L[M,\geq])$. A machine M' is a modification of M as follows; M' moves as in the same way as M for xc $(x \in \Sigma^*)$. After reading xc, M' continues to read ε, and changes sequentially memory configurations as in the same was as M reads y in $w\Delta^*$. M' has an autonmous finite state machine as a sub-machine which generates any y in $w\Delta^*$ nondeterministically. M' is a linear bounded automaton which recognizes $L_{GE}(f,\ell)$.

As for $L_G(f,\ell)$, the proofs are similar.

Corollary 1. Let L be a finite lattice, and let f: $\Sigma^* \to L$ be a fuzzy language. If f is f-recognized by a machine in T_i, then for any $\ell \in L$, each of $L_{GE}(f,\ell)$ and

$L_G(f,\ell)$ is recognized by a machine in T_i for i = 0,1,2 and 3.

(Proof) Assume that f is f-recognized by a machine in T_i, $M = \langle \Sigma \cup \Delta \cup \{c\}, \Gamma, \Psi, \theta, \delta, \Omega, \kappa, \gamma_0 \rangle$. Let $L = \{\ell_1, \ell_2, \ldots, \ell_s\}$. Then there exists w_i' s in Δ^* such that $r(\ell_i)$ is in $w_i\Delta^\infty$ but is not in $w_j\Delta^\infty$ for $i \neq j$ $(1 \leq i, j \leq s)$. From the definition of f-recognition, the set $L[M,\geq]$ is recognized by a machine in T_i, where

$$L[M,\geq] = \{xcy \mid x \in \Sigma^*, y \in \Delta^*, \gamma_0 \underset{xcy}{\Longrightarrow} \gamma \text{ such that } \kappa(\theta(\gamma)) \text{ is } (=) \text{ or } (>)\}$$

Clearly it holds that for $1 \leq i \leq s$

$$\{x \in \Sigma^* \mid f(x) \geq \ell_i\} = \{x \in \Sigma^* \mid xcy \text{ in } L[M,\geq] \text{ for some } y \text{ in } w_i\Delta^*\}.$$

The rest of the proof is the same as in the proof of Theorem 1.

Corollary 2. Let L be a lattice with minimum element 0. If an L-fuzzy language f is f-recgonized by a machine in T_i with a representation r, and for $\ell \in L$, $r(\ell)$ is generated by an autonomous finite automaton sequentially, then $L_G(f,\ell)$ and $L_{GE}(f,\ell)$ are recognized by a machine in T_i (i = 0,1,2 and 3).

Theorem 2. Let L be a lattice with minimum element 0 which has a representation r_0 over Δ_0. Let $f: \Sigma^* \to L$ be a fuzzy language such that $f(\Sigma^*) = \{f(x) \mid x \in \Sigma^*\}$ is finite. If for any $f(x)$ $(x \in \Sigma^*)$, $L_{GE}(f,f(x))$ is recognized by a machine in T_i, then f is f-recognized by a machine in T_i, for i = 0, 1,2 and 3.

(Proof) Assume that $f(\Sigma^*) = \{\ell_1, \ell_2, \ldots, \ell_s\}$. Let M_i be a machine recognizing $L_{GE}(f,\ell_i)$ for $1 \leq i \leq s$. A machine M which f-recognizes f with a representation r over Δ is given as follows; Let $\Delta = \Delta_0 \cup \Delta_1 \cup \Delta_2$ where $\Delta_1 = \{\ell_1', \ell_2', \ldots, \ell_s'\}$ (ℓ_i' is a new symbol corresponding uniquely to ℓ_i for $1 \leq$

$i \leq s)$ $\Delta_2 = 2^{\Delta 1} \times 2^{\Delta 1}$. We define r as

$$r(\ell_i) = \ell_i' r_0(\ell_i) \quad \text{for} \quad 1 \leq i \leq s$$

$$r(\ell) = (A_\ell, B_\ell) r_0(\ell) \quad \text{if} \quad \ell \overline{\in} f(\Sigma^*),$$

where $A_\ell = \{\ell_j' | \ell_j > \ell\}$ and $B_\ell = \{\ell_j' | \ell_j < \ell\}$.

M contains M_i for $1 \leq i \leq s$ as sub-machines. For any word xc with x in Σ^*, M reads first ε and chooses nondeterministically the initial configuration of any one of M_j's, say M_k, and hereafter M_k moves reading x as an input word. From any memory configuration γ^i corresponding to accepting configuration $(\gamma^i)'$ in some M_i $(1 \leq i \leq s)$, M moves into a memory configuration γ_c^i by the input symbol c, and M with the configuration γ_c^i, say $M(\gamma_c^i)$, moves in a deterministic way as; $M(\gamma_c^i)$ emits one of =, >, <, and ! as the output and halts according to the cases where it reads ℓ_i', ℓ_j' such that $\ell_j < \ell_i$, ℓ_k' such that $\ell_k > \ell_i$, and ℓ_ℓ' such that ℓ_i and ℓ_ℓ are incomparable. Reading (A_ℓ, B_ℓ), $M(\ell_c^i)$ emits one of >,< and ! according to the cases $\ell_i' \in A_\ell$, $\ell_i' \in B_\ell$ and $\ell_i' \overline{\in} A_\ell \cup B_\ell$, and halts. We set ν as

$$\nu(\gamma_c^i) = \ell_i \qquad (1 \leq i \leq s).$$

If $f(x) = \ell_k$, then $S_x = \{\gamma_c^i | \ell_i \leq \ell_k\}$. Thus $f(x) = \nu(\gamma_c^k) = \max\{\nu(\gamma) | \gamma \in S_x\}$.

We obtain the following corollaries directly from Theorem 1, Corollary 1 and Theorem 2.

Corollary 3. If L is a totally ordered set with minimum element or a finite lattice and f is an L-fuzzy language over some alphabet Σ such that $f(\Sigma^*)$ is finite, then for $0 \leq i \leq 3$, a necessary and sufficient condition for f to be f-recognized by a machine in T_i is that for any $f(x)$ $(x \in \Sigma^*)$ $L_{GE}(f, f(x))$ is recognized by a machine in T_i.

Corollary 4. Let L be a language over Σ and let f_L: $\Sigma^* \to B_1$ be the characteristic function of L, where B_1 is the Boolian lattice with two elements. For $0 \leq i \leq 3$ L is recognized by a machine in T_i if and only if f_L is f-recognized by a machine in T_i.

Corollary 4 shows that the recognition concept for fuzzy languages introduced in this paper is a fairly good extension of the one for ordinary languages.

Example 5. Let L be a lattice with the minimum element 0 and the maximum element 1. An L-fuzzy context-free grammar is defined as a quadruple $G = (V,\Sigma,P,S)$ where V is a finite set of symbols, $\Sigma \subset V$ is the set of terminal symbols, $V-\Sigma$ is the set of non-terminal symbols, S is in $V-\Sigma$ and P is a finite set of production rules of the form

$$A \overset{\ell}{\to} \alpha$$

with $A \in V-\Sigma$, $\alpha \in V^*$ and $\ell \in L$. For β, γ in V^*, we will write

$$\beta \overset{\ell}{\Longrightarrow} \gamma ,$$

if there exists δ, η in V^* and $A \overset{\ell}{\to} \gamma$ in P such that $\beta = \delta A \eta$ and $\gamma = \delta\alpha\eta$. We will write

$$\beta \underset{*}{\overset{1}{\Longrightarrow}} \beta$$

$$\beta \underset{*}{\overset{0}{\Longrightarrow}} \gamma$$

for all β, $\gamma \in V^*$, and

$$\beta \underset{*}{\overset{m}{\Longrightarrow}} \gamma$$

if and only if there exists a sequence of elements in V^*, $\beta_0, \beta_1,\cdots,\beta_t$ such that $\beta_0 = \beta$, $\beta_t = \gamma$, $\beta_{i-1} \overset{\ell_i}{\Longrightarrow} \beta_i$ for $1 \leq \ell \leq t$ and $\bigwedge_{i=1}^{t} \ell_i = m$. For any x in Σ^*, let ℓ_x the least upper bound of $\{\ell \in L \mid S \underset{*}{\overset{\ell}{\Longrightarrow}} x\}$. The L-fuzzy language f defined by

$$f(x) = \ell_x \quad \text{for all } x \in \Sigma^*$$

is said to be generated by G. An L-fuzzy language generated by some L-fuzzy context free grammar is called an L-fuzzy context free language. (L-fuzzy phrase structure, L-fuzzy context sensitive, L-fuzzy regular grammars and languages are similarly defined respectively.) $L_{[0,1]}$-fuzzy context-free languages were studied by Lee and Zadeh [3]. From Proposition 18 in [4] and Theorem 2, it follows that any $L_{[0,1]}$-fuzzy context free language is f-recognized by a pushdown automaton.

Now we consider B_n-fuzzy context free language, where B_n is the Boolian lattice with n atoms. (An example of B_2-fuzzy context-free language was given by [5].)

Proposition. The family of languages $\{L_{GE}(f,\lambda)\}$ with a B_n-fuzzy context-free language f and $\lambda \in B_n$ is exactly the same as the family of n-intersection languages introduced by Liu and Weiner [6]. (A language L is defined to be an n-intersection language if L is expressible as an intersection of n context-free languages.)

From Corollary 3 and the above Proposition, it follows that any B_n-fuzzy context-free language is f-recognized by a linear bounded automaton, but for $n \geq 2$, a B_n-fuzzy context-free language is not generally f-recognized by a pushdown automaton.

Example 6. Let $\Sigma = \{a,b,c\}$ and let f_2 be an $L_{[0,1]}$-fuzzy language over Σ defined by

$$f_2(a^i b^j c^k) = (\tfrac{1}{2})^{|i-j|} + (\tfrac{1}{2})^{|j-k|}$$

$$f_2(w) = 0 \quad \text{if} \quad w \overline{\in} a^*b^*c^*$$

Then 1 is an isolated cut-point of f_2 and $L_{GE}(f_2,1) = \{a^i b^i c^i \mid i \geq 0\}$ is not recognized by any pushdown automaton.

Hence from Theorem 1, f_2 is not f-recognized by any pushdown automaton. It can be seen that f_2 is f-recognized by some deterministic linear bounded automaton with representation r_1.

6. A FUZZY LANGUAGE WHICH IS NOT F-RECOGNIZED BY A MACHINE IN DT_2

Considering Theorem 1 and Corollary 1, it is easy to find fuzzy languages not f-recognized by a machine in T_i for $0 \leq i \leq 3$. But we can not use Theorem 1 and Corollary 1 to find an L-fuzzy language whose membership function-values distribute densely over L and is not f-recognized by a machine in T_i for $0 \leq i \leq 2$. Hence, it is interesting to find such a language. But only a following result has been obtained.

Example 7. Let $\Sigma = \{0,1\}$ and let f_3 be an $L_{[0,1]_R}$-fuzzy language over Σ such that for $a_i \in \Sigma$ $(1 \leq i \leq k)$,

$$f_3(a_1 a_2 \cdots a_k) = a_1 2^{-1} + a_2 2^{-2} + \cdots + a_k 2^{-k}$$

(binary expansion).

$$f_3(\varepsilon) = 0$$

Then f_2 is not f-recognized by any deterministic pushdown automaton.

(Proof) Assume that f_3 is f-recognized by a deterministic pushdown automaton $M = \langle \Sigma \cup \Delta \cup \{c\}, \Gamma, \{<, =, >\}, \theta, \delta, \Omega, \kappa, \gamma_0 \rangle$ with a representation r over Δ. Let

$$L_1 = \{xcy \mid \gamma_0 \underset{xcy}{\Longrightarrow} \gamma \text{ and } \kappa(\theta(\gamma)) = (=)\},$$

then L_1 is a context-free language included in $\Sigma^* c \Delta^*$. Let $L_2 = L_1 \cap 0^*1^*c\Delta^*$, L_2 is also a context-free language. Since M is deterministic, for any xcy in L_1, there exists only α in Δ^∞ such that $r(f_3(x)) = y\alpha$. Due to the pumping lemma of the theory of context free languages, it holds that there exists a constant K such that if $|z| \geq K$ and $z \in L_2$, then we can write $z = uvwxy$ such that $vx \neq \varepsilon$, $|vwx| \leq K$, and for all i, $uv^i wx^i y$ is in L_2. Let $m \geq K$ and let z_p be an element in L_2 of the form $0^p 1^m cg_p$ ($g_p \in \Delta^*$) for any $p \geq 0$. We can write $z_p = u_p v_p w_p x_p y_p$ such that $v_p x_p \neq \varepsilon$ $|v_p w_p x_p| \leq K$, and for all $i \geq 0$, $u_p v_p^i w_p x_p^i y_p$ in L_2. Since M is deterministic and halts immediately after it emits (=), there exist no x in Σ^* and y and y' in Δ^* with $y \neq y'$ such that both xcy and xcy' are L_2. Since $f_3(x) \neq f_3(x')$ for distinct x and x' in 0^*1^m, there exist no x and x' in 0^*1^m with $x \neq x'$ and y in Δ^* such that both xcy and x'cy are in L_2. Thus for all p neither u_p nor y_p contains c. Obviously for all p, c can not occur in either v_p or x_p. Hence, w_p contains c and both v_p and x_p are not ε for all $p \geq 0$, so that for all $p \geq 0$ we can write $v_p = 1^{s_p}$ for some $s_p \geq 1$ and $w_p = 1^{t_p} cW_p$ for some $t_p \geq 0$ and W_p in Δ^*. Since $|v_p w_p x_p| \leq K$ for all p, there exist nonnegative integersp and q, and W and x in Δ^* such that $p < q$, $W_p = W_q = W$, $s_p = s_q = s$ amd $x_p = x_q = x$. Hence

$$z_p = 0^p 1^m cWxy_p$$

$$z_q = 0^q 1^m cWxy_q$$

and for all $i \geq 0$

$$0^p 1^{m-s} 1^{si} cWx^i y_p \in L_2$$

$$0^q 1^{m-s} 1^{si} cWx^i y_q \in L_2$$

For some α_0 and α_1 in Δ^∞, it holds that

$$r^{-1}(Wy_p\alpha_0) = f_3(0^p 1^{m-s}) < f_3(0^p 1^m) = r^{-1}(Wxy_p\alpha_0)$$

Let d_0 be the minimum D-length of $Wy_p\alpha_0$ and $Wxy_p\alpha_1$, and let $j > d_0$, then for some α_2 in Δ^∞,

$$r^{-1}(Wx^j y_p \alpha_2) = f_3(0^p 1^{m-s} 1^{sj}) > f_3(0^p 1^m).$$

Hence the minimum D-length d_1 of $Wy_p\alpha_0$ and $Wx^j y_p \alpha_2$ is not greater than d_0. Since $j > d_0$, $j > d_1$. But for some α_3 in Δ^∞,

$$r^{-1}(Wx^j y_q \alpha_3) = f_3(0^q 1^{m-s} 1^{sj}) < f_3(0^p 1^{m-s}) = r^{-1}(Wy_p \alpha_0),$$

which contradicts the definition of the representation. Thus f_3 can not be f-recognized by any deterministic push-down automaton.

Note. $L_{GE}(f_3, \ell)$ and $L_G(f_3, \ell)$ are regular for any ℓ in $L_{[0,1]_R}$. This is a well-known result in the theory of probabilistic automata [9].

7. RECURSIVE FUZZY LANGUAGES

The relation between deterministic machines and non-deterministic machines with respect to the f-recognizability of fuzzy languages is somewhat different from that of ordinary languages. It will be shown that in the f-recognition of fuzzy languages nondeterministic Turing machines are more powerful than deterministic Turing machines. Let $\{t_0, t_1, t_2, \ldots\}$ be an enumeration of deterministic Turing machines. Let L_3 be a lattice with three elements 0, a and 1 such that $0 < a < 1$. Let $\Sigma = \{\sigma\}$. f_4 and f_5 are L_3-fuzzy languages over Σ defined as follows: For $n \geq 0$

$f_4(\sigma^n) = 1$ if t_n with the blank tape eventually halts,
$= 0$ otherwise.

$f_5(\sigma^n) = 1$ if t_n with the blank tape eventually halts,
$= a$ otherwise.

Then the following lemma holds.

Lemma. f_4 is f-recognized by a deterministic Turing machine. f_5 is f-recognized by a non-deterministic Turing machine, but it is not f-recognized by any deterministic Turing machine.

(Proof) It is easy to show a deterministic Turing machine which f-recognizes f_4. Since $L_{GE}(f_5,1)$ and $L_{GE}(f_5,a)$ are recursively enumerable languages, it follows from Theorem 2 that f_5 is f-recognized by a Turing machine. Assume that f_5 is f-recognized by a deterministic Turing machine, then it is easily shown that the halting problem of Turing machine is solvable. This cannot be valid. Thus f_5 is not f-recognized by any deterministic Turing machine.

Let $\mathcal{L}(T_i)$ and $\mathcal{L}(DT_i)$ be the families of fuzzy languages f-recognized by a machine in T_i and DT_i respectively for $i = 0, 1, 2$ and 3.

Theorem 3. (i) $\mathcal{L}(T_0) \supsetneq \mathcal{L}(DT_0)$ (ii) $\mathcal{L}(T_2) \supsetneq \mathcal{L}(DT_2)$ (iii) $\mathcal{L}(T_3) = \mathcal{L}(DT_3)$.

(i) is a direct consequence of Lemma. (ii) follows from Corollary 4. The proof of (iii) is easy. But it is not known whether $\mathcal{L}(T_1) \supsetneq \mathcal{L}(DT_1)$ or not.

Considering Lemma, it seems reasonable to define recursive fuzzy languages as follows: A fuzzy language f over Σ is recursive if and only if f is f-recognized by some machine $M = \langle \Sigma \cup \Delta \cup \{c\}, \Gamma, \Psi, \theta, \delta, \Omega, \kappa, \gamma_0 \rangle$ in DT_0 with some representation r over Δ with the condition that for any $x \in \Sigma^*$, $S_x \neq \phi$, where S_x is $\{\gamma \in \Gamma^* | \gamma_0 \underset{xc}{\Longrightarrow} \gamma\}$.

Obviously any fuzzy language in $\mathcal{L}(T_3)$ is recursive. And the following Proposition is easily proved.

Proposition. A fuzzy language in $\mathcal{L}(DT_2) \cup \mathcal{L}(T_1)$ is a recursive fuzzy language.

ACKNOWLEDGMENT

The authors wish to thank the members of their research group for their helpful discussions.

REFERENCES

1. L. A. Zadeh, "Fuzzy Sets", Information and Control, 8, pp. 338-353, 1965.
2. J. A. Goguen, "L-Fuzzy Sets", J. Math. Anal. Appl., 18, pp. 145-174, 1967.
3. E. T. Lee and L. A. Zadeh, "Note on Fuzzy Languages", Inf. Sci., 1, pp. 421-434, 1969.
4. L. A. Zadeh, "Fuzzy Languages and Their Relation to Human and Machine Intelligence", Man And Computer, Proc. Int. Conf. Bordeaux 1970, pp. 130-165, (Karger, Basel 1972).
5. M. Mizumoto, J. Toyoda and K. Tanaka, "Examples of Formal Grammars With Weights", Information Processing Letters, 2, pp. 74-78, 1973.
6. L. Y. Liu and P. Weiner, "An Infinite Hierarchy of Intersections of Context-Free Languages", Mathematical Systems Theory, Vol. 7, No. 2, pp. 185-192, 1973.
7. H. H. Kim, M. Mizumoto, J. Toyoda, and K. Tanaka, "Lattice Grammars", Trans. Inst. Elect. Commun. Engrs. Japan, Vol. 57-D, No. 5, pp. 253-260 (in Japanese).
8. A. V. Aho and J. D. Ullman, The Theory of Parsing, Translation, and Compiling, Vol. 1: Parsing, Prentice-Hall, Inc. 1972.
9. A. Paz, Introduction to Probabilistic Automata, Academic Press, 1971.

ON THE DESCRIPTION OF FUZZY MEANING OF CONTEXT-FREE LANGUAGE

Yasuyoshi Inagaki and Teruo Fukumura
Department of Information Science
Faculty of Engineering, Nagoya University
Furo-cho, Chikusa-ku, Nagoya 464 Japan

I. INTRODUCTION

The theory of fuzzy grammars and quantitative fuzzy semantics have been proposed by Prof. L. A. Zadeh (1, 2, 3) and other researchers (4,5). The theory gives us many interesting ideas and provides us a natural way to reduce the gap between formal languages and natural languages. On the other hand, if there is a connection between context-free grammars and grammars of natural languages, it is undoubtedly, as Prof. N. Chomsky proposes, through some stronger concept like that of transformational grammar. In this framework, it is not the context-free language itself that is of interest, but instead, the set of derivation trees (structural descriptions or P-markers). It is also true from the view point of the syntax directed description of fuzzy meanings that sets of trees are of prime importance as opposed to sets of strings.

These observations motivate interest in systems to manipulate fuzzy sets of trees. The purpose of this note is to offer three systems to manipulate fuzzy sets of trees, generators, acceptors, and transducers. Main results of this note are:

(1) The set of derivation trees of any fuzzy context-free grammar is shown to be a fuzzy set of trees generated by a fuzzy context-free dendro language generating system and

also to be a fuzzy set of trees recognizable by a fuzzy tree automaton.

(2) The fuzzy tree transducer is shown to be able to describe the fuzzy meanings of fuzzy context-free languages at the level of syntax structure in the sense that it can fuzzily associate each fuzzy derivation tree of the fuzzy language wwith a tree representation of the computation process of its fuzzy meaning.

2. TREES AND PSEUDOTERMS

We begin by describing fundamental concepts concerning trees.[†] Let N be the set of natural numbers and N* be the set of strings on N containing the null string ε. A finite closed subset U of N* is called a <u>finite tree domain</u> if

(i) $w \in U$ and $w = uv$ implies $u \in U$ $(u, v, w \in N^*)$

(ii) $wn \in U$ and $m \leq n$ implies $wm \in U$ $(w \in N^*, m, n \in N)$

For a finite tree domain U, the subset $\overline{U} = \{w | w \in U, w \cdot 1 \notin U\}$ is called the <u>leaf node set</u>. A <u>partially ranked alphabet</u> is a pair $(V;\Sigma)$ of finite alphabets V and Σ, where $V \cap \Sigma = \phi$. A <u>tree</u> t on a partially ranked alphabet $(V ; \Sigma)$ is a mapping from a finite tree domain U into $V \cup \Sigma$ such that $t\colon U \to (V;\Sigma)$, where

$t(w) \in V$ for $w \in U - \overline{U}$

$t(w) \in \Sigma$ for $w \in \overline{U}$.

As easily known, a finite tree $t\colon U \to (V;\Sigma)$ can be represented by a finite set of pairs $(w,t(w))$, i.e., $\{(w,t(w)) | w \in U\}$.

It should be fairly clear that trees on $(V;\Sigma)$ can be represented graphically by constructing a rooted tree (where the

---

[†] Concerning detailed description of these concepts, refer to the references (6), (7).

successors of each node are ordered), representing the domain of the mapping, and labeling the nodes with elements of $V \cup \Sigma$, representing the values of the function. Thus, in the following figures there are two examples; as a mapping, the left-hand tree has the domain $\{\varepsilon, 1, 2, 11, 12\}$ and the value at 11, for example, is a.

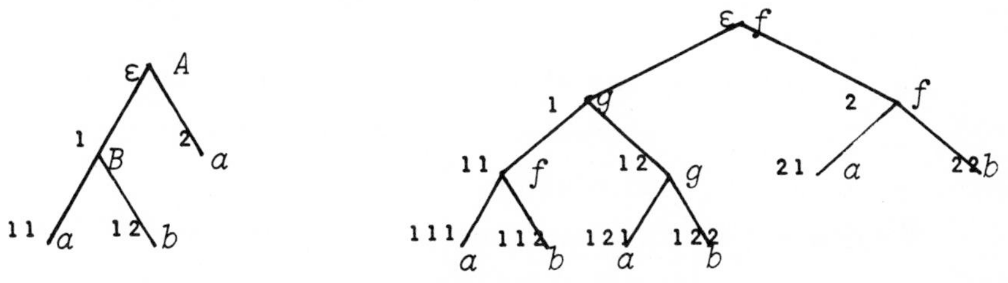

Figure 1

The definition of tree and the corresponding pictorial representation provide a good basis for intuition in considering tree manipulating systems. The development of the theory, however, is simpler if we consider the familiar linear representation of such trees. For this purpose we define the set $T^p_{(V;\Sigma)}$ of <u>pseudoterms</u> on $V \cup \Sigma$ as the smallest subset of $[V \cup \Sigma \cup \{\ (,)\}]^*$ satisfying †:

(i) $\Sigma \subset T^p_{(V;\Sigma)}$

(ii) If $n > 0$ and $A \in V$ and $t_1, t_2, \ldots, t_n \in T^p_{(V;\Sigma)}$, then

$$A(t_1 \ \ t_2 \cdots t_n) \in T^p_{(V;\Sigma)}$$

q We will consider trees and pseudoterms to be equivalent formalizations in the followings. The translation between the two is the usual one. By way of example, the trees of the above figures correspond to the following pseudoterms:

$A(B(a\ b)a)$, $f(g(f(a\ b)\ g(a\ b))\ f(a\ b))$

For completeness, we note that this correspondence can be

† It is assumed that the parentheses are not symbols of $V \cup \Sigma$.

made precise in the following way.

(i) If a pseudoterm $t^p \in T^p_{(V;\Sigma)}$ is atomic, i.e., $t^p = a \in \Sigma$, then the corresponding tree t has domain $\{\varepsilon\}$ and $t(\varepsilon) = a$.

(ii) If $t^p = A(t_1^p \cdots t_m^p)$, then t has domain $\bigcup_{i \leqq m}\{iw|w \in \text{domain } (t_i)\} \cup \{\varepsilon\}$, $t(\varepsilon) = A$, and for $w = iw'$ in the domain of t, $t(w) = t_i(w')$.

In the following, we will denote the set of trees on $(V;\Sigma)$ by $T_{(V;\Sigma)}$, its element by t, and the pseudoterm corresponding to a tree t by p(t) or t^p.

A <u>fuzzy set T of trees</u> is defined by a membership function $\mu_T\colon T_{(V;\Sigma)} \to [0, 1]$. The set of all fuzzy set of trees will be denoted by $F(T_{(V;\Sigma)})$.

3. FUZZY DENDROLANGUAGE GENERATING SYSTEMS

We introduce a fuzzy system which generates fuzzy sets of trees, as an extension of dendrolanguage generating system of authors (6).

<u>Definition 1</u>. A <u>fuzzy context-free dendrolanguage generating system</u> (F-CFDS) is 5-tuple,

$$S = (\Lambda, V, \Sigma, P, \lambda_0),$$

where (1) Λ: a finite set of symbols, of which elements are called <u>nonterminal node symbols</u>,

(2) V: a finite set of symbols, of which elements are called <u>node symbols</u>,

(3) Σ: a finite set of symbols, of which elements are called <u>leaf symbols</u>,

(4) P: a finite set of <u>fuzzy rewriting rules</u> of the form

$$\mu(\lambda \to t) = \rho$$

which are usually represented by

$\lambda \overset{\rho}{\to} t$; $t \in T_{(V;\Lambda \cup \Sigma)}$; $\rho \in [0,1]$

or equivalently by

$\lambda \overset{\rho}{\to} p(t)$; $p(t)$ is a pseudoterm corresponding to a tree t.

(5) $\lambda_0 \in \Lambda$: an <u>initial nonterminal node symbol</u>.

Let us define a fuzzy relation $\overset{\rho}{\Rightarrow}$ on the set $T_{(V;\Lambda \cup \Sigma)}$ of trees: For any two trees $\alpha, \beta \in T_{(V;\Lambda \cup \Sigma)}$

$$\alpha \overset{\rho}{\Longrightarrow} \beta$$

if and only if (i) $p(\alpha) = x\lambda y$, (ii) $p(\beta) = xp(t)y$, and (iii) $\lambda \overset{\rho}{\to} t$ is in P, where $x, y \in [V \cup \Lambda \cup \Sigma \cup \{(,)\}]^*$, $\lambda \in \Lambda$ and $t \in T_{(V;\Lambda \quad \Sigma)}$. Furthermore, we define the transitive closure $\overset{*\rho}{\Longrightarrow}$ of fuzzy relation $\overset{\rho}{\Longrightarrow}$ by:

(i) $\alpha \overset{*1}{\Longrightarrow} \alpha$ for all $\alpha \in T_{(V;\Lambda \cup \Sigma)}$

(ii) $\alpha \overset{*\rho}{\Longrightarrow} \beta$ iff $\rho = \underset{\gamma \in T_{(V;\Lambda \cup \Sigma)}}{\mathrm{Sup}} \{\min\{\rho', \rho'' | \alpha \overset{*\rho'}{\Longrightarrow} \gamma, \gamma \overset{\rho''}{\Longrightarrow} \beta\}\}$

<u>Definition 2</u>. The fuzzy set $T(S) = \{(t;\rho) | \lambda_0 \overset{*\rho}{\Longrightarrow} t \in T_{(V;\Sigma)}\}$ is called <u>fuzzy context-free dendrolanguage</u> (F-CFDL) generated by F-CFDS, S.

<u>Example 1</u>. Suppose that $\Lambda = \{\lambda, \xi, \eta\}$, $V = \{A\}$, $\Sigma = \{a\}$ and P is given by

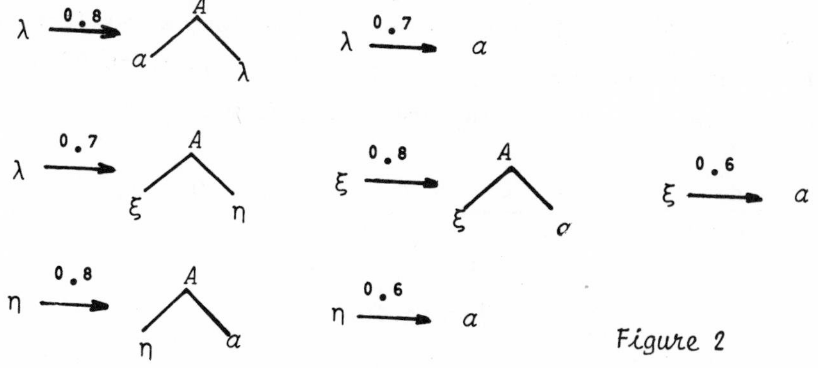

Figure 2

Then, F-CFDS, $S = (\Lambda, V, \Sigma, P, \lambda)$ generates F-CFDS, $T(S) =$

$$\{(t;\ 0.7) \mid p(t) = \overbrace{A(a\cdots A(a}^{n}A(aa)\overbrace{)\cdots)}^{n},\ n \geq 0\}$$

$$\cup\ \{(t;\ 0.6) \mid p(t) = \overbrace{A(\cdots A(}^{n}A(aa)\overbrace{a)\cdots a)}^{n},\ n \geq 1\}$$

For example, as a derivation, we have

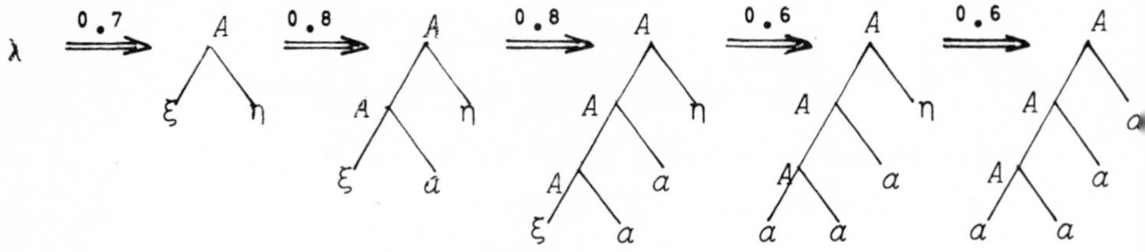

Figure 3

equivalently,

$$\lambda \overset{0.7}{\Longrightarrow} A(\xi\eta) \overset{0.8}{\Longrightarrow} A(A(\xi a)\eta) \overset{0.8}{\Longrightarrow} A(A(A(\xi a)a)\eta)$$

$$\overset{0.6}{\Longrightarrow} A(A(A(aa)a)\eta) \overset{0.6}{\Longrightarrow} A(A(A(aa)a)a)$$

4. NORMAL FORM OF F-CFDS

The <u>depth</u> of tree t with domain U_t is defined by

$$d(t) = \max\ \{lg(w) \mid w \in U_t\},$$

where $lg(w)$ is the length of w. The order of F-CFDS is defined as the maximum value of depths of trees appeared in the right-hand side of the rules.

Two F-CFDS's are said to be <u>equivalent</u>, if they generate the same fuzzy dendrolanguage. In this section, we will prove that for any F-CFDS we can construct an equivalent F-CFDS of order 1, i.e., of which rules are in the form of

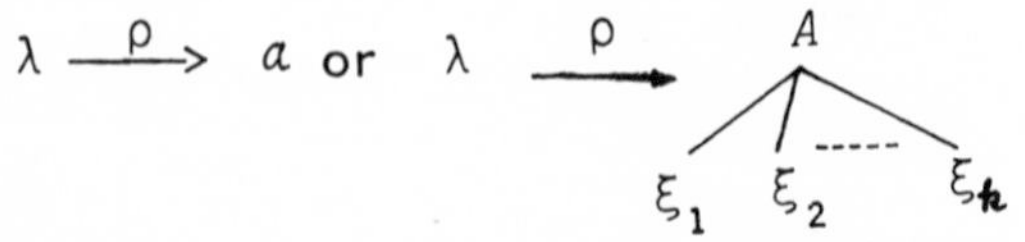

$(k \geq 1)$

Figure 4

where λ, $\xi_i \in \Lambda$ $(i = 1,\ldots,k)$, $A \in V$ and $a \in \Sigma$.

Lemma 1 Let $S = (\Lambda, V, \Sigma, P, \lambda_0)$ be a F-CFDS of order $n(n \geq 2)$. Then we can construct an equivalent F-CFDS of order $(n - 1)$.

Proof. Let us determine a new F-CFDS, $S' = (\Lambda', V, \Sigma, P', \lambda_0)$ from a given F-CFDS, S as follows:
P' is defined by: For each rule

$$\lambda \xrightarrow{\rho} t$$

in P, (i) If $d(t) < n$, then $\lambda \xrightarrow{\rho} t$ should be in P', (ii) If $d(t) = n$ and $p(t) = X(p(t_1)\ldots p(t_k))$ then

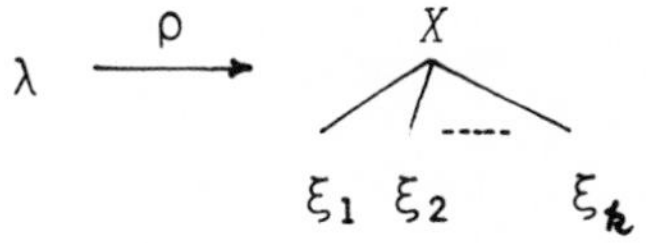

Figure 5

and

$$\xi_i \xrightarrow{1} t_i \qquad \text{for all } i \text{ such that } p(t_i) \notin \Lambda$$

should be in P', where ξ_i is a new distinct nonterminal node symbol if $p(t_i) \notin \Lambda$ and $\xi_i = p(t_i)$ if $p(t_i) \in \Lambda$.

It should be clear that Λ' is the union of Λ and the set of all new nonterminal node symbols introduced by applying the above rule (ii).

Suppose $\alpha \overset{\rho}{\Longrightarrow} \beta$ under S. Then $p(\alpha) = x\,\lambda\,y$, $p(\beta) = xp(t)y$ and $\lambda \xrightarrow{\rho} t$ is in P. If $d(t) < n$, the above construction asserts that $\alpha \overset{\rho}{\Longrightarrow} \beta$ under S' since $\lambda \xrightarrow{\rho} t$ is also contained in P'. If $d(t) = n$, by the above construction we can have

$$p(\alpha) = x\lambda y \overset{\rho}{\Longrightarrow} x\ X(\xi_1\cdots\xi_k)\ y \overset{*1}{\Longrightarrow} x\ X(p(t_1)\cdots p(t_k))y = p(\beta).$$

Conversely, if $x\ \lambda\ y \overset{\rho^k}{\Longrightarrow} x\ X(\xi_1\cdots\xi_k)\ y$ under S' then $\xi_i \xrightarrow{1} t_i$; $i = 1,\cdots,k$ should be applied since nonterminal symbols ξ_i's can be rewritten only by them. Thus,

$$x\ \lambda\ y \overset{\rho}{\Longrightarrow} x\ X(\xi_1\cdots\xi_k)\ y \overset{*1}{\Longrightarrow} x\ X(p(t_1)\cdots p(t_k))\ y.$$

For this derivation, we can have

$$x\ \lambda\ y \overset{\rho}{\Longrightarrow} x\ p(t)\ y$$

under S.

Thus we know that T(S) = T(S'). It should be clear from the construction procedure of S' that S' is of order (n - 1).

(Q.E.D.)

By repeating application of Lemma 1, we obtain:

Lemma 2 For any given F-CFDS, S, there exists an equivalent F-CFDS, S' of order 1.

Theorem 1 For any given F-CFDS, S, we can construct an equivalent F-CFDS, S' of which rules are in the form of

(i) $\lambda \xrightarrow{\rho} a$ or (ii) 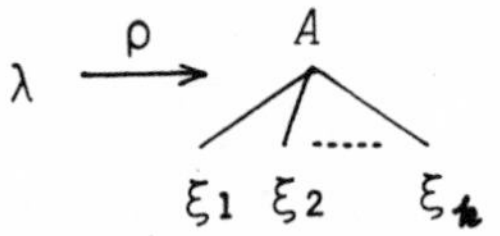

$(k \geq 1)$

Figure 6

where λ, ξ_i's are nonterminal node symbols, a is a leaf symbol, and A is a terminal node symbol.

Proof. By Lemma 2, for any given F-CFDS, S, we can construct an equivalent F-CFDS of which rules are in the form of

(i) $\lambda \xrightarrow{\rho} a$ or (ii) 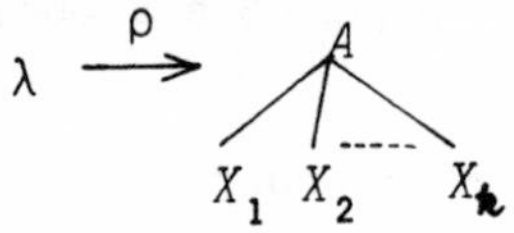

; $X_i \in \Sigma \cup \Lambda$

Figure 7

Here, if we replace the rule of type (ii) by a rule

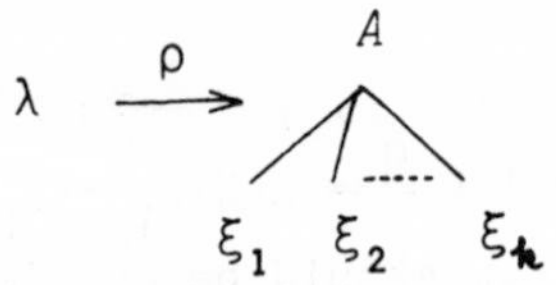

Figure 8

where $\xi_i = X_i$ if $X_i \in \Lambda$ and ξ_i is a new symbol if $X_i \in \Sigma$, and rules

$$\xi_i \xrightarrow{1} X_i \qquad \text{for all } X_i \in \Sigma,$$

then we can obtain the desired F-CFDS. (Q.E.D.)

In the following, a F-CFDS of which rules are in the form of (i) or (ii) of Theorem 1 will be said to be <u>normal</u>.

<u>Example 2</u>. Consider a set of rules:

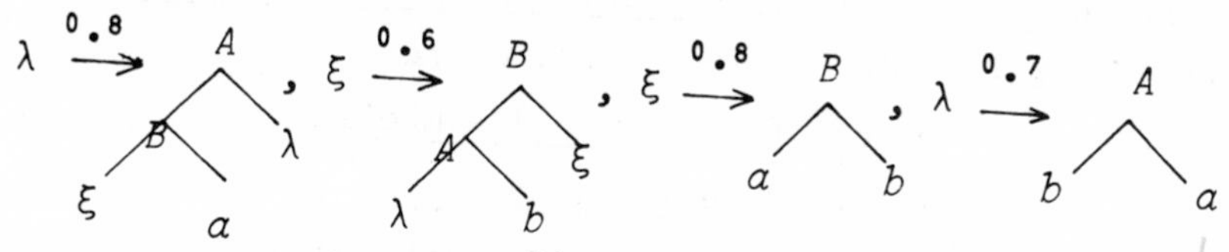

Figure 9

This gives a F-CDFS of order 2. The normal form for this F-CFDS is given by the following rules:

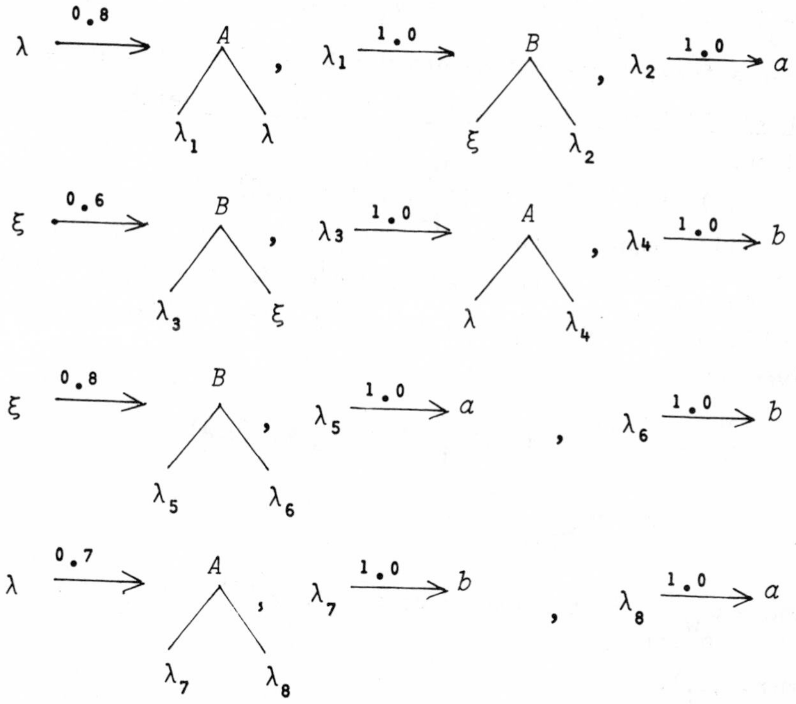

Figure 10

5. CHARACTERIZATION OF SETS OF DERIVATION TREES OF FUZZY CONTEXT-FREE GRAMMARS

We define the sets of derivation trees of fuzzy context-free grammars as fuzzy set of trees and we characterize them by F-CFDS's.

Definition 3. (1,2) A fuzzy context-free grammar (F-CFG) is a 4-tuple $G = (V,\Sigma,P,S)$, where

(i) V: a set of nonterminal symbols,

(ii) Σ: a set of terminal symbols,

(iii) P: a set of fuzzy production rules,

(iv) S: an initial nonterminal symbol.

For a derivation

$$w_0(=S) \overset{\rho_1}{\Longrightarrow} w_1 \overset{\rho_2}{\Longrightarrow} \cdots \overset{\rho_m}{\Longrightarrow} w_m(=w)$$

under a fuzzy context-free grammar G, we formally define a derivation tree with a value of degree of membership as follows:

(i) For w_0 (= S), $(\alpha^{w_0};1) = (\{(\varepsilon,S)\};1)$

(ii) Suppose that $(\alpha^{w_{i-1}};\rho)$ is given for some i and that $w_{i-1} \overset{\rho_i}{\Longrightarrow} w_i$ is realized by $A \overset{\rho_i}{\longrightarrow} Y_1Y_2\cdots Y_k$ $(Y_i \in V \cup \Sigma)$ with $w_{i-1} = xAy$ and $w_i = xY_1Y_2\cdots Y_ky$. Then $(\alpha^{w_i};\rho')$ is given by

$$\alpha^{w_i} = \alpha^{w_{i-1}} \cup \{(u\cdot i,Y_i) \mid 1 \le i \le k,\ (u,A) \in \alpha^{xAy},\ u \in \overline{U}_{\alpha^{w_{i-1}}}\}^{\dagger},$$

where $\overline{U}_{\alpha^{w_{i-1}}}$ is the leaf node set of $\alpha^{w_{i-1}}$ and by $\rho' = \min(\rho,\rho_i)$.

---

† It is assumed that the symbol A replaced by $Y_1Y_2\cdots Y_k$ corresponds to a leaf node u in $\overline{U}_{\alpha^{w_{i-1}}}$.

Let D_G be a fuzzy set of trees on $(V;\Sigma)$ defined by the above procedure for all possible derivations of a fuzzy context-free grammar G. The fuzzy set D_G will be called a <u>fuzzy set of fuzzy derivation trees</u> of G.

<u>Theorem 2</u> For any given F-CFG, $G = (V,\Sigma,P_G,S)$, there exists a F-CFDS, $S = (\Lambda,V_S,\Sigma_S,P_S,\lambda_0)$ which generates the fuzzy set D_G of fuzzy derivation trees of G.

<u>Proof</u>. Put $\Lambda = \{\lambda_X | X \in V\}$, $V_S = V$, $\Sigma_S = \Sigma$ and $\lambda_0 = \lambda_S$. Determine P_S as follows: If $X \overset{\rho}{\to} Y_1Y_2\cdots Y_k$ is in P, then

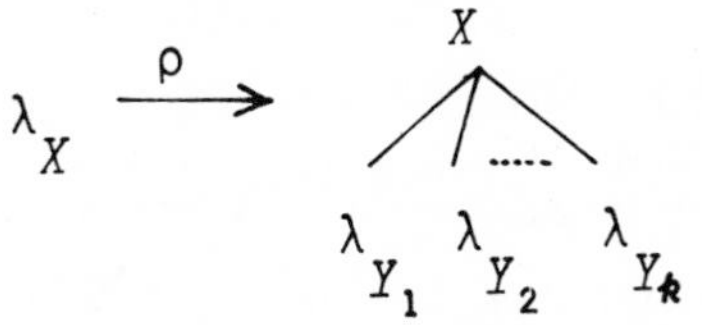

Figure 11

should be contained in P_S, where we understand that if $Y_i = a \in \Sigma$ then $\lambda_{Y_i} = a$.

By noting that the process obtaining (α^{w_i},ρ') from $(\alpha^{w_{i-1}},\rho)$ in the definition of D_G corresponds to the application of the rule

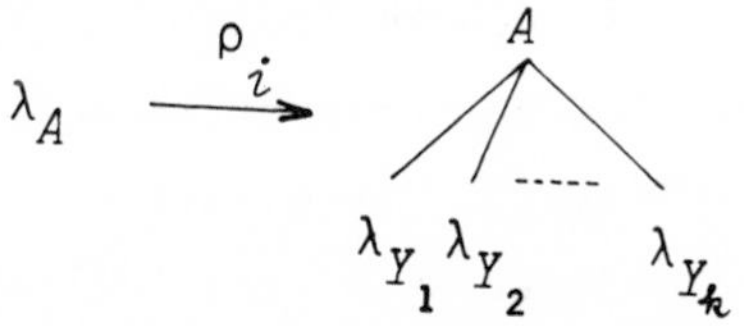

Figure 12

in F-CFDS, S, we can easily prove that $D_G = T(S)$. (Q.E.D.)

<u>Example 3</u>. Consider a F-CFG given by rules:

$S \xrightarrow{0.5} AB$, $A \xrightarrow{0.8} AB$, $B \xrightarrow{0.4} BA$, $A \xrightarrow{0.8} a$, $A \xrightarrow{0.3} b$, $B \xrightarrow{0.9} b$

For this F-CFG, construct a F-CFDS determined by rules:

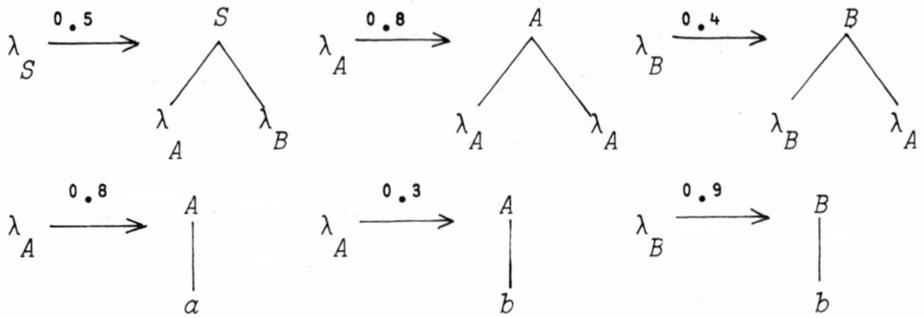

Figure 13

For a derivation of F-CFG,

$$S \xrightarrow{0.5} AB \xrightarrow{0.4} ABA \xrightarrow{0.8} aBA \xrightarrow{0.9} abA \xrightarrow{0.3} abb$$

its derivation tree is generated by the F-CFDS as follows:

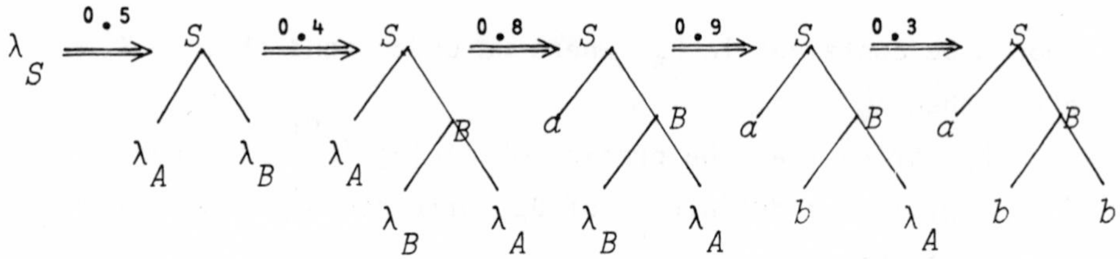

Figure 14

Next we consider the converse of Theorem 2: We will prove that for any F-CFDS, S, there exists a F-CFG, G corresponding to S in the sense of the following Theorem 3. Let h: $[V \cup \Sigma \cup \{(,)\}]^* \to \Sigma$ be a homomorphism defined by $h(a) = a$ for a in Σ and $h(X) = \varepsilon$ for $X \notin \Sigma$.

Lemma 3 Let S be a F-CFDS. Then, the fuzzy set p(T(S)) of pseudoterms of T(S) is a fuzzy context-free language.

Proof. Let $S = (\Lambda, V, \Sigma, P, \lambda_0)$ be a F-CFDS. Construct a F-CFG, $G = (V_G, \Sigma_G, P_G, S_G)$ as follows:
Put $V_G = \Lambda$, $\Sigma_G = V \cup \Sigma \cup \{(,)\}$, $S_G = \lambda_0$ and determine P_G by: if $\lambda \overset{\rho}{\to} t$ is in P then $\lambda \overset{\rho}{\to} p(t)$ should be in P_G. It is clear

from the above construction that $L(G) = p(T(S))$. (Q.E.D.)

<u>Theorem 3</u> For any F-CFDS, S, $h(p(T(S)))$ is a fuzzy context-free language on Σ.

<u>Proof</u>. By Lemma 3 with the fact that homomorphic image of a fuzzy context-free language is also a fuzzy context-free language. (Q.E.D.)

We can prove a stronger result:

<u>Theorem 4</u> Every F-CFDL is a projection of the fuzzy set of derivation trees of a F-CFG.

<u>Proof</u>. Let $S = (\Lambda, V, \Sigma, P, \lambda_0)$ be a normal F-CFDS. We define a F-CFG, $G = (V_G, \Sigma_G, P_G, S_G)$, where $V_G = \Lambda \times (V \cup \Sigma)$, $\Sigma_G = \{\langle \delta, a\rangle | a \in \Sigma\}$ (δ is a new symbol not in Λ), $S_G = \{\langle \lambda_0, X\rangle | X \in V \cup \Sigma\}$ and P_G is defined by:

(i) If 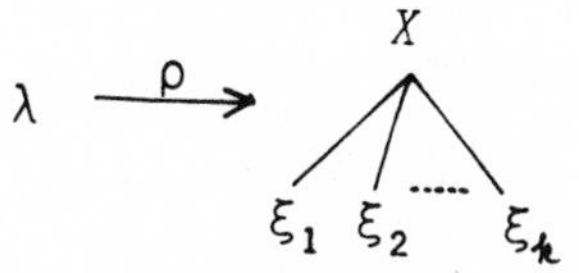 is in P, then rules

Figure 15

$$\langle \lambda, X\rangle \overset{\rho}{\to} \langle \xi_1, X_1\rangle \langle \xi_2, X_2\rangle \cdots \langle \xi_k, X_k\rangle$$

should be in P_G, where $X_i \in V \cup \Sigma$.

(ii) If $\lambda \overset{\rho}{\to} a$ is in P, then a rule

$$\langle \lambda, a\rangle \overset{\rho}{\to} \langle \delta, a\rangle$$

should be contained in P_G.

Again, it is easy to check that if $(t;\rho)$ is a fuzzy derivation tree of this grammar, $(\pi(t);\rho)$, a projection of $(t;\rho)$, is a fuzzy tree generated by the F-CFDS, S, where $\pi(t)$ is defined, in terms of pseudoterms, as follows:

(i) $\pi[\langle \lambda, a\rangle (\langle \delta, a\rangle)] = a$

(ii) $\pi[\langle \lambda, X\rangle (p(t_1) \cdots p(t_k))] = X(\pi[p(t_1)] \cdots \pi[p(t_k)])$.

(Q.E.D.)

Example 4. Consider a F-CFDS given by rules:

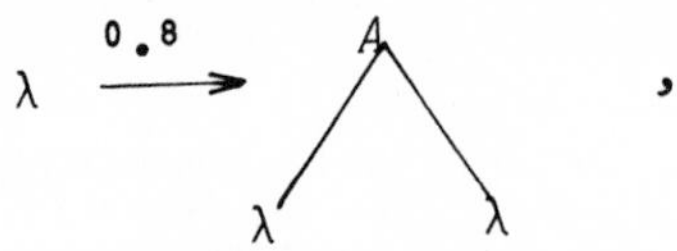

, $\lambda \xrightarrow{0.9} a$

Figure 16

For this F-CFDS, define a F-CFG by the following rules:

$$\langle \lambda, A\rangle \xrightarrow{0.8} \langle \lambda, A\rangle \ \langle \lambda, A\rangle$$

$$\langle \lambda, A\rangle \xrightarrow{0.8} \langle \lambda, A\rangle \ \langle \lambda, a\rangle$$

$$\langle \lambda, A\rangle \xrightarrow{0.8} \langle \lambda, a\rangle \ \langle \lambda, A\rangle$$

$$\langle \lambda, A\rangle \xrightarrow{0.8} \langle \lambda, a\rangle \ \langle \lambda, a\rangle$$

$$\langle \lambda, a\rangle \xrightarrow{0.9} \langle f, a\rangle$$

For example, a fuzzy derivation

$$\langle \lambda, A\rangle \xRightarrow{0.8} \langle \lambda, A\rangle \ \langle \lambda, a\rangle \xRightarrow{0.8} \langle \lambda, a\rangle \ \langle \lambda, a\rangle \ \langle \lambda, a\rangle$$
$$\xRightarrow{*0.9} \langle f, a\rangle \ \langle f, a\rangle \ \langle f, a\rangle$$

has a derivation tree

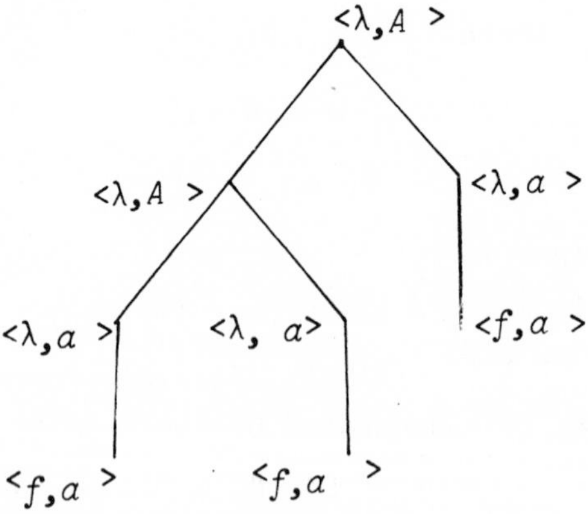

Figure 17

and its degree of membership is 0.8.

The projection of this tree defined in the proof of Theorem 4 is

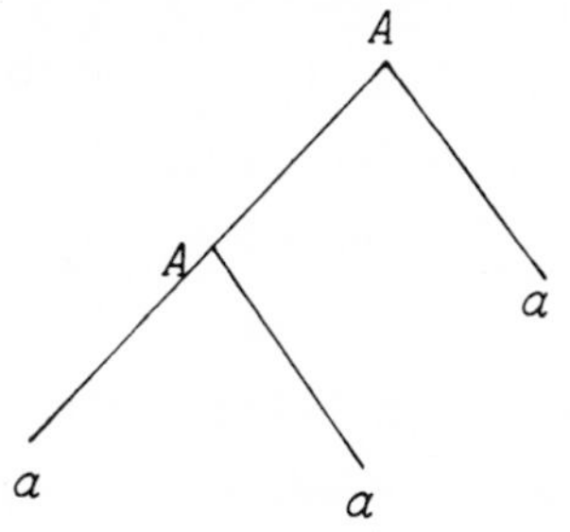

Figure 18

, which is contained in the F-CFDL generated by the given F-CFDS.

6. FUZZY TREE AUTOMATON

In the section 4, we have introduced a fuzzy dendrolanguage generating system, as a generator, which is used to characterize the fuzzy set of derivation trees of a fuzzy context-free grammar in the preceeding section 5. Here we define a fuzzy tree automaton as an acceptor of fuzzy dendrolanguage.

Definition 4. A fuzzy tree automaton (F-TA) is a 5-tuple

$A = (S, V, \Sigma, \alpha, F)$

where (i) S: a finite set of state symbols,

(ii) V: a finite set of terminal node symbols,

(iii) Σ: a finite set of leaf symbols, where $V \cap \Sigma = \phi$

(iv)† α: $(V \cup \Sigma) \to [\, \mathcal{S} \times S \to [0,1]]$, where $\mathcal{S}$ is a finite subset of S^* containing the null string ε. For $X \in V$, $\alpha(X) = \alpha_X$ is a mapping from $(\mathcal{S} - \{\varepsilon\}) \times S$ into $[0,1]$, i.e., a fuzzy relation from $\mathcal{S}$ to S, which will be called

†[A → B] denotes the set of all mappings from a set A to B.

a fuzzy direct transition function. For a ε Σ, $\alpha(a) = \alpha_a$ is a mapping from $\{\varepsilon\} \times S$ into $[0,1]$. α_a defines a fuzzy set on S which should be assigned to the node of a. $\alpha_X(S_1 S_2 \cdots S_k, S) = \rho$ means that when a node of X has k sons with states $S_1, S_2, \ldots, S_k$, the state S can be assigned to the node with degree ρ. This may graphically be represented by

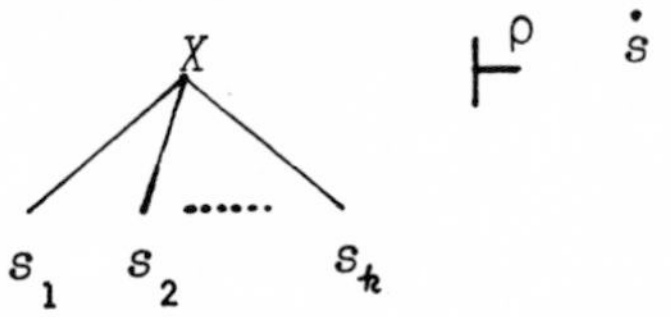

Figure 19

Finally, (V) F: a distinct subset of S, called a set of final states.

Now, for a tree $t \varepsilon T_{(V;\Sigma)}$, we define a fuzzy transition function

$$\alpha_t : [S^* \to S] \to [0,1]$$

Let t be $X(t_1 t_2 \cdots t_k)$ in terms of pseudoterm. Then

$$\alpha_t(s_1 s_2 \cdots s_n, s) = \alpha_{X(t_1 \cdots t_k)}(s_1 s_2 \cdots s_n, s)$$
$$= \sup_{\substack{\bar{s}_i \varepsilon S \\ i=1,\ldots,k}} \min\{\alpha_X(\bar{s}_1 \bar{s}_2 \cdots \bar{s}_k, s), \alpha_{t_1}(s_1 s_2 \cdots s_n, s_1), \cdots, \alpha_{t_k}(s_1 s_2 \cdots s_n, s_k)\}$$

If $t = a \; \varepsilon \; \Sigma$, then $\alpha_t = \alpha_a$.

By α_t, therefore, we can assign a fuzzy set on S to the root node of the tree t. The mapping α_t can also define a fuzzy set of trees, i.e., a fuzzy set on $T_{(V;\Sigma)}$, by

$$T(A) = \{(t;\rho) \mid \rho = \max_{s \in F} \{\alpha_t(\varepsilon, s)\}$$

which will be called a <u>fuzzy set of trees recognized by a fuzzy tree automaton</u> A.

<u>Example 5</u>. Put $S = \{s_\lambda, s_\xi, s_\eta\}$, $V = \{A\}$, $\Sigma = \{a\}$ and $F = \{s_\eta\}$. Suppose that α is defined by:

$\alpha(a)$	s_λ	s_ξ	s_η
ε	0.8	0.7	1.0

$\alpha(A)$	s_λ	s_ξ	s_η
s_λ	1.0	0	0
s_ξ	0	1.0	0
s_η	0	0	1.0
$s_\lambda s_\lambda$	0.2	0.8	0
$s_\lambda s_\xi s_\eta$	0	0	1.0

For the F-TA, $A = (S, V, \Sigma, \alpha, F)$, T(A) contains

$t =$ 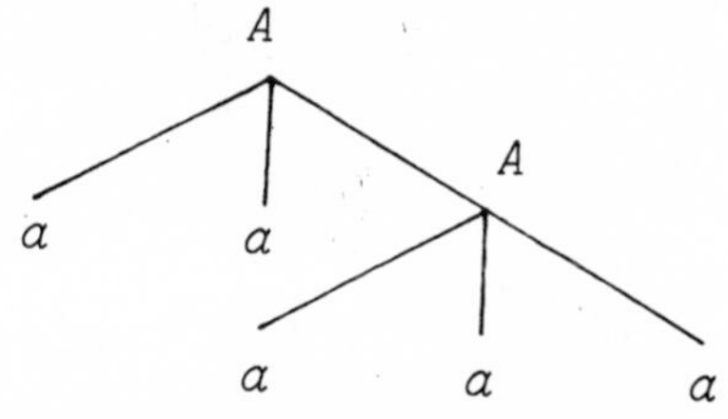

Figure 20

with the membership degree 0.7. The computation process of $\alpha_t(\varepsilon, s_\eta)$ can be graphically represented by:

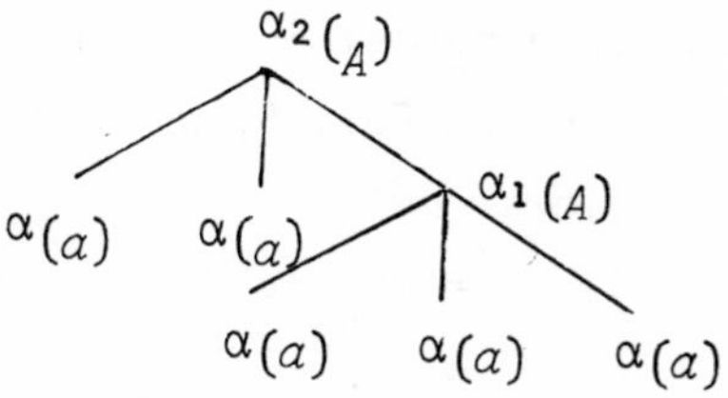

Figure 21

where

$$\alpha(a) = [0.8s_\lambda,\ 0.7s_\xi,\ 1.0s_\eta]$$

(i.e., $\alpha_a(s_\lambda) = 0.8$, $\alpha_a(s_\xi) = 0.7$, $\alpha_a(s_\eta) = 1.0$),

$$\alpha_1(A) = [0\ s_\lambda,\ 0\ s_\xi,\ 0.7\ s_\eta]$$

and

$$\alpha_2(A) = [\ 0\ s_\lambda,\ 0\ s_\xi,\ 0.7\ s_\eta]$$

Thus, we can know that $\alpha_t(\varepsilon, s_\eta) = 0.7$.

Alternatively, $\alpha_t(\varepsilon, s_\eta)$ can also be known by enumerating all the fuzzy reductions from t to s_η, such as following one:

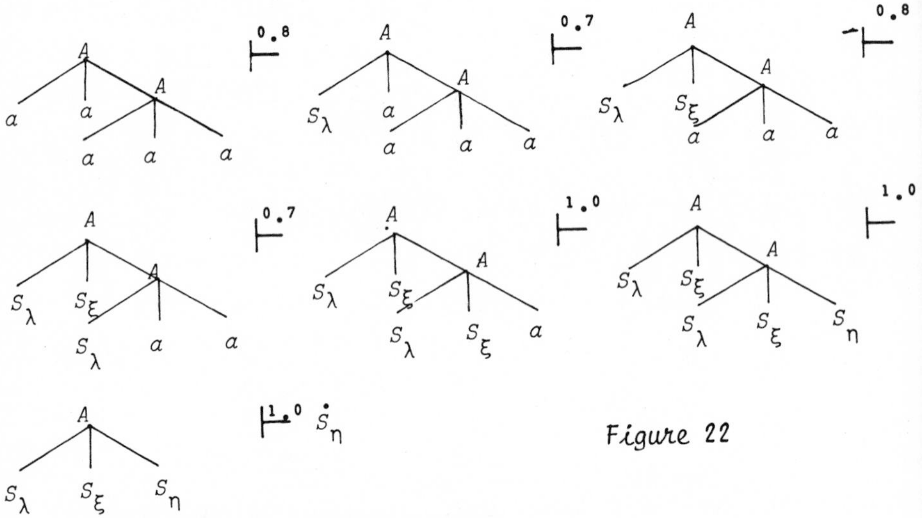

Figure 22

Theorem 5 Any F-CFDL is recognizable by a F-TA.

Proof. By Theorem 1, we can assume that for any given F-CFDL, T, there exists a normal F-CFDS, $S = (\Lambda, V, \Sigma, P, \lambda_0)$ with $T(S) = T$. From the F-CFDS, S, we construct a F-TA, $A = (S, V, \Sigma, \alpha, F)$ as follows:

Put $S = \{s_\lambda | \lambda \in \Lambda\}$, $F = \{s_{\lambda_0}\}$.

The mapping α is determined by:

(i) If $\lambda \overset{\rho}{\rightarrow} X(\lambda_1\lambda_2\cdots\lambda_k)$ is in P, then

$$\alpha(X)(s_{\lambda_1}s_{\lambda_2}\cdots s_{\lambda_k}, s_\lambda) = \rho$$

(ii) If $\lambda \overset{\rho}{\to} a$ is in P, then

$$\alpha(a)(\varepsilon, s_\lambda) = \rho$$

As easily known, the set S in the Definition 4 is

$$\{s_{\lambda_1}s_{\lambda_2}\cdots s_{\lambda_k} \mid \lambda \overset{\rho}{\to} X(\lambda_1\cdots\lambda_k) \text{ is in P}\} \cup \{\varepsilon\}.$$

The above construction of A assures that T(A) = T(S).

(Q.E.D.)

Theorem 6 Any fuzzy set of trees recognizable by F-TA is a F-CFDL.

Proof. Again the converse of the construction in the proof of Theorem 5 proves the theorem. (Q.E.D.)

Corollary 1 A fuzzy set of derivation trees of any F-CFG is a fuzzy set of trees recognizable by a F-TA.

Proof. By Theorems 2, 5 and 6.

It should be clear that for F-TA, the results corresponding to Lemma 3, Theorems 3 and 4 also hold.

7. FUZZY TREE TRANSDUCER

In the previous sections, we have introduced F-CFDS as generator of fuzzy set of trees and F-TA as acceptor. These fuzzy tree manipulating systems have been used to characterize a fuzzy set of derivation trees of a fuzzy context-free grammar. In this section, we introduce a fuzzy tree transducer which can define a fuzzy mapping from a set of trees to another one. These three tree manipulating systems will be able to be used to describe fuzzy meanings of context-free languages.

Let $(V_1;\Sigma_1)$ and $(V_2;\Sigma_2)$ be two finite partially ranked alphabets. A fuzzy tree translation from $T_{(V_1;\Sigma_1)}$ to $T_{(V_2;\Sigma_2)}$ is a fuzzy subset Φ of $T_{(V_1;\Sigma_1)} \times T_{(V_2;\Sigma_2)}$ in which the grade of membership of an element (t_1,t_2) of $T_{(V_1;\Sigma_1)} \times T_{(V_2;\Sigma_2)}$ is defined by

$$\mu_\Phi(t_1,t_2) = \rho \in [0,1].$$

We also denote it by a triple $(t_1,t_2;\rho)$ and then the fuzzy subset Φ can be considered to be a set of such triples. The domain of a fuzzy tree translation Φ is $\{t_1 | \text{for some } t_2 \text{ and some } \rho > 0, (t_1,t_2;\rho) \text{ is in } \Phi\} = \{t_1 | \text{for some } t_2, \mu_\Phi(t_1,t_2) > 0\}$, which will be denoted by dom Φ. The range of a fuzzy tree translation Φ is $\{t_2 | \text{for some } t_1 \text{ and some } \rho > 0, (t_1,t_2;\rho) \text{ is in } \Phi\} = \{t_2 | \text{for some } t_1, \mu_\Phi(t_1,t_2)>0\}$, which will be denoted by range Φ. Furthermore, we define two underlying fuzzy dendrolanguages; the one is that of domain which is defined to be a fuzzy subset ufd Φ of $T_{(V_1;\Sigma_1)}$ in which the grade of membership of an element t_1 of $T_{(V_1;\Sigma_1)}$ is given by

$$\mu_{ufd\Phi}(t_1) = \text{Sup}\{\rho | \mu_\Phi(t_1,t_2) = \rho,\ t_2 \in T_{(V_2;\Sigma_2)}\}$$

The other is that of range which is a fuzzy subset ufrΦ of $T_{(V_2;\Sigma_2)}$. The membership function is given by

$$\mu_{ufr\Phi}(t_2) = \text{Sup}\{\rho | \mu(t_1,t_2) = \rho,\ t_1 \in T_{(V_1;\Sigma_1)}\}.$$

Now, we introduce a relatively simple system, called a fuzzy tree transducer, to define a fuzzy tree translation:

Definition 5. A fuzzy simple tree transducer, (F-STT) is a 7-tuple

$$M = (\Lambda, V_1, \Sigma_1, V_2, \Sigma_2, R, \lambda_0)$$

where (i) Λ: a finite set of symbols, of which elements are called nonterminal node symbols

(ii) V_1, V_2: finite sets of symbols, called sets of node symbols

(iii) Σ_1, Σ_2: finite sets of symbols, called sets of leaf symbols.

(iv) R: $\Lambda \times T_{(V_1; \Lambda \cup \Sigma_1)} \times T_{(V_2; \Lambda \cup \Sigma_2)} \to [0,1]$, if $\mu(\lambda, t_1, t_2) = \rho$ then we write $\lambda \overset{\rho}{\to} (t_1, t_2)$, where t_1, t_2 contain the same nonterminal symbols in the same order. We call it a <u>fuzzy translation rule</u>.

(v) λ_0: initial nonterminal node symbol.

A form of M is a pair (t_1, t_2) where t_1 is in $T_{(V_1; \Lambda \cup \Sigma_1)}$ and t_2 is in $T_{(V_2; \Lambda \cup \Sigma_2)}$. If (i) $\lambda \overset{\rho}{\to} (t_1, t_2)$ is a fuzzy translation rule, (ii) (α_1, α_2) and (β_1, β_2) are forms such that $p(\alpha_1) = x_1 \lambda y_1$, $p(\alpha_2) = x_2 \lambda y_2$, $p(\beta_1) = x_1 p(t_1) y_1$, $p(\beta_2) = x_2 p(t_2) y_2$ and (iii) if λ is the k-th nonterminal node symbol in $p(\alpha_1)$ then λ of $p(\alpha_2) = x_2 \lambda y_2$ is also the k-th one in it, then we write

$$(\alpha_1, \alpha_2) \overset{\rho}{\Rightarrow} (\beta_1, \beta_2).$$

We also define the relation $\overset{*\rho}{\Longrightarrow}$ by:

$$(\alpha_1, \alpha_2) \overset{*1}{\Longrightarrow} (\alpha_1, \alpha_2)$$

and if $(\alpha_1, \alpha_2) \overset{*\rho_1}{\Longrightarrow} (\beta_1, \beta_2)$ and $(\beta_1, \beta_2) \overset{\rho_2}{\Longrightarrow} (\gamma_1, \gamma_2)$ then $(\alpha_1, \alpha_2) \overset{*\rho}{\Longrightarrow} (\gamma_1, \gamma_2)$, where $\rho = \underset{(\beta_1, \beta_2)}{\text{Sup}} \{\min(\rho_1, \rho_2)\}$.

The fuzzy tree translation defined by M, written as $\Phi(M)$, is

$$\{(t_1, t_2; \rho) \mid (\lambda, \lambda) \overset{*\rho}{\Longrightarrow} (t_1, t_2) \in T_{(V_1; \Sigma_1)} \times T_{(V_2; \Sigma_2)}\},$$

i.e., $\Phi(M)$ is a fuzzy set of $T_{(V_1; \Sigma_1)} \times T_{(V_2; \Sigma_2)}$ in which the grade of membership of an element (t_1, t_2) of $T_{(V_1; \Sigma_1)} \times T_{(V_2; \Sigma_2)}$ is given by

$$\mu_{\Phi(M)}(t_1, t_2) = \rho \text{ of } (\lambda, \lambda) \overset{*\rho}{\Longrightarrow} (t_1, t_2)$$

<u>Example 6.</u> Consider a set of rules:

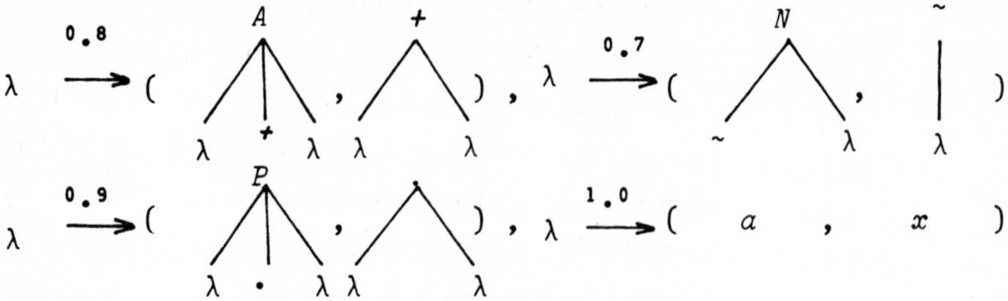

Figure 23

By these rules, we have a fuzzy tree translation of which elements can be obtained, for example, as follows:

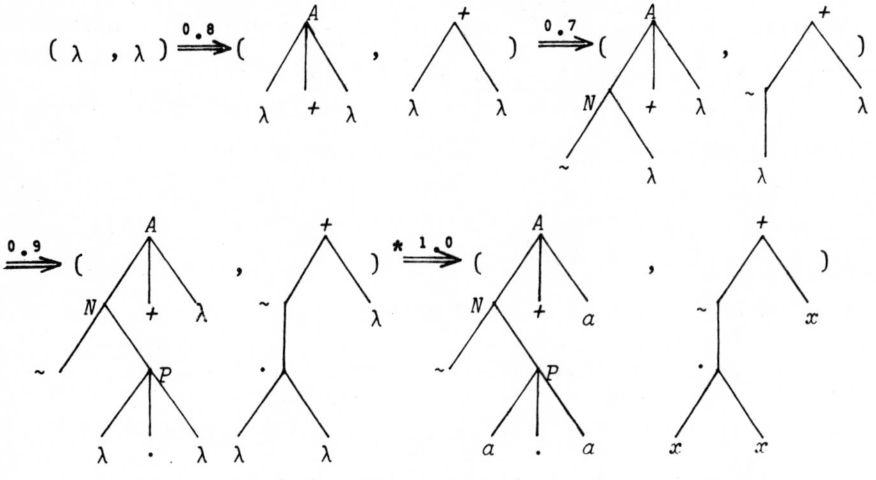

Figure 24

Theorem 7 For any F-STT, M, both of ufdΦ(M) and ufrΦ(M) are F-CFDL's.

Proof. From a given F-STT, $M = (\Lambda, V_1, \Sigma_1, V_2, \Sigma_2, R, \lambda_0)$, we can construct a F-CFDS, $S = (\Lambda, V_1, \Sigma_1, P, \lambda_0)$ by defining P as follows: If $\lambda \overset{\rho}{\to} (t_1, t_2)$ is in R, then P contains $\lambda \overset{\rho}{\to} t_1$.

By noting that $(\lambda_0, \lambda_0) \overset{*\rho}{\Longrightarrow} (t_1, t_2)$ if and only if $\lambda_0 \overset{*\rho}{\Longrightarrow} t_1$, we can prove ufd$\Phi$(M) = T(S).

The remaining part of the theorem can be proved similarly. (Q.E.D.)

We can also prove the following theorem:

Theorem 8 For any F-STT, M, both of dom $\Phi(M)$ and range $\Phi(M)$ are context-free dendrolanguages.

Discussions similar to Theorem 7 can prove this theorem.

8. FUZZY MEANING OF CONTEXT-FREE LANGUAGE

A central problem of semantics is that of specifying a set of semantic rules which can serve as an algorithm for computing the meaning of a composite term from the knowledge of the meanings of its components. But the complexity of natural languages is so great that it is not even clear what the form of the rules should be. In such circumstances, it is natural to start with a few relatively simple cases involving fragments of natural or artificial languages.

Prof. L. A. Zadeh has suggested a possible start to approach to the problem of the semantics by proposing a quantitative theory of semantics: The theory shows that the meaning of a term is defined to be a fuzzy subset of a universe of discourse and that an approach similar to that described by Prof. D. E. Knuth (8) can be used to compute the meaning of a composite term.

This method of assigning the meanings to a composite term is essentially considered to be a syntax-directed one. On the other hand, we consider that a fuzzy set to be assigned to a composite term is an image of some composite function of fuzzy sets on the universe of discourse. Then, it should be fairly natural to consider that we can define the semantic domain as the universe of discourse and (composite) functions. Here we can also recognize a syntax structure in an expression representing a function. In other words, we consider as the semantic domain only the set of all functions representable by using some syntax rules and we consider that assigning the meaning to a composite term is to assign a syntax tree representation of a corresponding function to it.

These discussions lead us to apply our fuzzy tree transducer to describing the fuzzy meaning of context-free language. This will be exemplified by the following discussions:

Example 7. Let us construct a fuzzy tree transducer for an example which is slightly modified one described by Prof. L. A. Zadeh in (2,3) as follows:

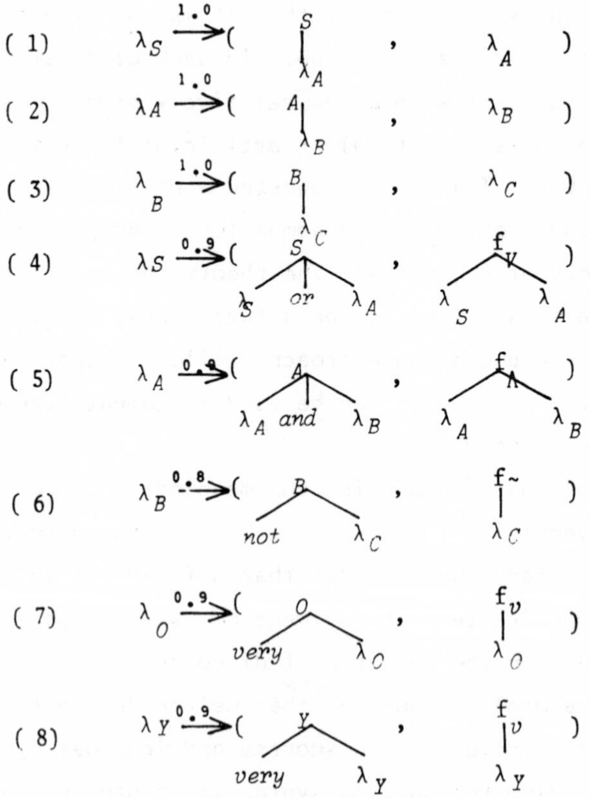

(9) $\lambda_C \xrightarrow{1.0} (\; C(\lambda_O),\; \lambda_O \;)$

(10) $\lambda_C \xrightarrow{1.0} (\; C(\lambda_Y),\; \lambda_Y \;)$

(11) $\lambda_C \xrightarrow{0.6} (\; C(\lambda_S),\; \lambda_S \;)$

(12) $\lambda_O \xrightarrow{1.0} (\; O(old),\; f_{old} \;)$

(13) $\lambda_Y \xrightarrow{1.0} (\; Y(young),\; f_{young} \;)$

Here we assume that f_V, f_Λ and $f_\sim$ are fuzzy set operations union, intersection and complement and f_v is the concentrating function defined by: if $f_v(A) = B$, then the membership function $\mu_B(x)$ of B is given by $\mu_B(x) = \mu_A^2(x)$. Furthermore we assume that f_{old} and f_{young} are constant functions of which values are, for example, the fuzzy subsets of the set of integers K = [1,100], characterized by the membership functions

$$\mu_N(\text{old},y) = 0\,, \quad \text{for } y < 50$$
$$= [1 + (\tfrac{y-50}{5})^{-2}]^{-1}\,, \quad \text{for } y \geqq 50$$

and

$$\mu_N(\text{young},y) = 1, \quad \text{for } y < 25$$
$$= [1 + (\tfrac{y-25}{5})^{2}]^{-1}, \quad \text{for } y \geqq 25,$$

respectively†.

Let us consider a composite term x = old or young and not very old. For the term x, the translation is given by

†Concerning the definitions of μ_N and related concepts, refer to the references (2,3).

(A) and (B):

(A) $(\lambda_S, \lambda_S) \xRightarrow{*0.6}$ (

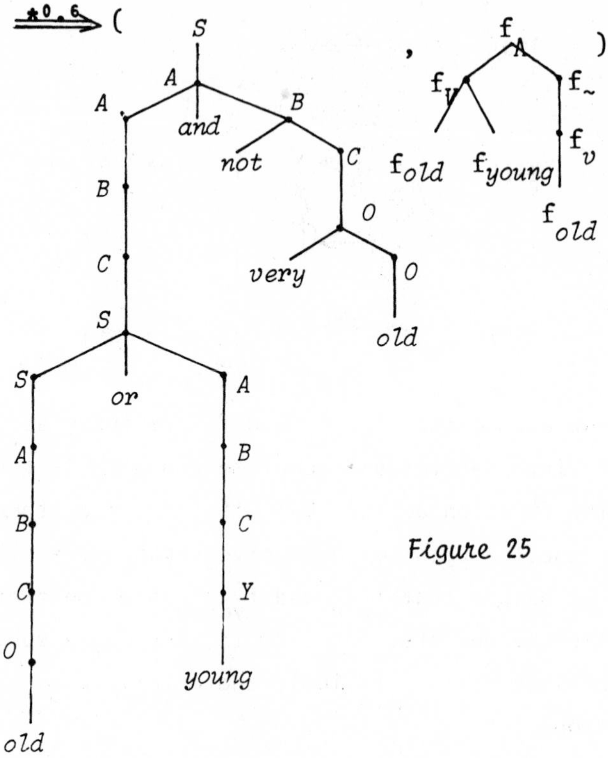

)

Figure 25

which is realized by applications of rules (1), (5), (2), (3), (11), (4), (1), (2), (3), (9), (12), (2), (3), (10), (13), (6), (9), (7) and (12) in this order.

(B) $(\lambda_S, \lambda_S) \overset{*0.8}{\Longrightarrow}$ (

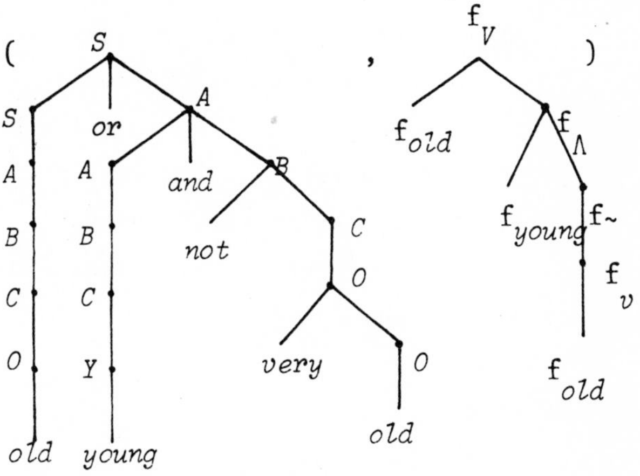

)

which is realized by applications of rules (4), (1), (2), (3), (9), (12), (5), (2), (3), (10), (13), (6), (9), (7) and (12).

Thus, we know that in the system under consideration the composite term x = old or young and not very old has two meanings $f_\Lambda(f_V(f_{old}, f_{young}), f_\sim(f_v(f_{old}))) = [\mu(old) \vee \mu(young)] \wedge [1 - \mu^2(old)]$ and $f_V(f_{old}, f_\Lambda(f_{young}, f_\sim(f_v(f_{old})))) = \mu(old) \vee [\mu(young) \wedge [1 - \mu^2(old)]]$ with degrees 0.6 and 0.8, respectively.†

As easily known from the above example, a fuzzy tree transducer can be a reasonable model to describe fuzzy meaning of fuzzy context-free language at the level of syntax structure in the sense that it can fuzzily associate each derivation tree of the fuzzy language with a tree representation of the computation process of its fuzzy meaning.

---

†Concerning the abbreviation of μ(old) and μ(young), refer to the references (2,3).

ACKNOWLEDGEMENT

The authors would like to acknowledge the continuing guidance and encouragement of Professor N. Honda of Tohoku University. They would like to thank Professor K. Tanaka of Osaka University for his kindness in giving them this opportunity of representing this research. They are indebted to Professor L. A. Zadeh for his papers to draw their attention to this problem.

REFERENCES

(1) Lee, E. T. and L. A. Zadeh, "Note on Fuzzy Languages," Information Sciences, 1, pp. 421-434, (1969).
(2) Zadeh, L. A., "Fuzzy Languages and Their Relation to Human and Machine Intelligence," Man and Computer, Proceedings Int. Conf. Bordeaux (1970) pp. 130-165, Ed. M. Marinos, Karger, Basel, 1972, Paris.
(3) Zadeh, L. A., "Quantitative Fuzzy Semantics," Information Sciences, 3, pp. 159-176, (1971).
(4) Mizumoto, M., J. Toyoda and K. Tanaka, "N-Fold Fuzzy Grammars," Information Sciences, 5, pp. 25-43, (1973).
(5) Kandel, A., "Codes Over Languages," IEEE Trans. on SMC, pp. 135-138, (Jan. 1974).
(6) Ito, H., Y. Inagaki and T. Fukumura, "Hierarchical Studies of Dendrolanguages with an Application to Characterizing Derivation Trees of Phrase Structure Grammars," Memoirs of the Faculty of Engineering, Nagoya University, 25 1, pp. 1-46, (1973).
(7) Thatcher, J. W., "Characterizing Derivation Trees of Context-Free Grammars through a Generalization of Finite Automata Theory," J.C.S.S., 1, pp. 317-322 (1967).
(8) Knuth, D. E., "Semantics of Context-Free Languages," Math. Sys. Theory, 2, 2, pp. 127-145, (1968).

FRACTIONALLY FUZZY GRAMMARS WITH APPLICATION TO PATTERN RECOGNITION

G. F. DePalma and S. S. Yau
Departments of Computer Sciences
and Electrical Engineering
Northwestern University
Evanston, Illinois 60201 U.S.A.

ABSTRACT

A new type of fuzzy grammar, called the fractionally fuzzy grammar, is introduced. These grammars are especially suitable for pattern recognition because they are powerful and can easily be parsed. It is shown that the languages produced by the class of type i (Chomsky) fractionally fuzzy grammars properly includes the set of languages generated by type i fuzzy grammars. It is also shown that the set of languages generated by all type 3 (regular) fractionally fuzzy grammars is not a subset of the set of languages produced by all unrestricted (type 0) fuzzy grammars. It is found that context-sensitive fractionally fuzzy grammars are recursive and can be parsed by most methods used for ordinary context-free grammars. Finally, a pattern recognition experiment which uses fractionally fuzzy grammars to recognize the script letters *i*, *e*, *t* and *ℓ* without the help of the dot on the *i* or the crossing of the *t* is given. The construction of a fractionally fuzzy grammar based on a training set and the experimental results are discussed.

I. INTRODUCTION

Formal language theory has been applied to pattern recognition problems in which the patterns contain most of their information in their structure rather than in their numeric values (1-5). In order to increase the generative power of grammars and to make grammars more powerful so that they become more suited to pattern recognition, the concept of a phrase structured grammar has been extended in several ways. One is to randomize the use of the production rules, resulting in stochastic grammars (2,6,7) and fuzzy grammars (3-5, 8-10). A fuzzy grammar produces a language which is a fuzzy set (11) of strings with each string's membership in the language measured on the interval [0,1], where 0 indicates no membership and 1 indicates full membership. These languages have shown some promise in dealing with pattern recognition problems, where the underlying concept may be probabilistic or fuzzy (3-5). A second way of extending the concept of a grammar is to restrict the use of the productions (10,12,13) resulting in programmed grammars and controlled grammars. These grammars can generate all recursively enumerable sets with a context-free core grammar. Programmed grammars have the added advantage that they are easily implemented on a computer.

Cursive script recognition experiments (16-21) have so far had the major emphasis on recognizing whole words, and none have used a syntactic approach. A typical method presented by Eden (18) decomposes the words to be recognized into sequences of strokes which are then combined into letters, and into words. Mermelstein and Eden (19) tried to distinguish words which were similar in appearance such as *fell*, *feel*, *foul*, etc. All of these experiments except that of Sayre (21) input their data on a graphics device and kept the sequence of the points as a part of the data. Sayre inputs pictures of writing and thus did not have the sequence information available.

In this paper, we will introduce a new type of fuzzy

grammar, called the fractionally fuzzy grammar. Fractionally fuzzy grammars are especially suited to pattern recognition because they are powerful and can easily be parsed. We will show that the languages generated by the class of type i (Chomsky) fractionally fuzzy grammars properly includes the set of languages generated by type i fuzzy grammars. We will also show that the set of languages generated by all type 3 (regular) fractionally fuzzy grammars is not a subset of the set of languages generated by all unrestricted (type 0) fuzzy grammars. We will find that context-sensitive fractionally fuzzy grammars are recursive and can be parsed by most methods used for ordinary context-free grammars. Finally, a pattern recognition experiment which uses fractionally fuzzy grammars to recognize the script letters $\mathscr{i}$, $\mathscr{e}$, $\mathscr{t}$ and ℓ without the help of the dot on the $\mathscr{i}$ or the crossing of the $\mathscr{t}$ will be described. The construction of a fractionally fuzzy grammar based on a given training set and the experimental results will also be discussed.

II. BACKGROUND AND NOTATION

An ordinary grammar is a four-tuple $G = (V_T, V_N, S, P)$, where V_T is a finite set of terminal symbols, V_N is a finite set of non-terminal symbols, $V = V_T \cup V_N$ is the vocabulary of the grammar, $S \in V_N$ is the starting symbol and P is a finite set of production rules of the form $\alpha \to \beta$, where α and β are members of the set V^* of all strings (including the null string ε) over the vocabulary V. A derivation within a grammar proceeds as follows: Starting with the current string being the starting symbol S, we search the productions for a rule whose left-hand side α is a substring of the current string. Any rule which is matched in this manner can be applied by replacing the substring with the right-hand side of the rule β. A derivation is complete when there are only terminal symbols in the current string.

Grammars are classified according to the form of the production rules used (Chomsky hierarchy). Unrestricted or type 0 grammars have no restrictions on the form of the rules. Context-sensitive or type 1 grammars require that if $\alpha \rightarrow \beta$ is a production, then the length of the string α, denoted by $|\alpha|$, must be not greater than the length $|\beta|$ of the string β. A grammar is said to be context-free or type 2 if every production is of the form $A \rightarrow \beta$, where A is in V_N. A grammar is said to be regular (left linear) if all its productions are of the form $A \rightarrow B$, $A \rightarrow b$, or $A \rightarrow bB$, where A and B are in V_N and b is in V_T. It is sometimes useful to write a grammar in a particular form. Two forms for context-free grammars have been commonly used: Chomsky normal form and Greibach normal form. A context-free grammar is said to be in Chomsky normal form if every production rule is of one of the forms $A \rightarrow BC$ or $A \rightarrow a$ where A, B, and C are in V_N and a is in V_T. Also, if ε is in the language L(G) generated by G, then $S \rightarrow \varepsilon$ is a production and S does not appear on the right-hand side of any production. A context-free grammar is said to be in Greiback normal from if every production rule is of the form $A \rightarrow a\theta$, where A is in V_N, a is in V_T, and θ is in V_N^*. Also, if ε is in L(G), then $S \rightarrow \varepsilon$ is a production and S does not appear on the right-hand side of any production.

Zadeh (11) first introduced the concept of a fuzzy set. A fuzzy set f is a mapping of the elements of the universe into [0,1]. If x is an element, f(x) is its membership in f with f(x) = 0 indicating no membership and f(x) = 1 indicating full membership. If f and g are fuzzy sets, then f = g if and only if f(x) = g(x) for all x. The union and intersection of fuzzy set f and g are given by the mappings:

$$f \cup g(x) = \max[f(x), g(x)]$$
$$f \cap g(x) = \min[f(x), g(x)]$$

An extension of the idea of an ordinary grammar is the fuzzy grammar based on the concept of fuzzy sets. Fuzziness is added to a language by modifying the grammar as follows: A fuzzy grammar FG is a six-tuple $FG = (V_T, V_N, P, S, J, f)$, where V_T, V_N, S and P are respectively the terminal alphabet, the non-terminal alphabet, the starting symbol and the set of production rules as with an ordinary grammar. $J = \{r_i | i = 1, 2, \ldots, n\}$ is a set of distinct labels for the productions in P, and f is a fuzzy membership function, $f: J \to [0,1]$. If rule r_i is $\alpha \to \beta$, when we write the application of r_i as:

$$\gamma\alpha\delta \xrightarrow[r_i]{f(r_i)} \gamma\beta\delta$$

where α, β, γ and δ are in V^*. If θ_i, $i = 0,1,2,\ldots,m$, are strings in V^* and

$$S = \theta_0 \xrightarrow[r_1]{f(r_1)} \theta_1 \xrightarrow[r_2]{f(r_2)} \theta_2 \xrightarrow[r_3]{f(r_3)} \cdots \xrightarrow[r_m]{f(r_m)} \theta_m = x$$

is a derivation of x in FG, we write

$$S = \theta_0 \xrightarrow[r_1 r_2 \ldots r_m]{f(r_1 r_2 \ldots r_m)} \theta_m = x$$

with $f(r_1 r_2 \ldots r_m) = \min[f(r_1), f(r_2), \ldots f(r_m)]$. The grade of membership of the string $x \in V_T^*$ is given by

$$f(x) = \sup[f(r_1 r_2 \ldots r_m)],$$

where the supremum is taken over all derivations of x in the language L(FG) generated by FG. It can be seen that ordinary grammars are a special case of fuzzy grammars, namely when $f(r_i) = 1$ for all r_i in J. Fuzzy grammars can be classified according to the form of the production rules. Lee and Zadeh (8) have shown that for every context-free fuzzy grammar G, there exists two context-free fuzzy grammars G_g and G_c such that $L(G) = L(G_g) = L(G_c)$ and G_g is in Greibach normal form and G_c is in Chomsky normal form.

EXAMPLE 1: Consider the fuzzy grammar $FG = (V_T, V_N, S, P, J, f)$, where $V_T = \{a, b\}$, $V_N = \{S, A, B, C\}$, and P, J, and f are given as follows:

r_1:	$S \to AB$	$f(r_1) = 1$
r_2:	$A \to a$	$f(r_2) = 1$
r_3:	$B \to b$	$f(r_3) = 1$
r_4:	$A \to aAB$	$f(r_4) = 0.9$
r_5:	$A \to aB$	$f(r_5) = 0.5$
r_6:	$A \to aC$	$f(r_6) = 0.5$
r_7:	$C \to a$	$f(r_7) = 0.5$
r_8:	$C \to aa$	$f(r_8) = 0.2$
r_9:	$A \to B$	$f(r_9) = 0.2$

The language generated by this fuzzy grammar consists of strings of the form $a^n f^m$ with $n, m > 0$. The fuzzy membership of these strings depends on n and m as follows: If $m = n = 1$, $f(x) = 1$; if $m = n \neq 1$, $f(x) = .9$; if $m = n \pm 1$, $f(x) = .5$; if $m = n \pm 2$, $f(x) = .2$; otherwise, $f(x) = 0$. Thus, it is seen that this grammar detects strings of the form $a^n b^m$, where $|n-m| \leq 2$.

A fuzzy grammar can be used to generate non-fuzzy languages by the use of thresholds. The language generated by a fuzzy grammar FG with a threshold λ is the set of strings

$$L(FG,\lambda) = \{x \in L(FG) \mid f(x) > \lambda\}$$

and is called the λ-fuzzy language. There are two other threshold languages defined by Mizumoto, Toyoda and Tanaka (9). They are the two-threshold language and the equal-threshold language defined as follows:

$$L(FG,\lambda_1,\lambda_2) = \{x \in L(FG) \mid \lambda_1 < f(x) \leq \lambda_2\}$$

and

$$L(FG,=,\lambda) = \{x \in L(FG) \mid f(x) = \lambda\}.$$

$L(FG,\lambda)$ is most often used to compare the generating power of a fuzzy grammar to that of an ordinary grammar. However, $L(FG,\lambda) = L(G)$, where G is the grammar obtained from the FG by removing all productions whose fuzzy membership is less than or equal to λ and then removing the fuzziness from the remaining rules. Therefore, for fuzzy grammars, the threshold language appears to be of limited use.

III. FRACTIONALLY FUZZY GRAMMARS

In syntactic pattern recognition, the patterns are strings over the terminal alphabet. These strings must be parsed in order to find the pattern classes to which they most likely belong. Many parsing algorithms (15) require backtracking. That is, after applying some rules, it is discovered that the input string cannot be parsed successfully by this sequence of rules. Rather than starting it from the beginning again, it is desirable to reverse the action of one or more of the most recently applied rules in order to try another sequence of productions. With non-fuzzy grammars, it is sufficient to keep track of the derivation tree as it is generated with each node being labeled with a symbol from V. However, with fuzzy grammars, this tree is not sufficient since the fuzzy value at the ith step is the minimum of the value at $(i-1)^{th}$ step and the fuzzy membership of the ith rule. If this minimum was the ith rule's membership, there is no way to know the fuzzy value at the $(i-1)^{th}$ step. Thus, the fuzzy value at each step must also be remembered at each node, and hence the memory requirements are greatly increased for many practical problems.

A second drawback of fuzzy grammars in pattern recognition is the fact that all strings in L(FG) can be classified into a finite number of subsets by their membership in the language. The number of such subsets is strictly limited by the number of productions in the grammar. This is so because if x is a string in L(FG) with a membership f(x), then there must be a rule in FG with the membership f(x) since $f(x) = \min_j[f(r_{i_j})]$ for some sequence of rules $r_{i_1} r_{i_2} \ldots r_{i_m}$ in P. This is also the reason that L(FG,λ) for some threshold λ is always a language generated by those rules in the grammar with a membership greater than λ.

To overcome these restrictions, we introduce a new method of computing the membership of a string x which can be derived by the m sequences of production rules, $r_1^k r_2^k \ldots r_{\ell_k}^k$, of lengths ℓ_k, where k = 1,2,...,m. This leads to the following definition.

<u>Definition 1</u>: <u>A fractionally fuzzy grammar</u> is a 7-tuple $FFG = (V_N, V_T, S, P, J, g, h)$, where V_N, V_T, S, P, and J are the non-terminal alphabet, the terminal alphabet, the starting symbol, the set of productions and a distinct set of labels on the productions as a fuzzy grammar. The functions g and h map the set of productions into the non-negative integers such that $g(r_k) \leq h(r_i)$ for all r_i in P. A string is generated in the same manner as that by a fuzzy grammar, except that the membership of the derived string is given by

$$f(x) = \sup_k \frac{\sum_{j=1}^{\ell_k} g(r_j^k)}{\sum_{j=1}^{\ell_k} h(r_j^k)},$$

where 0/0 is defined as 0.

Because 0/0 is defined as 0, it is clear that $0 \leq f(x) \leq 1$ for all x. It is also clear that backtracking over a rule r can now be accomplished by simply subtracting g(r) and h(r) from the respective running totals.

Let us interpret the above definition in a heuristic sense. As each rule r is applied, g(r) and h(r) are added to the respective running totals for the numerator and denominator of the fuzzy membership. Immediately, it is seen that the fuzzy membership of a string could be any rational number in [0,1]. Also, the number of fuzzy membership levels is not limited by the number of productions. With pattern recognition in mind, we see that the amount of impact a rule will have on the final membership level is proportional to the value of h(r). The value of g(r) will tend to increase the membership of the string if g(r) is approximately equal to h(r) and will tend to decrease the membership of the string if $g(r) << h(r)$ or if g(r) is close to 0. Rules for which g(r) and h(r) are both 0 will have no effect on the membership. Thus, we can divide the rules into three classes. Those which strongly indicate membership in the class, those which strongly indicate membership in another class, and those which serve little purpose in separating the classes but which are traits between different classes.

<u>EXAMPLE 2</u>: The following four rules define a fractionally fuzzy grammar which has $V = \{a,b\}$, $V_N = \{S\}$ and with P, J, g and h given by

$$\begin{array}{llll} r_1: & S \to ab & g(r_1) = 1 & h(r_1) = 1 \\ r_2: & S \to aSb & g(r_2) = 1 & h(r_2) = 1 \\ r_3: & S \to aS & g(r_3) = 0 & h(r_3) = 1 \\ r_4: & S \to Sb & g(r_4) = 0 & h(r_4) = 1 \end{array}$$

This grammar generates the fuzzy language $\{a^n b^m \mid n,m > 0\}$. The membership of the string $a^n b^m$ is given by:

$$f(a^n b^m) = \min(n,m)/\max(n,m)$$

This set of strings is the fuzzy set of strings which are

"almost" $a^n b^n$. That is, the first pair of rules generates the set of strings $a^n b^n$ for $n > 0$, and the second pair of rules allow for variations in the number of a's and the number of b's. The closer n and m are, the greater the membership of the string. This is because $g(r) = h(r)$ for the first pair of rules and $g(r) << h(r)$ for the second pair. However, unlike the grammar shown in Example 1, which could only measure finite differences between m and n, this grammar measures membership on percentage difference between m and n, which is similar to the way a person would judge whether or not the lengths of two lines were the same. We will now compare the relative generative power of fractionally fuzzy grammars to that of fuzzy grammars.

Theorem 1: The set of all languages generated by type i fractionally fuzzy grammars properly includes the set of all languages generated by type i fuzzy grammars, where i = 0,1, 2 and 3.

Proof: It is clear from the preceding remarks that the number of distinct levels of membership in a language generated by a fuzzy grammar FG is limited by the number of production rules in FG which must be finite. Thus, it is sufficient to show that for every fuzzy grammar of type i, there exists a fractionally fuzzy grammar of type i which generates the same language, and that there exists a fractionally fuzzy grammar of type i which generates a language with an infinite number of distinct membership levels.

Let $FG = (V_N, V_T, S, P, J, f)$ be a fuzzy grammar of type i. We can construct a fractionally fuzzy grammar $FFG = (V_N', V_T', S', P', J', g, h)$ as follows: Let f_i, $i = 1,2,\ldots,m$ be the distinct values of $f(r_i)$ and arrange them in the order $f_1 > f_2 > \cdots > f_m$. Let g_i and h_i be integers such that $f_i = g_i/h_i$, $i = 1,2,\ldots,m$. For each symbol A in V_N, we define m distinct symbols A_i, $i = 1,2,\ldots,m$. These symbols and the new starting symbol S' make up the set of non-terminals V_N' for the fractionally fuzzy grammar. Let P' initially be the set of rules

$$r_0^i\colon\ S' \to S_i \qquad g(r_0^i) = g_i \qquad h(r_0^i) = h_i, \quad i = 1,2,\ldots,m.$$

To these, for each rule r_j: $\alpha \to \beta$ in P and each f_i, $i = 1,2, \ldots,m$, if $f(r_j) \geq f_i$, we add the rule

$$r_j^i\colon\ \alpha_i \to \beta_i \qquad g(r_j^i) = g_i \qquad h(r_j^i) = h_i,$$

where α_i and β_i are the strings α and β with each non-terminal A replaced with the associated new symbol A_i. The FFG has, in effect, m sub-grammars reachable by the r_0^i rules. The ith sub-grammar will produce only those strings which would have been generated with a membership of at least f_i in L(FG). In the ith sub-grammar, these strings have membership $g_i/h_i = f_i$. Since their membership in the L(FFG) is defined as the supremum of all derivations, we find that each string has equal membership in both languages. In other words, L(FFG) = L(FG).

To show the proper inclusion, the following fractionally fuzzy grammar is considered:

$$r_1\colon\ S \to a \qquad g(r_1) = 1 \qquad h(r_1) = 1$$
$$r_2\colon\ S \to aS \qquad g(r_2) = 0 \qquad h(r_2) = 1$$

This grammar produces strings of the form a^n for $n > 0$. Since the fuzzy membership of the string a^n is $1/n$, the language certainly has an infinite number of distinct membership levels. Since the grammar is regular, it is also context-free, context-sensitive, and a member of the class of type-0 grammars. Q.E.D.

Fractionally fuzzy grammars are of no use in pattern recognition if it is not possible to determine whether a given string is a member of the language. In other words, in order to apply fractionally fuzzy grammars to pattern recognition, we need an algorithm which can compute the membership of a string in L(FFG) which is bounded in time. The following lemmas will lead to Theorem 2 which proves that context-sensitive fractionally fuzzy grammars are recursive and such an algorithm exists.

Lemma 1: Let FFG be a context-sensitive fractionally fuzzy grammar and let some derivation contain the sequence

$$\ldots \to \theta_i \to \theta_{i+1} \to \ldots \to \theta_{i+k} \to \ldots,$$

where $\theta_i = \theta_{i+k}$. Then, either $k \leq n^p$, where $n = |V|$ is the number of symbols in the vocabulary and $p = |\theta|$ is the length of the string θ_i, or $\theta_{i+j} = \theta_{i+m}$ for $0 \leq j < m < n^p$.

Proof: The lemma is obvious because of the non-contracting nature of context-sensitive grammars (i.e., $|\theta_u| \leq |\theta_v|$ for $u \leq v$) and because there are exactly n^p distinct strings over V of length p. Q.E.D.

Lemma 2: Let FFG be a fractionally fuzzy grammar, and $x \in L(FFG)$ which is derivable by the sequence

$$S = \theta_0 \to \theta_1 \to \theta_2 \to \ldots \to \theta_n = x.$$

If $\theta_j = \theta_k$ for $j < k$ and $0 \leq j < n$, then the membership of x in FFG is at least

$$\max \left[\frac{\sum_{m=1}^{j} g(r_{i_m}) + \sum_{m=k+1}^{n} g(r_{i_m})}{\sum_{m=1}^{j} h(r_{i_m}) + \sum_{m=k+1}^{n} h(r_{i_m})}, \frac{\sum_{m=j}^{k} g(r_{i_m})}{\sum_{m=j}^{k} h(r_{i_m})} \right]$$

Proof: The loop in the derivation sequence from θ_j to θ_k can be removed. The first argument of the maximum represents the membership given to x by this shortened derivation. The loop can also be repeated as many times as desired. In the limit, as the number of times the loop is repeated is increased, the membership of x approaches the second argument. There may be other derivations of x, and thus the membership of x in L(FFG) is at least the maximum of these two terms. Q.E.D.

Lemma 2 may be applied repeatedly in a derivation. Thus, if a derivation contains a loop nested within a loop, the loops can be considered separately and the membership of the string is the maximum given by the loop-free derivation, the inner loop and the outer loop.

<u>Lemma 3</u>: Let FFG be a context-sensitive fractionally fuzzy grammar with n symbols in V. Let $R_0^1 = \{S\}$. Let R_0^k be the set of all strings over V of length k which can be directly generated from a string of length less than k. Let R_j^k be the set of all strings of length k which can be directly generated from a string in R_{j-1}^k, $j = 1,2,\ldots$ Then, the set $R^k = R_0^k R_1^k \ldots R_n^k$ contains all strings over V of length k which can be generated by the FFG, and the derivation needed to generate R^k will contain all the simple derivation loops on strings of length k in L(FFG).

<u>Proof</u>: Assume that there exists a θ_m such that θ_m is not in R^k, $|\theta_m| = k$, and θ_m is derivable from $\theta_i \in R_0^k$. Let $\theta_i \rightarrow \theta_{i+1} \rightarrow \ldots \rightarrow \theta_{i+j} = \theta_m$ be the shortest sequence from θ_i to θ_m for FFG. But $j > n$ by assumption. Thus, by Lemma 1, this sequence must contain a loop and is therefore not the shortest sequence. Therefore, no such θ_m exists.

Assume that $\theta_j \rightarrow \theta_{j+1} \rightarrow \ldots \rightarrow \theta_{j+s} = \theta_j$ is a simple loop in FFG, where $|\theta_j| = k$. Assume this loop was not detected by the derivations which generated R^k and that loop can be detected from $\theta_i \in R_0^k$ by the shortest sequence

$$\theta_i \rightarrow \theta_{i+1} \rightarrow \ldots \rightarrow \theta_{i+r} = \theta_j$$

By assumption, $r + s > n$ since the loop was not detected by generating R^k. However, the r + s strings $\theta_i, \theta_{i+1}, \ldots, \theta_{j+s-1}$ cannot all be distinct. Since $\theta_i, \ldots, \theta_{i+r}$ are distinct by assumption and $\theta_j, \ldots, \theta_{j+s-1}$ are also all distinct by assumption, then $\theta_{i+u} = \theta_{j+v}$ for some $0 \leq u < r$ and $0 \leq v < s$. Therefore, the derivation sequence

$$\theta_i \rightarrow \theta_{i+1} \rightarrow \ldots \rightarrow \theta_{i+u} = \theta_{j+v} \rightarrow \theta_{j+v+1} \rightarrow \ldots \rightarrow \theta_{j+s} = \theta_j \rightarrow \ldots \rightarrow \theta_{j+v}$$

also detects the simple loop and is shorter than the original sequence. This contradicts the original assumption, and therefore no such loop can appear in a derivation. Q.E.D.

<u>Theorem 2</u>: If FFG is a context-sensitive fractionally fuzzy grammar, then FFG is recursive.

<u>Proof</u>: Let $n = |V|$. For any string x, we need only generate $R^1, R^2, \ldots, R^k$, where $k = |x|$. Let R_i be the set of all strings derivable from S in i steps. For any finite i, we can find this set. Since it takes n_j steps to generate R^j from R_0^j, clearly R_m contains R^j if $m \geq 1 + n + n^2 + \ldots + n^j$. Thus, in a finite number of steps, all strings of length k can be found and all loops in these derivations can be detected. Thus, we can determine whether or not x is in L(FFG) and its membership if it is.

The following theorem gives an interesting property of λ-fractionally fuzzy grammars:

<u>Theorem 3</u>: Let FFG be a regular fractionally fuzzy grammar. Then, the language L(FFG,λ) is not necessarily a regular language.

<u>Proof</u>: Assume that L(FFG,λ) is a regular language for all regular fractionally fuzzy grammars. Consider the two fractionally fuzzy grammars $FFG_1 = (V_T, V_N, X, P, J, f_1, g_1)$ and $FFG_2 = (V_T, V_N, S, P, J, f_2, g_2)$, where $V_T = \{0,1\}$, $V_N = \{S,A\}$, and P is given by

r_1: $S \rightarrow 0S$
r_2: $S \rightarrow 0$
r_3: $S \rightarrow A$
r_4: $A \rightarrow 1A$
r_5: $A \rightarrow 1$

Let $g_1(r_i) = g_2(r_i) = 1$ for $i = 1,2,3,4,5$, and let $f_1(r_i) = 1$ for $i = 1,2,3$, $f_1(r_i) = 0$ for $i = 4,5$. Let $f_2(r_i) = 0$ for $i = 1,2$ and let $f_2(r_i) = 1$ for $i = 3,4,5$. It is clear that both grammars produce strings of the form 0^n1^m, where $m, n \geq 0$ and $m + n > 0$. Further examination shows that the fuzzy membership of 0^n1^m is given by the following two equations:

$$f_1(0^n1^m) = \frac{n+1}{m+n+1}$$

and

$$f_2(0^n1^m) = \frac{m+1}{m+n+1} \text{ if } m > 0 \text{ or } f_2(0^n1^m) = 0 \text{ if } m = 0$$

The set $L(FFG_1,0.5) = \{0^n1^m \mid n \geq m\}$ and the set $L(FFG_2,0.5) = \{0^n1^m \mid n \leq m\}$ must be regular by assumption. Since regular sets are closed under intersection, the set $L(FFG_1,0.5) \cap L(FFG_2,0.5) = \{0^n1^n \mid n > 0\}$ must be regular. However, it is known that 0^n1^n, where $n \geq 0$, is context-free and not regular. Thus, the assumption is false and the theorem is proved.

Q.E.D.

IV. A PATTERN RECOGNITION EXPERIMENT

An experiment was conceived to test the usefulness of fractionally fuzzy grammars in pattern recognition. A pattern space which was a set of strings was needed. We chose to use script writing which was input to a computer on a graphics tablet. This data consisted of strings of points in the 2-dimensional space of the tablet. The data is a sample of seven persons' handwriting. Each person was given a list of 400 seven-letter words and was told approximately how large the person should write. The first three persons wrote all 400 words while the last four wrote only the first 100 words. The data was digitized in a continuous mode by the computer whenever the pen was down. Each point collected in this manner was compared to the previously stored point to see if the distance between them was greater than a given threshold (about 0.04 inch). If it was not, the new point was discarded and a new position of the pen was read. If the threshold had been exceeded, this point was added to the data and the process was repeated. This resulted in a record of 250 points in the X-Y plane (with zero fill-in) for each seven-letter word written. Figure 1 shows some examples of words input to the computer.

Figure 1

We needed to convert these points into a string of symbols which would comprise the terminal alphabet. This was accomplished by comparing each adjacent pair of points to see the relative direction traveled by the pen at that point and classifying the direction into one of eight directions, each separated by 45 degrees, with class 0 being centered at 0 degrees (the + X-direction) and the remaining classes being numbered 1 through 7 in a counter clockwise direction. Thus, the terminal alphabet consisted of the eight octal digits, i.e., $V_T = \{0,1,2,\ldots,7\}$. This quantization of directions introduced some distortion into the data as shown in Figure 2.

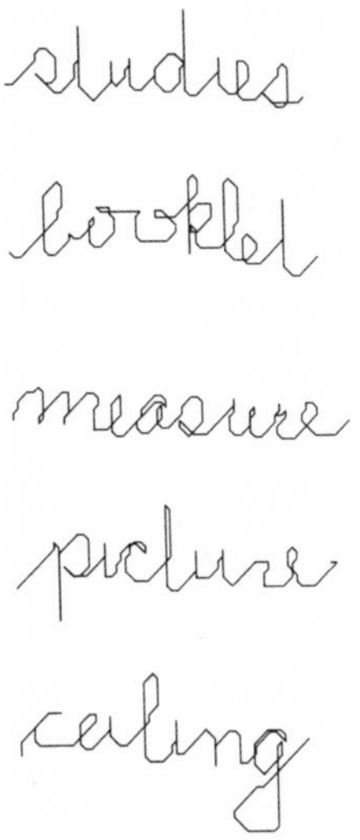

Figure 2

The individual letters were separated by an operator using an interactive graphics program. These letters then consisted of strings of octal digits whose lengths varied from 10 to about 70 characters in length. Because of the methods used, the crossings of the *t*'s and the dotting of the *i*'s was deleted since they did not necessarily follow the basic letter without other letters intervening. Thus, to keep the computer time down, only four letters were used in the test. The machine was asked to separate the *i*'s, *e*'s, *t*'s, and the *l*'s without the dots on the *i*'s and the crossings of the *t*'s. Because of Theorem 3, it was also decided to use only regular fractionally fuzzy grammars. The grammars listed in Figure 3 were generated by cut and try methods based on the following ideas.

Production Rule	g_e/h_e	g_i/h_i	g_l/h_l	g_t/h_t	Comments
S→0S	0/0	0/0	0/0	0/0	Allows horizontal initial stroke
S→1A	14/14	10/14	0/18	0/18	Beginning of a letter. Initial membership
S→2B	12/12	8/14	0/18	0/18	Ditto
A→0A	0/1	0/1	0/1	0/1	Small backtrack in direction (noise)
A→1A	0/0	0/0	0/0	0/0	Expected in allup strokes. No effect
A→2B	0/0	0/0	0/0	0/0	Ditto
A→2B	0/0	0/0	0/0	0/0	Ditto
A→3C	1/1	0/3	4/4	0/3	Top of a letter not pointed
A→4D	0/1	0/1	4/4	0/1	Ditto
A→5E	0/16	8/8	0/4	4/4	Top sharply pointed
A→6F	0/16	8/8	0/4	4/4	Ditto
B→0B	0/2	0/2	0/2	0/2	Noise on up stroke
B→1B	0/1	0/1	0/1	0/1	Noise on up stroke
B→2B	0/0	0/0	0/0	0/0	Expected on up stroke (no effect)
B→3C	1/1	0/3	4/4	0/3	Rounded top of a letter
B→3J	0/2	0/2	0/2	0/2	Noise sequence start
B→4D	0/0	0/0	4/4	0/0	Ditto
B→5E	0/5	5/5	3/3	5/5	Neutral top of a letter
B→6F	0/16	8/8	0/4	4/4	Pointed top of a letter
C→3C	2/2	0/7	4/4	0/7	Rounded top of a letter
C→4D	5/5	0/5	4/4	0/5	Ditto
C→5E	1/1	0/3	2/2	0/3	Neutral top of a letter
C→6F	0/3	3/3	1/1	3/3	Slightly pointed top
C→7G	0/7	6/6	0/3	6/6	Pointed top
D→4D	2/2	0/7	5/5	0/7	Very open loop in a letter
D→5E	1/1	0/3	2/2	0/3	Open loop or a highly slanted letter
D→6F	0/0	0/0	0/0	0/0	No effect
E→5E	0/1	0/1	2/2	0/1	Possible loop
E→6F	0/2	0/2	2/2	2/2	Expected part of down stroke
E→7F	0/1	0/1	0/1	0/1	Noise on down stroke
F→0H	0/0	0/0	0/0	0/0	End of a letter (tail)
F→0	0/0	0/0	0/0	0/0	Ditto
F→7G	0/2	0/2	2/2	2/2	Down stroke.
F→7	0/2	0/2	2/2	2/2	Down stroke. End of a letter
F→6F	0/1	0/2	2/2	2/2	Down stroke
F→6	0/2	0/2	2/2	2/2	Down stroke. End of a letter
F→5F	0/1	0/1	0/1	0/1	Noise on down stroke
G→0H	0/0	0/0	0/0	0/0	No effect (tail)
G→0	0/0	0/0	0/0	0/0	End of tail
G→7G	0/2	0/2	2/2	2/2	Down stroke
G→7	0/2	0/2	2/2	2/2	Ditto
G→6G	0/3	0/3	0/3	0/3	Noise
G→6	0/2	0/2	0/2	0/2	Noise or at end of a letter
H→0H	0/0	0/0	0/0	0/0	Tail of a letter (no effect)
H→0	0/0	0/0	0/0	0/0	Ditto
J→2B	0/0	0/0	0/0	0/0	Noise

Figure 3. The List of Production Rules of the Fractionally Fuzzy Grammar for the Experiment.

First, since all the letters under consideration started with a near horizontal, left to right stroke (octal direction 0) and continued in a counterclockwise direction (increasing octal direction) until returning to a near horizontal tail, the same set of production rules can be used for all classes. The productions used the non-terminal symbols A,B,...,G to represent the hightest octal direction thus far encountered, A representing 1, B representing 2, etc. From any non-terminal, only higher octal directions and higher non-terminal representations are reachable in the ideal case. However, to allow for noise in the less curved portions of the letters, we allow the terminal symbol generated to be one less than the highest thus far generated. A change of direction of more than 225 degrees counterclockwise was not allowed since this would never occur in these letters. The non-terminal symbol H was added to allow a tail of any length to be affixed to the ideal letter. The grammar was tested on a training set and was found to accept most of the strings. Minor modifications were made (e.g., J was added to the non-terminal alphabet to pick up an unusual noise condition) so that all strings in the training set were accepted. Now, came the task of generating the fractionally fuzzy membership functions. These were developed using the following criteria. First, a rule which could not help distinguish one class from another could be given the value 0/0 and would then have no effect on the final membership assuming some rule r, for which $h(r) \neq 0$, was also applied. Second, a rule for which h(r) was small would have little effect on the final membership of any string generated by that rule. Third, any rule for which h(r) was large would have a large effect on the final membership of any string generated by using that rule. Fourth, if rule r was used, the fuzzy membership of the string would be changed in the direction towards the value g(r)/h(r) by that application of rule r. Thus, if g(r)/h(r) was close to 1, the membership of the string would be increased and if g(r)/h(r) was close to 0, the membership of the string would be decreased. Finally, a rule which was used in all strings could be given a membership value which could serve as a starting point from which we could subtract by rules with $g(r)/h(r)=0$ and to which we could add by rules with $g(r)/h(r)=1$. In our case, there were two rules (second and third in Figure 3) one of which must be used in any valid derivation. Some comments are included in Figure 3 to give some insight into why the membership functions for that rule were chosen. Thus, the rule $B \rightarrow 6F$ is used when a vertical line changes direction abruptly from up to down. This would indicate a sharply pointed crown and the letters *i* and *t* are reinforced while *e* and *ℓ* are reduced in membership when this rule is used. After adjustment on the training set to allow a threshold of 0.5 or more to indicate in the class and less than 0.5 to indicate not in the class, the grammars were used on a random sampling of 121 letters from the remainder of patterns. The strings were parsed in a top down (left to right) manner by a program written in the SNOBOL 4 programming language. The results of this test are summarized in Figure 4.

CLASS	E	I	L	T
METHOD 1 % error	10%	16%	28%	74%
METHOD 2 % error	10%	4%	5%	27%

Figure 4. Results of the Experiment.

Two methods of categorizing were tested. The first classified the letter into any class for which the pattern had a fuzzy membership of 0.5 or more. This method left some letters unclassified and classified others into more than one class. The method was considered successful if the correct class was included possibly among other classes. This was done since a contextual post-processor could be used to find the correct letter. The second method classified the pattern into the class which had the highest fuzzy membership. As expected, the second method had better results, with 90% of the e's, 96% of the i's, 95% of the ℓ's and 73% of the t's correctly classified. The only distinction between a t and an ℓ is the width of the loop. Many of the t's were quite wide and were thus incorrectly classified as ℓ's. If the presence of one or more t's was detected by the presence or absence of a horizontal line written directly above some portion of the word, most of these incorrect classifications could be corrected by a contextual post-processor such as described by Ehrich (16). The distinction between the e's and ℓ's could have been improved if the data was prescaled to eliminate differences in the average height of the letters generated by the different subjects. All in all, considering the similarities in the four letters tested, the results are quite good.

V. CONCLUSION

In this paper we have presented a new type of fuzzy grammar, called the fractionally fuzzy grammar. It has been shown that fractionally fuzzy grammars are especially suitable to deal with pattern recognition problems, and a cursive script recognition experiment was given to demonstrate its capability. Further research along this line will be to establish an _algorithm_ for constructing the fractionally fuzzy grammars for a given training set in order to make them practical to solve pattern recognition problems.

REFERENCES

1. K. S. Fu and P. H. Swain, "On Syntactic Pattern Recognition", _Software Engineering_, Julius Tou (ed.), pp. 155-182, Academic Press, 1971.
2. P. H. Swain and K. S. Fu, "Stochastic Programmed Grammars for Syntactic Pattern Recognition", _Pattern Recognition_, Vol. 4, pp. 83-100, 1972.

3. W. G. Wee, "A Formulation of Fuzzy Automata and Its Application as a Model of Learning Systems", IEEE Trans. on Systems Science and Cybernetics, Vo. SSC-5, pp. 215-223, July 1969.
4. M. G. Thomason, "Finite Fuzzy Automata, Regular Fuzzy Languages, and Pattern Reocgnition", Pattern Recognition, Vol. 5, pp. 383-390, 1973.
5. S. Tamura and K. Tanaka, "Learning of Fuzzy Formal Language", IEEE Trans. on Systems, Man and Cybernetics, Vol. SMC-3, pp. 98-102, January, 1973.
6. K. S. Fu and T. J. Li, "On Stochastic Automata and Languages," Information Sciences, Vol. 1, pp. 403-419, 1969.
7. R. Kanst, "Finite State Probabilistic Languages", Information Sciences, Vol. 1, pp. 403-419, 1969.
8. E. T. Lee and L. A. Zadeh, "Note on Fuzzy Languages", Information Sciences, Vol. 1, pp. 421-434, 1969.
9. M. Mizumoto, J. Toyoda and K. Tanaka, "N-Fold Fuzzy Grammars", Information Sciences, Vol 5, pp. 25-43, 1973.
10. A. Kandel, "Codes Over Languages", IEEE Trans. on Systems, Man, and Cybernetics, Vol. SMC-4, pp. 135-138, January 1974.
11. L. A. Zadeh, "Fuzzy Sets", Information and Control, Vol. 8, pp. 338-353, 1965.
12. A. Salomaa, "On Grammars with Restricted Use of Productions", Ann. Acad. Sci. Fennicae, Vol. A.I. 454, pp. 3-32, November, 1969.
13. A. Salomaa, "On Some Families of Formal Languages", Ann. Acad. Sci. Fennicae, Vol A.I. 479, pp. 3-18, November, 1970.
14. S. Ginsburg,and E. H. Spanier, "Control Sets on Grammars", Math. Systems Theory, Vol. 2, pp. 159-177, 1968.
15. A. Aho and J. Ullman, The Theory of Parsing, Translation and Compiling, Vol. 1, Prentice-Hall, Englewood, N.J., 1972.
16. R. Ehrich, "A Contextual Post-Processor for Cursive Script Recognition", Proc. 1st International Joint Conf. on Pattern Recognition, pp. 169-171, October 30, 1973.
17. L. D. Earnest, "Machine Recognition of Cursive Writing", Information Processing (IFIP), C. M. Popplewell (ed.), North-Holland Amsterdam, 1962.
18. M. Eden, "Handwriting and Pattern Recognition", IRE Trans. on Information Theory, Vol. IT 8, pp. 160-166, 1962.
19. P. Mermelstein and M. Eden, "Experiments on Computer Recognition of Connected Handwritten Words", Information and Control, Vol. 7, pp. 255-270, 1964.

20. N. Lindgren, "Machine Recognition of Human Language, Part III--Cursive Script Recognition", IEEE Spectrum, Vol. 2, pp. 104-116, May, 1965.
21. K. M. Sayre, "Machine Recognition of Handwritten Words: A Project Report", Pattern Recognition, Vol. 5, pp. 213-228, 1973.

TOWARD INTEGRATED COGNITIVE SYSTEMS, WHICH MUST MAKE FUZZY DECISIONS ABOUT FUZZY PROBLEMS†

Leonard Uhr
Computer Sciences Department
The University of Wisconsin
Madison, Wisconsin 53706 U.S.A.

INTRODUCTION

When the separate cognitive processes of perception, thinking, remembering, language "understanding", acting, and learning are integrated into a single system, a variety of fuzzy problems inevitably arise. This paper examines such a computer-programmed system, one that is a first attempt to model the integrated, wholistic mind/brain.

The Search for Well-Formed Problems has Focussed "AI" Research on a Few Non-Fuzzy Problems

Artificial intelligence (AI) Research has to a great extent legislated that things be well-formed and non-fuzzy by developing separate systems to handle clear-cut problems with correct answers (e.g. the proof of a theorem, or the answer to a factual question) that can be deduced in a deterministic way. But as soon as we attempt the kinds of problems that human beings can handle (e.g. the description of a scene, or an on-going conversation in which several people exchange information) the system must constantly operate in a fuzzy domain.

AI research has attempted to reduce all cognitive

†This research has been partially supported by grants from the National Institute of Mental Health (MH-12266), the National Science Foundation (GJ-36312), (NGR-50-002-160) and the University of Wisconsin Graduate School.

processes to a search for a path from a set of Givens to a Goal, using only a set of legal Transforms. But it is rare that Givens, Goals, and Transforms are known (or, as in a game like chess or Go, easily computable). Rather, the crucial problem may be to find the set of transforms that humans, or some other intelligent entities use (as in pattern recognition), or to determine what are relevant givens (as in describing scenes or answering questions) or worthwhile goals (as in conversation, or in finding interesting new theorems).

Each area of AI uses slightly different basic transforms, overall networks, and search techniques. These are generalized in this paper, so that a single system (called a "SEER") can handle them all. This makes each separated function a good bit richer in its powers, and also a good bit fuzzier in the kinds of things that it does. It further raises new problems of fuzziness, at several ever-higher levels of integration of processes. Relevance must constantly be assessed, as a function of a wide variety of contextually interacting influences. The system must make fuzzy decisions as to the types of things perceived (e.g. objects vs. words), internal processes needed, and external acts suggested. And learning consists in a variety of types of fuzzy conjectures as to general hypotheses posited from particular pieces of experience, and how to use these to build, unbuild, and restructure the cognitive memory network.

The Ill-Formed and Fuzzy Nature of Everyday Thinking

Let's look at the kinds of simple everyday thinking that all of us spend most of our time doing. They are characterized by an intimate interaction between perceiving, deducing, searching, remembering, and acting, with constant monitoring and guidance from feedback. Rarely do we make deep or difficult deductions, or remember surprising or profound concepts. Rather, we conduct a kind of shallow and diffuse on-going "conversation" with our environment, one that finds and uses the relevant aspects of an impossibly large set of possibilities, in order to help us muddle along, often with surprising success.

Examples are the way we decide what to do on a holiday, and actually carry out all the steps needed to do it. Or choosing a dinner, or a restaurant; opening the refrigerator, getting food, baking a cake; deciding what to collect; building a bookcase.

It may feel like belaboring the obvious, but consider the subtle interactions between all the cognitive processes that go on in even the most mundane of acts:

I perceive a can which, in the context of the table it sits on and the conference I am attending, suggests it might contain a liquid to drink; so I glance about and move my head and then walk around it, to look for a spigot and an indicator - whether a sign or a telltale stain or smell - of whether it contains coffee. This arouses vague hunger needs for food, and I further look around for trays of soft objects and, remembering the time of day, for donuts. I sense that I like cream, deduce that it might be in a pitcher or, if powdered, a deep dish, and look about some more. I pick up a noticed cup, and move it under the spigot, pushing its lever with my other hand, and carefully monitor the drip. And so on.

Note that a 1-year-old infant engages in much the same interacting set of processes, albeit with fewer and simpler possibilities, e.g. when it babbles, cries, crawls, grabs, flails, pushes, bites, drools, etc., in order to get food.

But I must emphasize that despite its surface simplicity, such a process is far more difficult than anything we have approached with our artificial intelligence programs.

This paper describes a model for such processes, and shows some of their many inevitably fuzzy aspects.

ARTIFICIAL INTELLIGENCE HAS SIMPLIFIED THE PSYCHOLOGICAL PROCESSES TO PIECEMEAL FUNCTIONS FOR WELL-FORMED PATH-SEARCHING

Almost all AI research has concentrated on a single, separated cognitive function, and attempted to simplify the problem being attacked to the point where it is "well-formed."

The Separate Functions of Perception, Thinking, Remembering, Acting, Language, and Learning

The separate functions being attacked are closely related to, but usually simplifications of, the traditional cognitive processes that have always interested pshychologists (Figure 1):

Psychology	Artificial Intelligence		
	General Focus of Interest	Well-Formed	Ill-Formed and Fuzzy
Perception	Pattern Recognition	Name	Describe and Extract the relevant
Thinking	Deductive Problem-Solving	Path-find	Heuristics; interesting goals, everyday thinking
Remembering	Question-Answering	Parse, Path-find	Converse
Language-Understanding	Syntactic Analysis	Parse	Understand semantically
Motor Behavior	Robots	Touch, Move	"Conversational" interaction with environment
Learning	Learning	Add assertions, reweight	Induction, discovery, generalization

Traditional Problem-Areas of Psychology and Artificial Intelligence: Well-Formed and Fuzzy

Figure 1

Perception, which studies the absorbing and understanding of pertinent information from the cognitive system's external environment, has been attacked as the recognition and naming of an isolated patterned object input to the system (see e.g. Duda & Hart, 1973; Uhr, 1973a).

Thinking, which involves a variety of little-understood and only partially identified processes for deciding what are the most relevant and functional things to do in order to cope with external presses (perceived objects, imports, suggestions, commands, etc.) and internal presses (needs, desires, expectations, goals, etc.), has become deductive problem-solving in deterministic domains like games, puzzles, and logical systems (see e.g. Nilsson, 1971; Newell & Simon, 1973).

Remembering, which accesses information in the system's memory model of its world that is most relevant to its present situation and its attendant problems, becomes the search for "correct" answers to clear-cut questions that indeed have answers that can clearly be deduced to be correct (see e.g. Minsky, 1968; Simmons, 1965, 1970).

Acting, which includes a variety of things that the organism can choose to effect upon its world (e.g., touch, grasp, mix, melt, combine, heat, ingest; move its glance, eye, head, body), as a function of its percepts, thoughts and memories, becomes a very conventionalized set of actions, e.g., "move-self" or "push-object" (a specified distance in a specified direction) quite similar to the moves of the different pieces in a game like checkers or chess (see e.g. Nilsson, 1969; Uhr & Kochen, 1969; Winograd, 1971).

Language understanding, which is an especially mysterious process that involves symbolic reference to the import of percepts, feelings, ideas and acts, has been reduced to syntactic parsing and simple question-answering.

Learning, which must build the memory model and the system's ways of making use of that model in the first place, becomes largely inductive reweighting of the strength of already-existing connections (see e.g. Duda & Hart, 1973) and, to a slight extent, the adding of new connections and the extracting and inferring of new things to be connected (see e.g. Sauvain and Uhr, 1969; Quillian, 1969, Winston, 1970). But for work on learning that moves into ill-formed problems see Kochen, 1961; Uhr and Vossler, 1961; Hunt, 1962; Jordan, 1971; Uhr, 1964, 1973a).

The Deterministic Search for a Solution-Path Between Givens and Goal, Using Only Legal Transforms

Science must simplify, and this is especially true when it studies so complex a phenomenon as the intelligent mind. And the simplified problems that Artificial Intelligence examines are important, reasonably representative, and still extremely difficult. But there seems to be an increasingly strong tendency to simplify by choosing problems that are "well-formed" in the following sense: Three sets of things are specified when a problem is posed - "Givens" (e.g. the axioms of a system of logic, the starting board in a game), "Goals" (e.g. the theorem to be proved, the winning boards) and "Legal Transforms" (e.g. the rules of inference, the moves of the game). The problem then becomes one of finding a sequence of legal transforms that forms a "Solution Path" from givens to goals. Thus we have a deductive deterministic search for a legal path between two clearly specified sets of nodes in a graph.

This may be an adequate conceptualization (see Figure 2) for deductive problem-solving, including theorem-proving, game-playing, and puzzle-solving (see Nilsson, 1971) - and these have been the major areas of interest for researchers in "artificial intelligence." It is also adequate for syntactic parsing that insists upon a tree of paths connecting the given sentence with the goal "sentence" node that roots it in a parse (Chomsky, 1957, 1965; Feldman and Gries, 1968).

There is a great deal of effort today to absorb other cognitive processes into the same framework (or strait-jacket). "Syntactic Pattern recognition" (e.g. Narashimyan, 1964; Shaw,

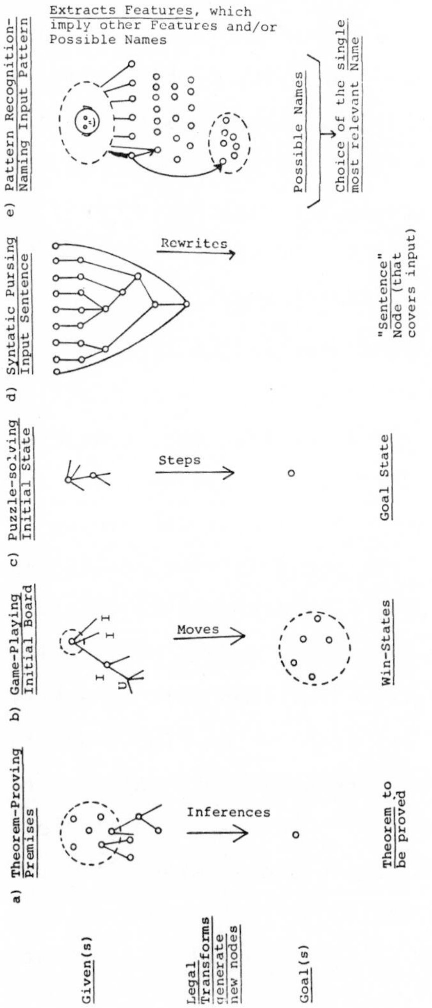

Well-Formed Problems of Deterministic Search for a Solution-Path; and One Ill-Formed and Fuzzy Problem

Figure 2

1969; see Swain & Fu, 1970; and Uhr, 1971) attempts to use parsing techniques to transform an input pattern into a well-formed tree, and "robot vision" (e.g. Guzman, 1968; Waltz, 1972; see Duda & Hart, 1973) attempts something quite similar, since it uses algorithms to match stored network models of objects with the input object. Question-answerers (e.g. Thompson, 1966; Shapiro, 1971; Quillian, 1969; see Simmons, 1965, 1970) typically put two such well-formed systems side by side: First a parser extracts information from the input query that is used to access the nodes in a memory network that the query is about. Then a deductive problem-solver searches for a path to answer nodes. Robot systems (e.g. Feldman et al., 1971; Nilsson, 1969; Winston, 1972; Winograd, 1971) interface four such systems, for vision, command-parsing, deductive problem-solving, and generating actions-sequences that will effect the deduced solution.

The Fuzzy Search for Relevance

Network models seem natural and attractive. The brain is a network of neurons connected at synapses that appear to compute complex threshold functions. The mind/brain as a cognitive network that models its world is an appealing conception that has been elaborated by Peirce (1931), Craik (1952) and many others. And almost all models of intelligence seem to be network models. (This may well be a trivial consequence of the fact that any complex function - and the functions that the intelligent mind/brain must compute are nothing if not complex - is best broken down into simpler steps, and any organization of these simple steps into the complex function forms a network.)

But there is far more to using and building networks

than the finding of legal solution paths between well-specified givens and goals. Rather than trying to simplify all our problems until they reduce to such a well-formed process, we should be examining the crucial - and much fuzzier - problems that arise. For example, in pattern recognition there are really no "legal transforms." Instead, any structuring of any set of feature-detectors or characterizers can be used to map the input pattern into its name (Figure 2e). The crucial problem for pattern recognition, and for perception in general, is to find from among the infinitely large set of possible transforms not the "correct" set but an adequate set, one that assigns names more or less the way we humans assign them, using our still-unknown set of transforms. In language processing the input is usually not a complete, perfectly grammatical sentence, and the sentence is always embedded in some larger perceived scene; yet we are able to extract its meaning and import, even though we cannot "parse" it. We typically must describe a scene, and converse about remembered concepts, rather than name objects and give correct answers (Figure 1).

In all these cases, givens, goals and transforms can only be inferred. Worse, there is no clear-cut search for a solution-path; rather, there is an interacting set of searches for relevant nodes and nodes relevant to these, etc.

Even deductive problem-solving, which gives us our only well-formed paradigms, is basically fuzzy. For it is only when we are laboring in an idealized world that a theorem-goal or a game-win-state is posed. And even there we must use "heuristics" to try to direct what now becomes an essentially fuzzy search. In everyday thinking, and in the logician's and mathematician's creative work, the goal to be reached or the theorem to be proved must be judged "valuable" or "interesting" - as important and relevant enough to be worth achieving. And in general the goals of everyday thinking must be achieved by subtle and complex assessments of "relevance."

A BRIEF DESCRIPTION OF AN INTEGRATED WHOLISTIC COGNITIVE SYSTEM (SEER)

Today's robots lash together several separated systems for relatively well-formed processes. In sharp contrast, I have been trying to develop cognitive "SEER" (Semantic Learner*) systems that do a variety of cognitive processes in as integrated and wholistic a manner as possible, using as simple and general a structure as possible (Uhr, 1973b,d, 1974). This seems to me mandatory from the points of view of scientific model-building, whose canons include simplicity and elegance as well as power and fruitfulness, and of evolution-learning, where each new (and inevitably small) change must be functional and serve some purpose.

General Transforms, to Give a Unified Memory Structure

Such an integrated system needs a unified memory structure that is built up from a single general kind of transform, one that can perform the entire gamut of cognitive functions. Figure 3a surveys the kinds of transforms that have been traditionally used in separate AI systems. A parser makes the built-in assumption that nodes are concatenated (i.e., touching, in order). A deductive system has built into it the specific relation among nodes - usually co-occurrence, or ordered. Transforms for associative memory searches and feature extraction imply a whole set of unordered

*or Sensed Environment Encoder, Recognizer and Responder

a) Specific Types of Transforms Used in Artificial Intelligence Systems

i) Parsing a sentence string

$\text{Node}_1\ \text{Node}_2 \ldots \text{Node}_N \Rightarrow$ Replacement

e.g.: Article Adjective Noun ⇒ Nounphrase

ii) Deducing a move, inference or other transform

$\text{Subexpression}_1\ \text{Subexpression}_2 \ldots \text{Subexpression}_n$

⇒ Transform

e.g.: ... P + Q ... ⇒ ... Q + P ...

iii) Searching through an associative memory net

$\text{Node}_1 \Rightarrow \text{Node}_{11}, \text{Node}_{12} \ldots \text{Node}_{IN}$

e.g.: Robin ⇒ Bird, Fly, Red-Breast, Animal, Harbinger-of-spring

iv) Extraction of Simple Features for Pattern Recognition

$\text{Feature} \Rightarrow \text{Name}_1, \text{Name}_2 \ldots \text{Name}_N$

e.g.: Small-closed-loop ⇒ P, B, R, Eye, Dumbbell, Scissors, Eyeglasses

b) A General Characterizing and Compounding Transform

$\text{Relation}_1(\text{Feature}_{11}, \text{Feature}_{12}, \ldots \text{Feature}_{1N}),$
$\text{Relation}_2(\ldots \Rightarrow$ (Names), (Compounds), (Characterizers)

e.g.: Top-right (vertical, closed-loop), Bottom-right (Vertical, closed-loop) ⇒ B, Dumbbell, Eyeglasses, Characterizers (Face)

(NOTE that weights, fuzzy values, relative locations and other attributes will also be expressed in the actual characterizing transform)

A General Transform, and the Ways it Handles Parsing, Deducing, Searching, Feature-Extraction, Characterizing and Compounding

Figure 3

things (the nearby nodes in memory, the possible names to be assigned). Often they have weights or other kinds of fuzzy values associated with them. The configurational characterizer used by SEERs (Figure 3b) is general enough, since relations, weights, and other attributes can be expressed explicitly, to represent all of these, (see the memory network in Appendix B for examples). Even more important, the assignment of fuzzy values to things, features, relations and implications which is essential for fuzzy problems also allows for a natural deepening of the methods for handling problems that have been treated as though well-formed. For examples, the heuristic search for a solution path can be directed by fuzzy inferences and fuzzy contexts; parsing need not insist upon a perfectly grammatical and noise-free sentence.

Interactive Merging of Implications Across Processes

Intimate interaction among the various processes is achieved by using these general transforms and by merging the implications of transforms into a very small number of common lists. Any transform can imply implications into any of these lists and any transform can require any number of conditions. This means that any kind of contextual interaction can occur. For example, a simple feature like a curve might imply that additional curve-features be looked for, to try to build a closed loop, and also that a face be looked for, which in turn implies looking for its other features (e.g. nose, hair, ears). Similarly, an internal need, e.g. for food, might imply particular food objects which in turn imply specific features that would characterize them.

Overall Architecture, Processes and Behavior

SEERs are memory networks that model their "world" (including themselves) in a usable fashion. They structure transforms into several major sub-systems, for a) perceptual recognition, b) thinking, including both deduction and association, and c) generation of actions-sequences. Several successively more powerful SEERs have been coded (see Uhr, 1974). Figure 4 sketches the structure of SEER-2, for 2-dimensional environments.

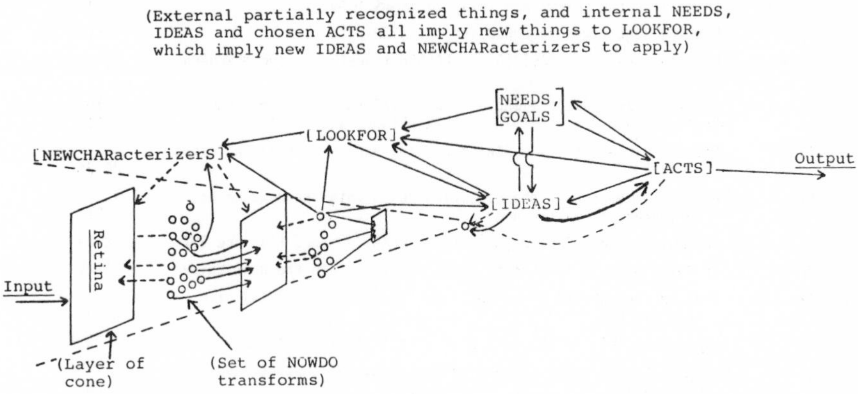

Overall Architecture of SEER-2, a First Attempt at an Integrated, Wholistic Cognitive System
Figure 4

Information flows into the system from the external environment, which impinges on the retina, and starts an outer-inner flow of processes. Successive sets of characterizers (called NOWDO in the SEER program) transform and coalesce information back into the next layer of the cone. But internal NEEDs and GOALs are simultaneously impelling inner-directed flows of processes. Similarly, partially recognized things can imply additional things to LOOKFOR and NEW CHARacterizerS to apply, in glancing about to gather information. Perception is thus a complex back-and forth many-layered parallel-serial process. "Thinking" consists in the serial application of transforms from the IDEAS list, to make associative memory searches and simple deduction. A certain amount of direction is got by having the system choose the single most highly implied transform on the IDEAS list to apply next. Thus the strengths of implications, that is, of connections between one node and another (where a node can be a transforming procedure, as well as a representation of an object, class, attribute or compound) serve to heuristically direct

TREPEAT	Initialize the LAYERS and CHARacterizerS TODO	10
T6	Get the next layer to NOWDO and its STEP-size	11
	Put the ERASER, NORMALIZER and PASSON transforms, and NEWCHARS, onto NOWDO	14
T5	Get the next TRANSform and attendant information from NOWDO.	15
	Get the bounds for applying the TRANSform, and its TYPE	16-18
T3	Get the pieces of its DESCRiption, and handle them as indicated by TYPE	20-24
A1	MERGE all members of the specified CLASS into the next Layer, to average or difference.	21
T1	Accumulate the TOTAL weights of the CLASS members that are GOT	22-24
T2	If TOTAL exceeds THRESHold, MERGE IMPLIEDS into the next Layer	25-26
A2	Iterate through the array until the bounds (CMAX, RMAX) have been reached.	27-28
E1	erase each cell in the next Layer, to initialize it.	29
N1	NORMALIZE each cell, to keep weights roughly constant despite convergence.	30
ITER	Go to process the next Layer, shrinking Row and Column by STEP-size, until the apex is reached	31-33 34
THINK	Initialize and go through up to 7 CYCLES	35-37
	Get NEWCHARS that POINTAT LOOKFOR and set them on IDEAS	38-40
	CHOOSE the most highly weighted THING on IDEAS, up to 100	41-45

(Cycles through IDEAS until an ACT is the most highly weighted and therefore chosen)

I	If TYPE is I, this is an Internal transform to apply to FOUND (the apex). If the TOTAL weight of CLASS members GOT exceeds THRESHold, MERGE IMPLIEDS.	46-54
ACT	If TYPE is ACT, get the next ACTION and its ARGumentS and other ARGuments from HISTory	55-58
OUT	output the result of the act (after having completed it with routines below)	59-60
SEARCH	If the act couldn't be completed, set up a SEARCH for things that POINTAT the NEEDED PARTS, and return to apply more IDEAS.	61-66
FAIL	output that have "FAILED" to execute the chosen act, and go to the next input.	67

(The routines for the different types of acts follow.)
(Note how Describe uses name, and Move uses Find.)

D	Describe the scene, giving the objects and their parts	68-76
T	Name the single most highly weighted thing of the class specified in ARGument.	69-71
F	Find the first THING that is a member of the specified ARGument and bracket it.	77-79
M	Move all Found THINGs FROM or TO (as indicated) the TARGET thing	77-85
R	Reply to a query by CHOOSEing all THINGs belonging to the specified ARGument whose weight exceeds half the MAXWeighTed THING (i.e. the first) chosen	86-90
SEARCHR	If no Reply was chosen, MERGE associations from what already found into IDEAS and return to think some more, by applying more IDEAS.	91
C	Compute, using the specified operators (ADD, subtract and divide are shown)	92-99
SEARCHC	If the needed numbers were not recognized, return to search for INTEGERS	96
ADD	ADD, subtract or divide, as indicated by computation commanded	97-99
G	In a Game, make a move by replacing the old board configuration with the move	100-104

(The following are the functions used by the main program.)

the processes of thinking. When an act is chosen (because it is the most highly implied thing on IDEAS) the system will start generating the specific actions-sequence needed to carry out that act. This will usually entail further calls for needed objects to LOOKFOR and transforming IDEAS to remember and deduce what might be usable objects, which are then looked for.

SEER thus carries out a rather complex set of parallel and serial operations. We might make the loose metaphorical comment that it widens and narrows its "conscious" "attention" as a function of its problems, tending to be more parallel in its perceptual processes, more serial in its central cognitive processes. This means that (serial) time must pass, and a new program (SEER-T) handles situations in which it interacts over time with a changing environment (see Uhr, 1973c).

For fuller descriptions, see Uhr, 1974, and in preparation.

Overview of the SEER-2 Program, and its Flow of Processes

The following is a succinct description of SEER-2. The actual program is given in the Appendix, along with a brief description of the programming language, EASEy (Uhr, 1973f), an English-like variant of SNOBOL (Griswold et al., 1968) in which it is coded. Note that caps, underlines and numbers refer to the program.

(Overview of the SEER-2 program.		Program Statement No.
START	Initialize Memory (Includes type-name to refer to lists and the Transforms repeatedly used - ERASER, NORMALIZE, PASSON.	
IN	input Memory, including NEW lists and ADDitions (The MEMory conversion program, which inputs the transforms needed (Appendix B) goes here.) (The environments - scenes and problems - to be sensed are also input here.)	1
INITialize	for the next input problem, by erasing all temporary lists.	
SENSE	Store the input as an array, in the first Layer of the recognition cone.	2-7
TRANSform	MERGE the things that POINTAT NEEDSGOALS and LOOKFOR into NEWCHARacterizers	8,9,12

MERGE	Get the THINGs on LISTA that belong to the specified CLASS and MERGE them, combining TOTAL WeighTs and HISTories, into LISTB.	ME1-11
POINTAT	Get the TOTHINGS that POINTAT each THING specified, and MERGE them into the lists specified.	PA1-4
NORMALIZE	Divide the WeighT of each thing TONORMalize by NORM, erasing it if WT is below 1.	NO1-5
ABS	Get the ABSsolute value of the argument (fails if the argument is not an integer).	ABS1-2
CHOOSE	CHOOSEs the MAXimum (or if TYPE specifies MIN the MINimum weighted thing in LISTA of the specified CLASS, getting the THING, TOTAL, LOC and HIST.	CH1-13

Some Simple Examples of the Variety of Behavior of SEER-2

SEER must be given a set of transforms, which form the system's "cognitive memory network model" of its world. Future systems will learn these transforms from experience. But for now we can start the system with a set of transforms (see Appendix B) already in its memory, and immediately begin to examine its behavior. The following examples show the kinds of problems that SEER can handle:

Naming Objects and Describing Scenes. When input an array of information that contains one or more objects, and also verbal statements, like "NAME THIS" or "WHAT IS THIS" or "DESCRIBE", SEER will successively apply feature-extracting transforms, and then configurational characterizers that were implied by these prior transforms, and also be tentatively implied things that would be further confirmed or denied by these transforms. This process continues until high-level compounds, like words, phrases, objects, and collections of objects are got. A recognized command or suggestion will imply whether the system should name or describe (or do something else), and the system will choose a particular type of act as a function of such perceived utterances, and also any internal needs and presses.

Thus when given the inputs shown in Figure 5a, SEER-2 will output:

(1) "CHAIR" (3) "TABLE"

(2) "FACE" (4) "FACE"

When given the inputs shown in Figure 5b, SEER-2 will output:

(1) "CHAIR (WITH BACK; SEAT; LEGS;)"

(2) "FACE (WITH VISAGE; EYE; EYE; NOSE; MOUTH; LEFTEAR;)"

(3) "BALLOON (WITH CIRCLE; STICK;) FACE (WITH EYE; NOSE; RIGHTEAR;)"

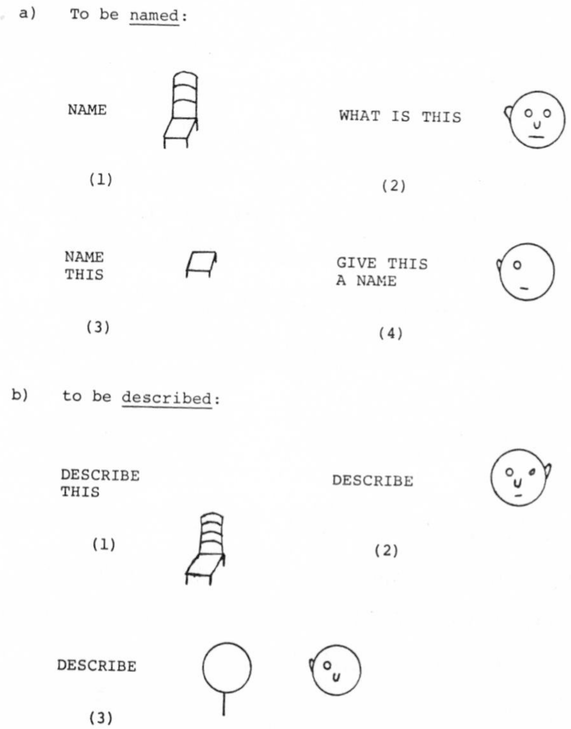

Simple Scenes of Mixed Objects and Words
Figure 5

See Uhr, 1973e for a similar system that gives a wider variety of descriptive information, where a stylized interaction allows the human recipient to direct the description.

For simplicity the rest of the examples of SEER's

behavior will be shown for 1-dimensional inputs, which can be handled by either SEER-1 or SEER-2 (see Uhr, 1974, for details). The first stages of the perceptual process handled by SEER-2's recognition cone become trivial, and a far simpler memory network is needed. For these examples we will use the convention @(thing name) to indicate objects, object parts and qualities (e.g. @PEAR is a symbol for an object whose name is "pear"), as a shorthand for the actual picture.

When input:
"NAME THIS @PEAR" or
"WHAT IS @PEAR THIS"
SEER will respond:
"PEAR"

When input a set of lower-level qualities, e.g.:
"SAY WHAT @STEM @YELLOW @TEARDROP YOU SEE" and
"@OVAL @STEM @RED WHAT DO YOU SEE"
SEER will respond:
"PEAR" AND
"APPLE" in turn.

Retrieving Information, Answering Questions, and Conversing.

If the input scene contains a question like:
"NAME THE FIRST PRESIDENT" or
"WHAT DOES WISCONSIN PRODUCE"
SEER will output:
"WASHINGTON;" and
"BEER; CHEESE; GARBAGE;"

Verbal inputs that are not simple questions, but rather act in a more conversational way, will lead to more variable, but more or less relevant, responses.

Deducing responses. Transforms can similarly lead to

simple deductions, e.g. as to arithmetic, logic, or the moves (not necessarily brilliant, or even good) in a game. For example,

"ADD 3 + 2" will give:

"5"

This result is achieved by first recognizing the parts (letters, words, phrases), which triggers the following of the command "ADD," which points to a transform that combines the numbers to be added according to the rules of addition.

Finding and Manipulating Objects, Driven by External and/or Interval Presses. When it recognizes some part of the external scene as a command or suggestion (e.g. "TOUCH" or "FIND" or "WHY DON'T YOU TOUCH"), or some internal need as a press (e.g. a high level of HUNGER will imply EAT any perceived food object), it may choose to act (e.g. touch the indicated object). Note that "touching" is simulated inside the computer, by the placing of agreed-upon symbols around the touched objects. The scene's mixture of words and objects remains, simulating a static visual scene with written words.

Thus a command like:

"TOUCH THE @BOX PAIL @PEAR @PAIL"

(where @object (e.g. @BOX) indicates the object, not the word) will lead to the result:

"A PAIL IS FOUND-" "TOUCH THE @BOX PAIL @PEAR @PAIL:"

(colons indicate the act that found and touched).

A command like:

"MOVE THE @PEAR @BOX FRUIT @APPLE TO THE @PAIL PAIL"

will lead to the result:

"A PEAR IS FOUND-"

"MOVE THE @BOX FRUIT @APPLE TO THE (@PAIL @PEAR) PAIL"

(where parentheses indicate that the objects within

them have been moved together).

An act can also be implied from internal needs as well as external objects. E.g.

@OBX @PAIL @APPLE @SEER-MOUTH" (along with an internal need-state of hunger)

will lead to the response:

"A APPLE IS FOUND-"

"@BOX @PAIL (@SEER-MOUTH @APPLE)"

because the act of putting the @APPLE to the MOUTH is now highly implied, by both the hunger need and the apple itself.

When conflicting commands and presses are perceived, SEER will make a fuzzy choice, not necessarily a "correct" choice, among them.

Choosing Among Different Types of Behavior

These examples illustrate the range of problems that SEER can handle. But they make little use of fuzzy values and give very little feeling for the complex set of fuzzy choices it must make when the possible set of alternatives grows larger and more contextual determinants become relevant. With simple problems and appropriate transforms there is little chance for ambiguity. But when there are many distortions of possible objects in a scene and much information in memory, decisions become multi-determined. This is handled by the use of fuzzy implications which are merged together into common lists where choices as to what to do next are constantly being made. Thus a variety of partial implications, as from different characteristics of a scene, and also from different sources of information such as externally sensed scenes and internally felt needs all merge together, are chosen among and determine when to choose.

It is not clear how well such a system will work as the size and complexity of its problems increase. The only way to find out is to test it on much larger sets of more difficult problems. This is a problem common to all AI systems. But we need not expect it to be perfect, or even to be exceptionally good - just as people do, it can exhibit its share of mistaken, stupid and rigid behavior.

HIGHER-LEVEL FUZZY PROBLEMS

The combining of functions into a single cognitive system forces us to eliminate any rigid flow of processes. The system must now choose among alternate possibilities, and these choices confront it with inevitably fuzzy situations, including the following (see Uhr, in preparation).

Relevance Must Constantly be Assessed, in an On-Going Conversational Description

Nothing is built into the system that compels it to do something like "assign a single name"; rather, the system must develop a relevant description of its environment, in the sense that it must notice those things that will help it do what it decides to do. But "description" is an extremely fuzzy concept. We can think of a "complete description," which is far too long to be of use, or a conventionalized description, which would make such a system impossibly rigid (see Uhr, 1973c,e for examinations of "description"). Instead, the system must develop a pertinent description, as a function of external and internal presses, and of acts it has decided to try to effect, and of deductions and memories that suggest what would be helpful and relevant parts of descriptions, toward effecting those acts.

Nothing can be built-in that says "find a path from this to that," or "choose the single most highly implied name." Rather, the system must constantly assess the relevance of nodes, transforms, and flows of procedures. Nor will it be clear how to assess relevance, which will be a fuzzy function of a variety of things. It seems best to think of the system as engaging in an on-going "conversation with its external world" (within which may be objects and other systems that converse verbally). There will rarely be a "correct response," and the system will often behave in a mediocre way. But it will always be trying to act relevantly and, especially with the aid of friendly elements in its environment (e.g. parents, teachers, ripe berries and other manna from heaven), will often muddle through, and even act reasonably well.

Relevant Imports Must be Got from Mixed Words and Things

Words, phrases, statements, suggestions, questions, commands, and any kind of verbal utterance must be input through the same perceptual channels that sense, recognize and understand objects and their relations and qualitites. There may indeed be several input channels, such as the two eyes, and also sensors for sound, touch, smell, or other type of impinging energy. But none can be exclusively reserved for verbal inputs, in the way that the robot's teletype inputs are known by the robot to be verbal commands, as opposed to its television inputs, which are sensed scenes of objects. Rather, the system must recognize the relevent things in its input scenes, and further recognize which combine into words and symbolic referential utterances, and what are the things to which they refer (Uhr, 1973b,d).

The System Must Choose the Type of Act to Effect

At the lowest level, any AI system must make fuzzy choices, using "heuristics" or "characterizers," in order to search through an overly-large network of possibilities. Unless the problem has been cut down to uninteresting toy size it cannot try everything.

But once we begin to ask a single system to handle a variety of different types of thing we force it to make higher-level fuzzy decisions. It must decide what is the appropriate type of act. E.g. it must decide whether to name, describe, draw, or touch an object; whether to treat an input recognized as a verbal utterance as a command to be followed or a statement to be responded to. This may entail several levels of fuzzy decisions. E.g. after it decides to name it must further decide whether to speak, write or print the name, and which language, e.g. French, English or American and, sometimes, which of several possible synonymous names, to use.

Certain Inputs Must be Recognized as Feedback

Just as there can be no special built-in channel for verbal utterances, there can be no special channel or signal for feedback. Rather, the system must recognize that something perceived (e.g. candy, a smile) or felt internally (e.g. pain) is feedback that refers to some previously perceived stimuli and consequent actions by the system. This entails a complex combination of hypotheses the system makes as to expected feedback consequences, which focus its attention for confirming or denying evidence, and also an ability to relate any identified feedback, whether anticipated or unanticipated, to percepts, thoughts and acts that the system must once again fuzzily infer are relevant.

The Learning of Things, Transforms and Hypotheses is Essentially Fuzzy

Learning is fuzzy in several ways.

First, the system must generalize from one, or at most a relatively small number of experiences. Such generalizations are guaranteed to be wrong a good deal if not most of the time, since the world is wondrously complex, diverse, and accidental. So the system must make an on-going experimental assessment of each tentatively-learned thing.

Second, it must generate new hypotheses in the first place, whether from experience, or by combining, refining, or in some other ways restructuring previously entertained hypotheses. Once again, there is no assurance that any such hypothesis will prove to be correct. Rather, the system must accrue evidence, through future experience, for each one of them, sifting and choosing among them as this evidence confirms and denies.

Third, a complex structure of hypotheses must be built. Once again, we have essentially fuzzy decisions as to which nodes to connect to which - e.g. what things to put into what classes, or compound into higher-level structures.

THE STEP-BY-STEP DEVELOPMENT OF MODELS OF INTELLIGENT MIND/BRAINS

A number of very difficult steps must be taken before we can hope to achieve intelligent systems. First, we must model the separate intellectual functions - perception, thinking, remembering, acting, language, and learning. But because these are so interdependent, we cannot expect to model them separately; rather, we must combine them into whole integrated systems. This paper describes a first attempt at such a system, one that begins to do a variety of cognitive tasks, where the different subsystems must interact smoothly and coherently. Such a system is just a beginning, and must be extended in a variety of ways.

It must be made to handle more things. To some extent this can be done by giving it more transforms, that is, by giving it a bigger memory network. At that point extensive tests must be made, since its behavior will be far too complex to predict. This further means that we must develop some conception about the range of behavior that such a system must model, so that we can say something meaningful as to how well we are sampling, and examining. We still need to develop the basic canons for experimental test and evaluation in this new science of complex entities.

Even more important, the system must be made more powerful, with more powerful transforms and overall structure.

At the same time, it must be simplified. We must strive for the simplest possible system, at the same time that we strive for the most powerful possible system. Simplicity is not only desirable from the point of view of efficiency and of the canons for building good models, but also because the simpler the system the more likely it may resemble living systems that have evolved under nature's canons of simplicity.

In addition to generality, integration, power and simplicity, we must also worry about the fit of the model's behavior. At first we can be quite satisfied with rather general fits, for we are modelling such a wide domain of behavior - describing the whole elephant, rather than a few hairs in the left fore-legpit. But at some point - I think only after we have developed models that exhibit a good deal of generality and power - we must begin to worry about fitting details. This completes the hypothetico-deductive enterprise of predicting human behavior, comparing with experimental results and thus testing our model, and hopefully, finding new disconfirming evidence that leads to changes that improve the model.

Finally, learning must be added. We can never hope to pre-program into such a system all the knowledge that it might need about the external world with which it must interact, that it must know about. In fact this is impossible in principle, because that external world is itself open-ended and constantly changing: New things, organisms and mutants are born, new words, concepts and other man-made things (e.g. bicycles, transistors, computers, poems) are created.

SUMMARY AND CONCLUSIONS

It is attractive to model the mind/brain as a network of neuron-like threshold elements that itself serves as a "cognitive model" of its world, including itself. Such a model must have a general kind of transform capability, one that can handle the specific types of transforms used in the typically separated Artificial Intelligence systems for perception, deductive problem-solving, language handling, remembering, acting and learning. And the different cognitive functions must be able to interact intimately, forming a well-integrated wholistic system with rich contextual influences on a constant stream of interacting decisions.

This paper examines a first step toward such a system, a programmed model called SEER that attempts to handle the various cognitive functions in as simple and integrated a way as possible.

A network model is general in that it is simply a

framework for computing structures of functions of any sort. But today's AI research has over-narrowed the use of networks to the point where they make a simple well-formed search for a path, using legal transforms (e.g. rules of inference, moves in a game) from a set of givens (e.g. premises, the initial board of the game) to a goal (e.g. a theorem, a win). It may be possible to handle a few cognitive tasks in such a well-formed way. But even that seems unlikely, for when the problem of path-searching becomes difficult (as it does the minute the problem becomes difficult enough to be interesting) fuzzy and conjectural "heuristics" must be used as hunches to guide the search.

Most cognitive problems are ill-formed and fuzzy. The mathematician doesn't prove theorems posed to him in an MIT exam; rather he must find theorems worthy of proof. A perceiver does not name single isolated easily discriminable objects (and even that is a fuzzy problem); rather he makes note of the relevant aspects of a scene of interacting objects. We cannot expect to be given only simple verbal questions, for which there is a "correct" answer; rather, we must usually engage in a conversation, to which we respond with hopefully relevant comments.

In just about every separated area of cognition the real problems are ill-formed and fuzzy, and it is over-simplifying to reduce them to the point where they can be handled by a search for a solution path.

Even more important, the moment we ask a system to handle a variety of cognitive functions at the same time - which is the typical process for human adults, and even for infants and higher animals, e.g. when we find, stalk, capture and prepare food - we force it to make a constant interacting stream of fuzzy decisions. Externally perceived objects and internally perceived needs fuzzily suggest acts that might be effected and memories and deductions that might help carry out these acts. Conversely, remembered objects, qualities, and procedures suggest objects that should be looked for and actions (e.g. glancing about, crawling around) that might help in gathering more information and testing the potential value of possible acts. In general, each cognitive function calls upon and affects all the others. There is no "correct" sequence of procedures. Rather, multi-determined fuzzy decisions are constantly being made to guide a fuzzy interacting set of sets of fuzzy processes.

Finally, several higher-level fuzzy problems emerge. Relevance must constantly be assessed. All perceived things are mixed together in the sensory input channels, and must be sorted out. Thus verbal utterances must be recognized as structures over perceived things, and then as having symbolic referential import. Feedback must also be recognized as feedback, and as probably relevant to fuzzily conjectured previous stimuli, acts, and hypotheses. And learning must make fuzzy decisions as to what to learn, and how to build, unbuild, or restructure the cognitive network memory.

APPENDIX A: THE SEER-2 PROGRAM

The following program (coded in EASEy, see Appendix C and Uhr, 1973f, and therefore able to run on any computer that has a translator for SNOBOL4, see Griswold et al., 1968) handles the examples given in the paper, among many others. It also handles far more complex problems e.g. perception of scenes of distorted objects, since it merges together multiple fuzzy implications, and chooses among them.

AI programs are too complex to describe fully, accurately

and fairly. The typical paper usually doesn't give many details, but rather extols a program's virtues. But we can not really begin to understand one another's programs, so that we can begin to borrow from and build upon one another, until we can observe them clearly. Think what monographs on mathematics would be without proofs, or books on architecture without pictures and diagrams.

The EASEy programming language was developed as a first attempt to bridge this communication gap. A program coded in EASEy is still cumbersome and hard to read and understand. It is like a complicated and messy proof in a peculiar and fuzzy notation. But it is precise and complete. For the reader who wants to dig in and see exactly what is happening, it is the thing itself. This is still difficult - but because of the intrinsic complexity of the model, not the peculiarities of the programming language.

Following the program is a set of characterizing transforms sufficient to handle all the examples given in both this paper and Uhr (1974), along with many others. The overview and description given in the paper should help when digging into the program.

```
(Program SEER-2)                                                        RECODES-1       SEER-2
                                                                        (see Uhr, 1973e)


START    set  ERASER = 'E 0 0 R C E '
              NORMALIZE = 'N 0 0 R C N '
              PASSON = 'P 0 0 R C P1 '
              P1 = 'A%D=  0 0 ]%I=%'
         set  SPOTSIZE = 1
              F = 'FOUND'
              N = 'NEWCHARS'
              L = 'LOOKFOR'
              I = 'IDEAS'
(INput and go to TYPE (initialize memory, SENSE or TRANSform)
IN       input TYPE DESCR % [ +to $('M' TYPE -end) ]                         1              1
(MEMory  input and format routine goes here-only the beginning is shown.    .1              M
(input      NEW lists or ADDed information on old lists
MNEW     from DESCR get NAME =
         set  $NAME = CONTENTS [to IN]                                      .2              M
MADD     from DESCR get NAME =                                              .3              M
         on $NAME set CONTENTS [IN]                                         .4              M
MINIT    erase R, L, EXTERNAL, TOOUT, NEWCHARS, LOOKFOR, IDEAS, ACTIVEWT   M(N+2)           M
(Input and SENSE a new scene into Layer 1 (the retina)
MSENSE   erase C                                                             2              2
         on EXTERNAL list DESCR                                             .1              3
S1       from DESCR get and call SPOTSIZE symbols SPOT erase [-to S2]      3.V              4
         list  $(L;R;C) = 'BRIGHT QUAL :BRIGHT ' SPOT ]                    4.V              5
```

```
                                                                              RECODES-1   SEER-2
         C = C + 1 [to S1]                                                      5          6
S2       R = R + 1 [to IN]                                                      6          7
MTRANS   output DESCR ' SCENE HAS BEEN INPUT, IS BEING TRANSFORMED'             7          8
(NEEDS and GOALS imply things to LOOKFOR that POINTAT them
         MERGE(POINTAT(NEEDSGOALS),L)                                           .1         9
TREPEAT  TODO = LAYERS CHARS                                                    8          10
T6       from TODO get STEP NOWDO % = [-TREPEAT ]                               9          11
(LOOKFOR implies NEWCHARacterizers that point at them.
         MERGE(POINTAT(LOOKFOR),N)                                              .1         12
         erase LOOKFOR                                                          .2         13
(To ERASE next LAYER,NORMALIZE, apply NEWly implied CHARacterizers, and PASSON
(found things
   at start of NOWDO set ':ERASER X ]:NORMALIZE X ]' NEWCHARS ':PASSON X ]'     10.V       14
(Get each TRANSform from NOWDO (to be applied at this layer)
T5       from NOWDO get CLASSES : TRANS WT HIST ]= [-ITER]                      11.V       15
         from $TRANS get RA CAA RMAX CMAX DO                                    12         16
T7       CA = CAA                                                               13         17
T4       from $DO get TYPE THRESH '%D=' DESCR '%I=' IMPLIEDS %                  14.V       18
         erase GOT TOTAL                                                        15         19
T3       from DESCR get : CLASS TYAL DR DC ]= [ +$(TYPE L) - ($(TYPE 2) ]       16.V       20
(TYPE A transform Averages or differences.
A1       MERGE($(L;RA+DR;CA+DC),'$(L+1;RA/STEP; CA/STEP),,TVAL,CLASS) [T3]      17.V       21

(TYPE I transform characterizes
I1       from $(L;RA+DR;CA+DC) get # that CLASS # REST : THING VAL [-T3]        18.V       22
         is TVAL lessthan VAL ? yes. TOTAL = TOTAL + 10 [T3 ]                   19.V       23
         on GOT list CLASS THING  [T3]                                          20.V       24
(the characterizer succeeds if the TOTAL weight reaches THRESHold.
I2       is THRESH greaterthan TOTAL ? [ + A2 ]                                 21.V       25
         MERGE(IMPLIEDS, '$(L+1;RA/STEP;CA/STEP)', GOT)                         22         26
(keeps applying this TRANSform till its upper bounds are reached.
A2       is CA lessthan $CMAX? yes - CA = CA + 1  [+T4 ]                        23         27
         is RA lessthan $RMAX? yes - RA = RA + 1  [+T7-T5]                      24         28
(Erases the next layer, to re-initialize.
E1       erase $(L+1; RA/STEP ; CA/STEP) [A2]                                   25         29
(Normalizes, dividing by 5 - roughly assuming a STEP-size of 2 plus a bit more.
N1       $(L;RA;CA) = NORMALIZE($(L;RA;CA),5) [ A2 ]                            .1         30
(Converge to the next layer.
ITER     L = L + 1                                                              26         31
         R = R / STEP                                                           27         32
         C = C / STEP                                                           28         33
(If the apex has been reached, start to "THINK" (apply IDEAS)
         is R lessthan 1 ? is C lessthan 1 ?[ -T6]                              29         34

(CYCLES 7 times applying 100 transform from IDEAS to FOUND (the apex)
THINK    CYCLES = 7                                                                        35
TCYCLE   CYCLES = CYCLES - 1                                                               36
         is CYCLES lessthan 1 ? [ + FAIL ]                                                 37
         MERGE(POINTAT(LOOKFOR),N)                                                         38
            on IDEAS set NEWCHARS                                                          39
         erase LOOKFOR, NEWCHARS                                                           40
         set TRIES = 100                                                                   41
TMORE    TRIES = TRIES - 1                                                                 42
         is tries lessthan 1 ? [+ TCYCLE ]                                                 43
(CHOOSES most highly weighted TRANSform from IDEAS
         from IDEAS get CHOOSE(IDEAS) = [ -TCYCLE]                              9.V        44
(Applies the TRANSform to the FOUND in the apex
         from $THING get TYPE THRESH '%D=' DESCR '%I=' IMPLIEDS % [$TYPE]       11.V       45
I        set FOUND = $(L;0;0)                                                   10.V       46
(Loops if All indicated (should use LOCs and get nearest.)
TH1      from DESCR get : CLASS WT ] = [-EVAL]                                  12.V       47
         from CLASS get '$' ALL = [ +THA ]                                                 48
           ALL = 1                                                                         49
THA      from FOUND get LEFT # that CLASS # RIGHT : THING WTF = [-TH1]          15.V       50

         on GOT list CLASS THING                                                16.V       51
         TOTAL = TOTAL + WT * WTF [ $('TH' ALL ]                                17.V       52
EVAL     is THRESH greaterthan TOTAL ? [-TMORE ]                                18.V       53
         MERGE(IMPLIEDS,F,GOT) [TMORE]                                                     54
(ACTs, because an act was chosen from IDEAS.
ACT      from IMPLIEDS get ACTION ARGS ; = [- OUT]                                         55
ACT2     from HIST get CLASS ARG = [+ $ACTION ]                                            56
           ARG = ARGS  [ $ACTION]                                                          57
(OUTputs its actions-sequence
OUT      is TOOUT sameas EMPTY ? [+ SEARCH ]                                               58
         output        TOOUT                                                               59
```

```
                                                                   RECODES-1  SEER-2
          output        EXTERNAL [IN]                                 44.V      60
(Initiates a SEARCH to help in completing the frustrated act.
SEARCH    from $ARG get '%I=' IMPLIEDS % [- FAIL ]                              61
SEARCH2   from IMPLIEDS get IMPLIED WT ; = [- FAIL ]                            62
          from IMPLIED get NEEDED '$' [-SEARCH2 ]                               63
          from $NEEDED get '%D=" PARTS %                                        64
          MERGE(POINTAT(PARTS),N)                                               65
RETURNACT  on NEWCHARS set CHOOSE [ TMORE ]                                     66
FAIL       output 'FAILED' EXTERNAL  [IN ]                                      67
D          [T]                                                                  68
(Names the most highly implied object of class specified in ARGument
T          from $(L;0;0) get CHOOSE($(L;0;0),ARG) = [ -ACT ]           36-43    69
(TOTAL weight must be above 5.
           is TOTAL greaterthan 5 ? [-ACT]                                      70
           on TOOUT list THING         [$('AC' ACTION) ]                        71
ACD        on TOOUT set '( WITH '                                               72
(Describes the scene
D3         from HISTC get CLASS THINGH =   [-D2 ]                               73
           from  $(L;0;0) get : that THINGH REST = [-D3]              46.V      74
           on TOOUT list THINGH ; [ D3 ]                              49.V      75
D2         on TOOUT set ' )  '  [T ]                                            76
(Finds the first THING for each ARGument.
F          from  $(L;0;0) get # that ARG # REST : THING MORE ] = [ -ACT2 ]    52.V      77
(Finds THING only if without variations in EXTERNAL input
           from EXTERNAL get that THING = : THING : [ -ACT2 ]         53.V      78
           on TOOUT list  'A ' THING ' IS FOUND- '  [ ACT2]           54.V      79
(Moves all Found things as PREPosition (TO or FROM) indicates
M          is CLASS sameas  'PREP' ? [ -F ]                                     80
           from HIST get CLASS ARGT     [ -ACT ]                                81
           from  $(L;0;0) get # that ARGT # REST : TARGET  [-SEARCH ]  56.V     82
M2         from EXTERNAL get : THINGA : = [+ $('M' ARG) ]             57.V      83
MTO        from EXTERNAL get that TARGET = '(' TARGET THINGA ')' [M2 ]          84
MFROM      from EXTERNAL get LEFT that TARGET RIGHT =                 58.V      85
+              : THINGA : LEFT RIGHT TARGET  [M2 ]
(Replies. If nothing about ARG, SEARCH associates out some more.
(needs more directed and conscious search
R          from $(L;0;0) get CHOOSE($(L;0;0), ARG) = [-SEARCHR]       60.V      86
              MAXWT = TOTAL                                                     87
R2         on TOOUT list THING ;                                                88
           from $(L;0;0) get CHOOSE($(L;0;0),ARG) = [ -ACT ]                    89
           is TOTAL lessthan MAXWT / 2 ? [+ACT - R2 ]                           90
SEARCHR    MERGE($(L;0;0),IDEAS) [ RETURNACT ]                                  91
(Computes
C          from ARGS get OP CLASSA CLASSB                                       92
           from $(L;0;0) get # that OP # REST : OP                              93
           from $(L;0;0) get # that CLASSA  # REST : ARGA [ - SEARCHC ]         94
           from $(L;0;0) get # that CLASSB # REST : ARGB [ + $OP ]              95
SEARCHC    MERGE(POINTAT(INTEGERS) [ RETURNACT ]                                96
(ADD, - etc. are OPS.
ADD        on TOOUT list ARGA + ARGB  [ACT]                                     97
-          on TOOUT list ARGA - ARGB  [ACT]                                     98
/          on TOOUT list ARGA / ARGB  [ACT]                                     99
(Game move (Needs to look deeper, choose with parallel heuristics)
G          from ARGS get OLD NEW                                                100
           from $(L;0;0) get # that OLD # REST : OLD [ -FAIL ]                  101
           from $(L;0;0) get # that NEW # REST : NEW [ -FAIL ]                  103
           from EXTERNAL get that OLD = NEW [ ACT ]                             104
(Functions used by the main program
(MERGE two lists, combining weights and HIStories
MERGE      (DEFINE: MERGE(LISTA,LISTB,HISTA,WT,CLASS)                 34.V      ME1
           is CLASS sameas EMPTY ? [-ME1]
           from LISTA get LEFT : THING WTA HIST ]  =  [ - return + ME3 ]
ME1        from LISTA get LEFT # that CLASS # REST : THING WTA HIST ] = [ -return]  35.V   ME2
ME3        from WTA get '$' = TOTAL                                   36.V      ME3
(Can optionally specify LISTB as part of THING
           from THING get '$' LISTB =                                  .1       ME4
           is LISTB sameas 'CH' ? [ + CH ]                            37.V      ME5
           from $LISTB get : that THING TOTAL HISTB =                 38.V      ME6
+            NAME TOTAL + (WT+1) * WTL HISTA HISTB ;    [+ME1 ]
           from $THING get '%C=' CLASSES %  [+ME2 ]                    .1       ME7
             erase CLASSES                                                      ME8
ME2        on $LISTB list THING CLASSES : THING (WT+1) * WTL LISTB HISTA ] [ME1]  39.V  ME9
CH         at start of CENTRAL list CHOOSE($(L;RA;CA),THING)  [ME1 ]  40.V      ME10
```

RECODES-1 SEER-2

```
(Back-links only to transforms with names                                          PA1
POINTAT    DEFINE: POINTAT(THINGS)
PA1        from THINGS get CLASSES : THING HIST ] = [- return ]
           from $THING get '%D=' TOTHINGS % [- PA1]
           MERGE(TOTHINGS;N)          [PA1 ]
(NORMALIZE to keep weights roughly constant even though converging passed-on things
                                                                                   NO1
NORMALIZE  DEFINE: NORMALIZE(TONORM,NORM)
NORM1      from TONORM get LEFT : THING WT RIGHT ] = [- return]
           WT = WT / NORM
           is WT lessthan 1 ? [+ NORM1]
           on NORMALIZE list LEFT : THING WT RIGHT ] [ NORM1 ]
(Get ABSolute value                                                                AB1
ABS        DEFINE: ABS(ABS)
ABS1       at start of ABS get '-' = [return]
(CHOOSE MAX or MIN weighted (MAX if no type is specified)                   41     CH1
CHOOSE     DEFINE: CHOOSE(LISTA,CLASS,TYPE)
(CLASS can be a specific thing, or empty (in which case all things are chosen
(among)                                                                     42.V
CH1        from LISTA get CLASSES : FIRST THWT HIHIST ] = [- -return ]      43
           is CLASS sameas EMPTY ? [+ CH2]                                  44.V
           from CLASSES get # that CLASS # [-CH1 ]                          50.V
           list CHOOSE = CLASSES: FIRST THWT L ; RA ; CA HIHIST             45.V
CH2        from LISTA get ORCLASSES : ORTHING ORWT ORHIST ] = [+CH4]        45.V
           from CHOOSE get CLASSES : THING TOTAL LOC HIST ] [return ]       46
CH4        is CLASS sameas EMPTY ? [+ $('CH'TYPE)                           47.V
           from ORCLASSES get # that CLASS # [+$('CH' TYPE)  -CH2 ]         48.V
CHMIN      is ORWT lessthan THWT ? yes - THWT = ORWT [+CH3 - CH2]
CH         [CHMAX ]                                                           .1
CHMAX      is THWT lessthan ORWT ? yes- THWT = ORWT  [+CH3 - CH2 ]          49.V
CH2        list CHOOSE = ORCLASSES : ORTHING ORWT LISTA HIHIST ] [CH2]
```

APPENDIX B: CHARACTERIZING TRANSFORMS THAT FORM SEER's MEMORY NETWORK

Memory networks input in the following form are automatically input and converted (see Uhr, 1974 for the conversion routines and a further description of details of transforms).

Section 1 gives the memory nodes needed to handle the 2-dimensional pattern recognition problems. Several layers of the recognition cone are set up, giving averaging, differencing and transforming characterizers rich enough to begin to handle a variety of distorted and shaded inputs. Many more such transforms would be needed to handle a wider range of objects but these examples should indicate how they can rather routinely be described.

Section 2 shows the additional memory nodes needed to handle all other problems. With 1-dimensional inputs perceptual recognition becomes much easier (although difficulties can arise with misspellings), and either SEER-1 or SEER-2 can handle these problems.

Figures 6 and 7 give a sketchy idea of two portions of the memory network into which the conversion routine transforms these inputs.

```
1)          Memory 1.  For 2-Dimensional Inputs

Type    Description          Implieds                              Classes                Name

L       L1:3 ]
MEM A:  121:BRIGHT ]
Y       242 ]
Y       121 ]

L       L2:3 ]
MEM A:  AAA:BRIGHT ]
Y       A8A ]
Y       AAA ]

L       L3:3 ]
MEM     00000(3;7):          HOR*2;ACROSS$N;:                      STROKE;:               HOR(3154) ]
Y       11111 ]
Y       00000 ]

MEM     01(2;6):             VERT*2;UP$N*3;LEGS$N;:                EDGE;:                 VERT ]
Y       01 ]
Y       01 ]
Y       01 ]

MEM     0001(4;7):           DIAGA*2;UP$N;                         STROKE;: ]
Y       0010 ]
Y       0100 ]
Y       1000 ]

MEM     1000(4;5):           DIAGB*2;:                             STROKE;: ]
Y       1100 ]
Y       0110 ]
Y       0011 ]

MEM     00111100(3;6):       CURVA*2;ACROSS$N;ENCLOSURE$N;:        STROKE;: ]
Y       01000010 ]
Y       10000001 ]

MEM     110(3;9):            RLOOP;LEFTEAR;FACE$N;:                STROKE;: ]
Y       001 ]
Y       110 ]

MEM     011(3;3):            LLOOP;RIGHTEAR;FACE$N;:               STROKE;: ]
Y       100 ]
Y       111 ]

MEM     101(1;7):            DLOOP;NOSE$2;:                        STROKE;: ]
Y       101 ]
Y       111 ]
Y       010 ]

MEM     010(3;9):            CIRCLE;EYE*3;FACE$N;:                 OBJECT;: EYE ]
Y       101 ]
Y       010 ]

MEM     10001(2;7):          DISH;MOUTH;:                          STROKE;: ]
Y       01110 ]

MEM     10000001(3;6):       SAUCER;ENCLOSURE$N;:                  STROKE;OBJECT;:SAUCER ]
Y       01000010 ]
Y       00111100 ]
```

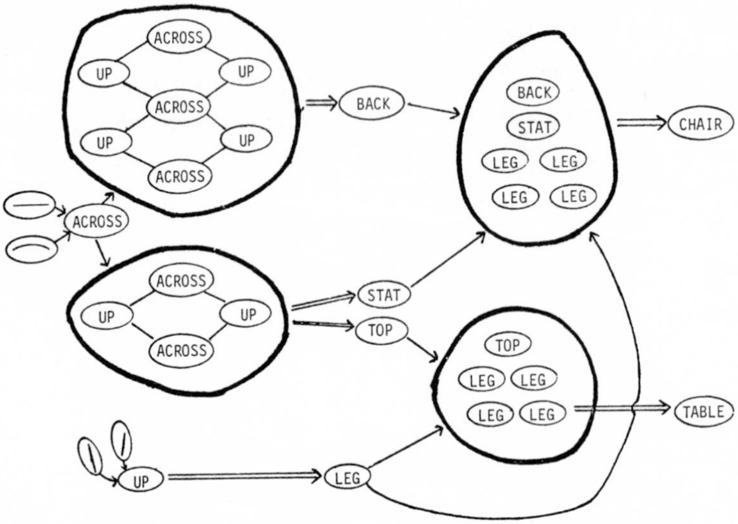

A Graphic Representation of the Part of Memory That is Needed to Recognize Chairs and Tables. Note That Many Details, Like Weights, Thresholds, and How Pointers are Used, are Not Shown.

Figure 6

```
Type  Description            Implieds                                     Classes              Name

L     CHARS:]
MEM   HOR;CURVEA;(1):        ACROSS*6;BACK$N*4;:                                                ACROSS]
MEM   VERT;DIAGA;(1):        UP*3BACK$N*4;SEATLEVEL$N*6;STICK;:                                 UP]
L     IDEAS]
MEM   :  !ACROSS(7):         BACK*9;CHAIR$N*9;:                                                 BACK]
Y     !!UP!     !UP!]
Y     !!UP!     !UP!]
Y     :   !ACROSS!]
Y     !!UP!     !UP!]
Y     !!ACROSS!]
MEM   !O!ACROSS!(7):         SEAT*14;CHAIR$N*8;TOP*13;TABLE$N*8;:                               LEVEL]
Y     !O!UP!    !UP!]
Y     !O!HOR!]
MEM   VERT;:VERT;VERT;VERT;]
        (7):]                LEGS*8;CHAIR$N*6;:                                                 LEGS]
MEM   !!CURVEA!:             VISAGE;FACE$N*3;BALLOON$N;
                               CIRCLE;:                                    FIGURE;:             ENCLOSURE]
Y     !!SAUCER!]             ((in IDEAS))
( If FACE succeeds, BALLOON is implied with a negative weight, to negate it.]
MEM   !!VISAGE*5!(6):        FACE*35;PERSON$N*4;BALLOON*-7;: OBJECT;:                           FACE]
      !! RIGHTEAR!!LEFTEAR!]
      :   !EYE! !EYE!]
      :     !NOSE!]
      :   !MOUTH!]
MEM   !!CIRCLE!              BALLOON*16;:                                  OBJECT;:             BALLOON]
      !!STICK!]
MEM   !!BACK!(9)             CHAIR*25;COUCH$L;:                            FURNITURE,OBJECT;:   CHAIR]
Y     !!SEAT!]
Y     !!LEGS!]
MEM   !!TOP!(9)              TABLE*21;CHAIR$L;:                            FURNITURE;OBJECT;:   TABLE]
Y     !!LEGS!]
```

1) MEMORY 2. For 1-Dimensional Inputs. (to be inserted, as indicated, into the proper layer, ordered to fit examples in the text.)

Type	Description	Implieds	Classes	Name
(The following go in Layer 1 (right after L L1:3]) (For Pattern Recognition)				
(The word PEAR implies "look for a pear (@PEAR)" and "Internally associate about pear)				
MEM	' PEAR '(1;7):	PEAR;@PEAR$N;IPEAR$N;:	WORD;TWORD;:	PEAR]
MEM	' @PEAR '(1;7):	@PEAR;NACT$N;IPEAR$N*2;PEAR;:	OBJECT;FRUIT;:	@PEAR]
MEM	NAME(1;(7):	NAME;TONAME$N*8;TACT$N*2;:	WORD;COMMAND;:	NAME]
MEM	WHAT:	TONAME$N;TOFIND$N;:	WORD;:	THIS]
MEM	NAME;SAY;WHAT;(1)	TACT$N*35;:	COMMAND;:	TONAME]
(The following goes in IDEAS (right after L IDEAS:])				
MEM	ACT:	T OBJECT;:		TACT]
(The following go in Layer 1)				
MEM	@TEARDROP(1;7):	@PEAR2$N;@TEARDROP;:	QUAL;:	@TEARDROP]
MEM	@STEM(1;5):	@APPLE2$N;@PEAR2$N;@STEM;STEM;	QUAL;:	@STEM]
(A 2d characterizer of @PEAR. Note variations (e.g. @YELLOW didn't point to it).)				
MEM	@YELLOW;@TEARDROP*2; @STEM;(4):	@PEAR*3;NACT$N;IPEAR$N;:	OBJECT;FRUIT;:	@PEAR2]
MEM	@OVAL(1;6):	@APPLE2$N;@OVAL;:	QUAL;:	
MEM	@RED;OVAL*2;STEM;(4)	@APPLE*4;NACT$N;IAPPLE$N;:	OBJECT;FRUIT;:	@APPLE2]
(For Describing (The following go in L1)				
MEM	DESCRIBE(1;6):	DACT$N*99;:	COMMAND;:	DESCRIBE]
MEM	ALL;:	DACT$N3;:	ADJ;:	ALL]
(The following goes in IDEAS)				
MEM	ACT:	D OBJECT;:		DACT]
(For retrieving information) (The following go in Layer 1)				
MEM	PRESIDENT(1;7):	PRESIDENT;IPRESIDENT$N*9; PNOW$N*5;PFIRST$N*5;:	NOUN;VIP;:	PRESIDENT]
MEM	BEER;:	BEER;:	PROD;:	BEER]
MEM	CHEESE;:	CHEESE;:	FOOD;PROD;:	CHEESE]
MEM	GARBAGE;:	GARBAGE;:	PROD;:	GARBAGE]
MEM	PROD;:	PROD;:	CLASS;:	PROD]
(The following go in IDEAS)				
MEM	PRESIDENT;FIRST;(9):	WASHINGTON*99;RACT$N*50;:	COMMAND;:	PFIRST]
MEM	ACT:	R ;:		RACT]
MEM	PRESIDENT;:	FORD*48;RACT$N*23;:	COMMAND;:	PNOW]
MEM	WISCONSIN(1;7):	WISCONSIN;IWIS$N*5;IWIS2$N*4;:	WORD;STATE;:	WISCONSIN]
MEM	WISCONSIN;CITIES;:	MILWAUKEE*23;MADISON*18; RACINE*15;RACT$N*25;:	COMMAND;:	IWIS]
MEM	PRODUCE(1;6):	PROD;IWIS2$N;GARBAGE*23;:	:	
MEM	WISCONSIN*3;PROD;:	BEER*27;CHEESE*27;RACT$N*25;:	COMMAND;:	IWIS2]

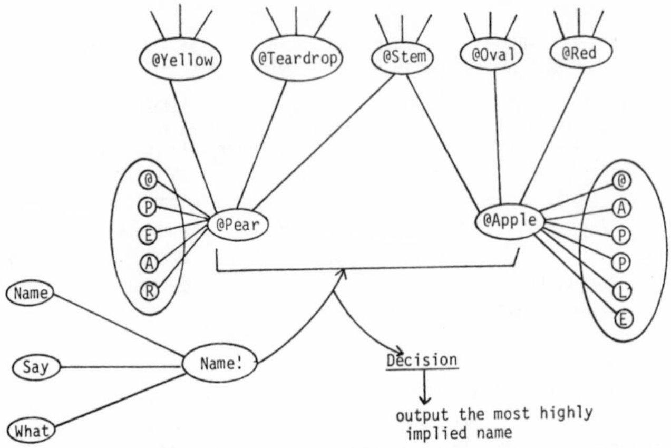

A Graphic Representation of Part of the Memory for Characterizing and Naming a Few Objects
Figure 7

Type	Description	Implieds	Classes	Name
(For deducing responses) (The following go in Layer 1)				
MEM	2;:	2*5;CACT$N;:	NUMBER;:	2]
MEM	TWO;:	2*5;:	NUMBER;:	TWO]
MEM	3 ;:	3*5;:	NUMBER;:	3]
MEM	ADD ;:	CACT$N*99;ADD*8;:	COMMAND;:	ADD]
(The following goes in IDEAS)				
MEM	ACT:	C COMMAND NUMBER NUMBER ;:		CACT]
(For Finding and Moving) (The following go in L1)				
MEM	@BOX(1;7):	@BOX;BOX;:	CONTAINER;:	@BOX]
MEM	BOX(1;6):	BOX;@BOX$N;:	WORD;TWORD;:	BOX]
MEM	@PAIL(1;6):	@PAIL;PAIL;:	CONTAINER;:	@PAIL]
MEM	PAIL(1;6):	PAIL;@PAIL$N;:	WORD;TWORD;:	PAIL]
MEM	@APPLE(;7):	@APPLE;APPLE;EAT$N;:	OBJECT;FRUIT;:	@APPLE]
MEM	APPLE(1;6):	APPLE;@APPLE$N;:	WORD;TWORD;:	APPLE]
MEM	FRUIT(1;6):	FRUIT;:	WORD;TWORD;CLASS;:	FRUIT]
MEM	TOUCH(;;7):	TOUCH;FACT$N*37;FIND;:	COMMAND;:	TOUCH]
MEM	FIND(1;7):	FIND;FACT$N*37;:	COMMAND;:	FIND]
MEM	' TO ':	TO;:	PREP:	TO]
NEW L	NEEDSGOALS NEEDS]	:N1 10]:N2 10]%		
MEM	FOOD;:	@HUNGER*9;EAT$N;:	OBJECT;:	N1]
MEM	FRUIT;CANDY;(1)	FOOD;@HUNGER;EAT$N;:		FOOD]
MEM	MOVE(1;7):	MOVE;TOMOVE$N*30;:	COMMAND;:	MOVE]
MEM	@SEER-MOUTH(1;7):	@SEER-MOUTH;EAT$N;TO;:	SELF;:	@SEER-MOUTH]
(The following go in IDEAS)				
MEM	OBJECT;TO;OBJECT;:	MACT$N*99;		TO MOVE]
MEM	ACT:	F OBJECT,:		FACT]
MEM	ACT:	M ;:		MACT]
MEM	@HUNGER;FRUIT;CANDY; TO;@SEER-MOUTH;(6):	EACT*99;:	COMMAND;:	EAT]
MEM	ACT:	M ;		EACT]

APPENDIX C: A NOTE ON EASEy PROGRAMS (See Uhr, 1973f for details)

1. Numbering at the right identifies statements, and allows for comparisons between programs. M indicates initializing Memory statements: I indicates cards that are Input by the program. .V indicates a Variant, .1 an additional statement.
2. A program consists of a sequence of statements, an end card, and any data cards for input. (Statements that start with a parenthesis are comments, and are ignored.) Statement labels start at the left; gotos are at the right, within brackets (+ means branch on success; - on failure; otherwise it is an unconditional branch). + signifies a continuation card.
3. Strings on capitals are programmer-defined. Strings in underlined lower-case are system commands that must be present (they would be keypunched in caps to run the program). These include input, output, erase, set, list, get, start, call, that and the inequalities. Other lower-case strings merely serve to help make the program understandable; they could be eliminated.
4. EASEy automatically treats a space following a string as though it were a delimiter; it thus automatically extracts a sequence of strings and treats them as names.].:,;, and % act similarly as a delimiter, but the programmer must specify it. The symbol # is used to stand for any delimiter (a space,], :, ;, % or #).
5. The symbol $stringI is used to indicate "get the contents of string I, and treat it as a name and get its contents" (as in SNOBOL).
6. Pattern-matching statements work just as in SNOBOL

statement: there are a) a name, b) a sequence of objects to be found in the named string in the order specified, c) the equal sign (meaning replace), and d) a replacement sequence of objects (b, c, and/or d can be absent). that string I means "get that particular object" - otherwise a new string is defined as the contents of stringI, which is taken to be a variable name.

7. size(...) is a built-in function that counts the symbols in the string (s) named within parentheses (its argument). integer(...) succeeds if its argument is an integer.

8. DEFINE: defines a programer-coded function. The function is executed whenever it is specified, FUNCTIONNAME (ARBUMENTS), in the program. It ends in success or failure when it reaches a [return] or [return] goto.

REFERENCES

Chomsky, N., Syntactic Structures, the Hague: Mouton, 1957.

Chomsky, N., Aspects of the Theory of Syntax, Cambridge: MIT, 1965.

Craik, K. J. W., The Nature of Explanation, New York: Cambridge, 1952.

Duda, R. O. and Hart, P. E., Pattern Classification and Scene Analysis, New York: Wiley, 1973.

Feldman, J. and Gries, D.,"Translator Writing Systems," Comm. ACM, 1968, 11, 77-113.

Feldman, J. A. et al.,"The Stanford Hand-Eye Project," Proc. 2nd Int. Joint Conf. on Artificial Intell., 1971, 521-526.

Griswold, R. E. et al. The SNOBOL4 Programming Language, Englewood-Cliffs: Prentice-Hall, 1968.

Guzman, A., "Decomposition of Visual Scenes into Three-Dimensional Bodies," Proc. AFIPS EJCC, 1968, 33, 291-304.

Hunt, E. B., Concept Formation: an Information Processing Problem, New York: Wiley, 1962.

Jordan, S. R., "Learning to use Contextual Patterns in Language Processing," Unpubl. Ph.D. Thesis, University of Wisconsin, Madison, 1971.

Kochen, M., "An Experimental Program for the Selection of Disjunctive Hypotheses," Proc, AFIPS WJCC, 1961, 19, 571-578.

Minsky, M. (Editor), Semantic Infomaation Processing, Cambridge: MIT, 1968.

Narsimhan, R., "Labelling Schemata and Syntactic Descriptions of Pictures, Information and Control, 1964, 9, 151-179.

Newell, A. and Simon, J., Human Problem Solving, Englewood-Cliffs: Prentice-Hall, 1972.

Nilsson, N., Problem-Solving Methods in Artificial Intelligence, New York: McGraw-Hill, 1971.

Nilsson, N. J., "A Mobile Automaton: an Application of A.I. Techniques," Proc. 1st Int. Joint Conf. on Artificial Intell., 1969, 509-520.

Peirce, C. S., Collected Papers, Cambridge: Harvard, 1931-1958.

Quillian, M. R., "The Teachable Language Comprehender: A Simulation Program and Theory of Language," Comm. ACM, 1969, 12, 459-476.

Sauvain, R. and Uhr, L., "A Teachable Pattern Describing and Recognizing Program, Pattern Recognition, 1969, 1, 219-232.

Shapiro, S., "A Net Structure for Semantic Information, Storage Deduction and Retrieval," Proc. 2nd Int. Joint Conf. on Artificial Intell., London, 1971.

Shaw, A. C., "A Formal Picture Description Scheme as a Basis for Picture Processing Systems, Information and Control, 1969, 14, 9-52.

Simmons, R. F., "Answering English Question by Computer: A Survey," Comm. ACM, 1965, 8, 53-70.

Simmons, R. F., "Natural Language Question-Answering Systems: 1969," Comm. ACM, 1970, 13, 15-30.

Swain, P. H. and Fu, K. S., Nonparametric and Linguistic Approaches to Pattern Recognition," Elect. Engin. Tech. Rept. Purdue University, Lafayette, 1970.

Thompson, F., "English for Computers, Proc. AFIPS FJCC, 1966, 28, 349-356.

Uhr, L. and Vossler, C., "A Pattern Recognition Program that Generates, Evaluates and Adjusts it Own Operators," Proc. AFIPS WJCC, 1961, 19, 555-570. (Reprinted, with additional results, in E. Feigenbaum and J. Feldman, Eds., Computers and Thought, New York: McGraw-Hill, 1963.)

Uhr, L. "Pattern-String Learning Programs, Behavioral Science, 1964, 9, 258-270.

Uhr, L., and Kochen, M., "MIKROKOSMs and Robots", Proc. 1st Int. Joint Conf. on Artificial Intell., 1969, 541-556.

Uhr, L. "Flexiable Linguistic Pattern Recognition," Pattern Recognition, 1971, 3, 363-384.

Uhr, L., "Layered 'Recognition Cone' Networks That Preprocess, Classify and Describe, IEEE Trans. Computer, 1972, 21, 758-768.

Uhr, L., Pattern Recognition Learning and Thought, Englewood-Cliffs: Prentice-Hall, 1973(a).

Uhr, L., "Recognizing, 'Understanding,' Deciding Whether to Obey, and Executing Commands," Computer Sci, Dept. Tech. Rept. 173, Univ. of Wisconsin, 1973(b).

Uhr, L., "The Description of Scenes Over Time and Space, Proc. AFIPS NCC, 1973, 42, 509-517(c).

Uhr, L. "DECIDER-1: A System That Chooses Among Different Types of Acts," Proc. 3rd Int. Joint Conf. on Artificial Intell., 1973, 396-401(d).

Uhr, L. "Describing, Using 'Recognition Cones'," Proc. 1st Int. Joint Conf. on Pattern Recognition, 1973(e). (Also Computer Sci. Dept. Tech. Rept. 176, Univ. of Wisconsin, Madison, 1973.

Uhr, L. "EASEy-2: An English-Like Program Language," Computer Sciences Dept. Tech. Rept. 178, Univ. of Wisonnsin, Madison, 1973(f).

Uhr, L., "A Wholistic Cognitive System (SEER-1) for Integrated Perception, Action and Thought," Computer Sci. Dept. Tech. Rept., Univ. of Wisconsin, 1974.

Uhr, L., Semantic Learning, in preparation.

Uhr, L., "An Integrated Cognitive System (SEER-2) That Interacts With Scenes That Change Over Time," in preparation.

Waltz, D. L., "Gene Rating Semantic Descriptions From Drawings of Scenes With Shadows," Unpubl. Ph.D. Diss., MIT, Cambridge, 1972 (AI TR-271).

Winograd, T., "Procedures as a Representation for Data in a Computer Program for Understanding Natural Language," Unpublished Ph.D. Diss., MIT, Cambridge, 1971. (Also Understanding Natural Language, New York: Academic, 1972.)

Winston, P. H., "Learning Structural Descriptions From Examples," Unpubl. Ph.D. Diss., MIT, Cambridge, 1970.

Winston, P. H., "The MIT Robot," in Machine Intelligence 7, (B. Melzer and D. Michie, Eds.), Edinburgh Univ. Press, 1972.

APPLICATIONS OF FUZZY SETS IN PSYCHOLOGY

Manfred Kochen
Mental Health Research Institute
University of Michigan
Ann Arbor, Michigan 48014 U.S.A.

1. INTRODUCTION

Since its inception, fuzzy set theory was guided by the assumption that classical sets were not natural, appropriate or useful notions in describing human behavior. The design of a control system that parks a car more in the manner of a human driver than in the manner of a "smart bomb" was one of the original challenges that motivated fuzzy set theorists. Another was the creation of a computer program that could respond appropriately to the same kind of instructions that are given to a human pilot by another human advisor or instructions for tying a knot.

Fuzzy set theory can offer psychology new concepts to use as building blocks for improved theories. In return, psychology can offer fuzzy set theory not only continuing challenges and test problems but methods of experimentation as well. It is more fruitful to introduce the notions of fuzzy set theory when the need for them arises in the development of psychological conceptualizations than to seek out psychological problems for potential applications of fuzzy set theory. The ideas and work reported here originated with the recognition, during the course of developing a new model of cognitive learning, that fuzzy sets are relevant, useful and possibly necessary to explain certain psychological findings.

2. BACKGROUND: SUMMARY OF PREVIOUS RESULTS ON THE USE OF FUZZY SET THEORY IN PSYCHOLOGY

Cognitive learning has been viewed (Kochen, 1974) as an algorithm which forms, revises and uses a system of representation for recognizing and coping with an increasing variety of opportunities and traps. This view led to such an hypothesis as: "if a problem-solver practices with tasks requiring shifts of representation, he is likely to perform better in solving an ill-defined problem than one who has no prior practice or one who has prior practice with well-defined problems not requiring representational shifting" (Badre, 1973).

To test such an hypothesis, a new experimental technique was developed in which human subjects (college students) were instructed only to ask questions that would help them recognize, formulate and perform a task that the experimenter had in mind and created for them. Certain words and actions are prespecified but not known to the subject. The subject's use of these is interpreted as indicative of representational shifting.

This technique was also applied in showing that 4th and 5th grade children can be taught to formulate realistic mathematical "story-problems" for themselves (Kochen, Badre and Badre, 1974), and in showing that if people ask better questions they tend to perform better on certain tasks that require information-seeking (Kochen and Badre, 1974). In ranking questions according to quality, a question that is judged to be similar to a second one along the generic-specific and the relevance-irrelevance dimensions but which is more precise[1] is considered better. To measure the precision of questions, such as "Is it very expensive?", "Is it

[1] Footnote on the bottom of the next page.

very far?", it was necessary to conceptualize precision more clearly, and the "fuzzy set" concept seemed to be the most appropriate foundation (Kochen and Badre, 1974). We presented subjects with samples of numbers and asked them to mark on a scale, such as

Agree Strongly	X	Disagree Strongly

, the strength of their belief in statements about a sample number being <u>much larger</u> than 5. We also repeated this with weights instead of numbers as stimuli. We found, for example, that in such a test people behave as if they considered "very much greater" more precise than "much greater."

This, of course, has long been regarded to be a plausible assumption by fuzzy set theorists. This finding is more important for the <u>method</u> of ranking precision than for its content. If the degree of agreement, measured by how far from the right a subject places his mark on the above scale, is plotted against the sample number, a grade-of-membership or characteristic curve is obtained. The precision with which the subject used "much larger than" is taken to be an estimate of the slope of this characteristic curve at inflection point.

We showed that, if $f_{ML}(x)$ denotes a subject's degree of agreement with "x is much larger than 5," and $f_{ML}(x)$ is assumed to be continuous and differentiable in x, with

$$f'_{ML}(x) = b\, f_{ML}(x)\ [1-f_{ML}(x)] \qquad (1)$$

then $f_L(x)$ is the logistic curve with an inflection point at $x = a/b$, where the (maximum) slope is $b/4$. The assumption in Equation (1) is plausible; it states that a subject's

[1] For simplicity of exposition, assume that the degree of precision is appropriate for the context. That is not always so. When it is not, the more precise question is not better than the fuzzier one.

strength of belief increases as x increases in proportion to how strongly he believes it already and also in proportion to how strongly he disagrees with the statement. If he disagrees strongly, it is probably because x is not much larger than 5.

We also found that anchoring increases the degree of precision, in line with analogous findings in psychophysics. A higher degree of response consistency over trials was found to occur if the subject is allowed to give an imprecise verbal response about a fuzzy set than if he is forced to give a precise "grade-of-membership" answer. This supports the assumption that the notion of a "linguistic variable" (Zadeh, 1973) is more realistic than that of "grade-of-membership."

This work revealed clearly the critical importance of context. In a subsequent study, using the above technique, we (Dreyfuss-Raimi, et al., 1974) presented sentences such as "25° C is cold for January," "25° C is cold for Miami," "25° C is cold for January in Montreal," "60 miles is a long distance to walk." We found that:

1. Characteristic curves were generally steeper for temperature than for distance, probably because temperature is an inherently less fuzzy concept than distance, since temperatures are interpreted by most people in the context of climate, while the interpretation of distances is much more sensitive to individual experiences.
2. For temperature, few contexts resulted in a steeper curve than the one for no context. But for distance, the curves for <u>appropriate</u> contexts resulted in a steeper curve than the one with no context.
3. Different contexts produced curves with strikingly

different degrees of steepness.

4. For both distance and temperature, the judgments were more precise for no context than for nonsensical contexts.

These results seem complementary to the only other (to my knowledge) behavioral approach to fuzzy set theory (Rosch, 1973, 1974), and to recent work connecting fuzzy set theory with linguistics (Lakoff, 1971; Zadeh, 1972).

3. A CONCEPTUAL ISSUE

When some persons are asked how strongly they believe that "x is a large number" they behave as if they had, for a time, fixed a threshold or decision criterion d that enables them to say, consistently, that if $x \geq d$ they agree, and if $x < d$ they disagree. The mark they place on the agreement-disagreement scale might be distributed uniformly over the right half of the scale whenever $x \geq d$ and it might be distributed uniformly over the left half if $x < d$. Let us call such people "thresholders."

Another kind of person might as best he can, try to place his mark close to the agreement side of the scale according to how large he thinks x is. This depends critically on the sample that is presented, for if he has "used up" the scale by placing a mark close to "agree," his response to the largest number just presented, and now an even larger one is presented, he will be "squeezed." The characteristic curve for such people will look like the continuous $f_L(x)$ discussed in the previous section. Let us call such people "estimators," because their strength of belief in "x>>5" resembles their estimate of x.

Yet another kind of person might place his mark near

"agree strongly" or "disagree strongly" when he feels strongly and place no mark when he is uncertain, if he has that option. The set of numbers is then partitioned into three subsets: those that such a subject considers not large; those he considers large; and those he will not call either large or not large. (People do not have to dichotomize, as prescribed by Aristotle, that x is either large or not large; the above remark is less objectionable to those who see Aristotlelian logic as a normative guide to cognition if we replaced "not large" by "small.") Let us call such people "reliables" or conservatives, as in the psychological literature.

Undoubtedly there are other possible kinds of people. <u>Fuzzy set theory applied to psychology might be interpreted to suggest the general hypothesis that most people are "estimators" rather than "thresholders" or "reliables."</u> If enough people in a sample behave as if their strength of belief varies nearly continuously with the stimulus variable in the statement to be believed, then this hypothesis would be supported, and the psychological reality of fuzzy sets would be made more evident.

If, on the other hand, too many people in a sample behaved as if they use a threshold-decision criterion - as would be implied by most models of decision-making used in decision theory and mathematical statistics (i.e., critical regions are classical sets) - then the psychological reality of fuzzy sets would be in doubt and other concepts more plausible.

Asking a person to respond to a statement like "25' is far to the right" on an agreement-disagreement scale, though direct, seems an unrealistic way to measure how he would <u>behave</u> in response to an instruction such as "move far to the right." The most obvious and natural thing to do is to ask him to move and observe where he goes. This should be done repeatedly with the same subject, with detractors between repeated instructions to avoid the effect of the subject's recalling or trying to be consistent. If he distributes the distances that he moves uniformly between some minimum distance and some upper limit of possible distances that he could move, we infer that he used a threshold: the minimum distance. The cumulative frequency would then be a straight line from the threshold to the maximum. The resulting curve might be viewed as his characteristic or grade-of-membership curve.

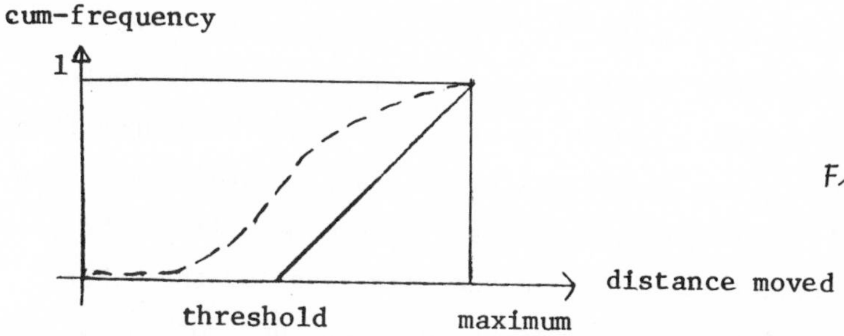

Figure 1

If he were an estimator, he might distribute the distances he moves according to a skew or Bell-shaped curve. The cumulative frequency would have a typical S-shape as shown by the dashed line in Figure 1.

It is common in psychological experiments to regard the subjects in a random sample of people as interchangeable. If the dependent variable that is observed has a bimodal distribution, that might indicate that the sample was drawn from two populations, such as thresholders and estimators. Figure 2 indicates what such a bimodal distribution would look like. If the population of estimators dominates, the general hypotheses about the psychological reality of fuzzy sets appears to be supported.

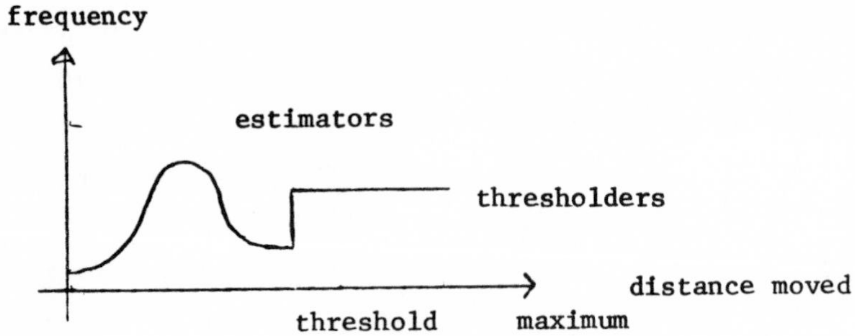

Figure 2

4. AN EXPERIMENT[1]

Twenty-four college freshmen were given the following instructions: "People interpret words like 'far,' 'hard,' etc. in different ways. In this experiment, I am trying to find out how they interpret such words in specific context. I will be asking a random sample of people like yourself the same thing to see if there are some laws according to which all people behave in such tasks. Your name will not be used and your responses will be mixed with those of the others. Just follow the instructions on each page and work as quickly and as accurately as you can.

In the first section, you are to put an X above the point on the line which is ________________ to the right of the 0.

For example: Put an X above the point on the line which is a little to the right of the 0.

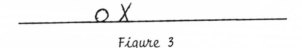

Figure 3

Any questions?"

---

[1]The help of Suzanne Brumer in conducting this experiment is gratefully acknowledged.

The subjects were then given 42 instructions as in the above example for the first part of the experiment. The blank was filled in by "far," "very far," "not so far," "not so close," "very close," or nothing. Two different line lengths were used to explore the effect of context in this sense. Two positions of the 0 were used with the smaller line to see if that made any difference. Each instruction was presented twice and the order of presenting the 42 was randomized.

In part 2 of the experiment, each subject was given 42 pictures in random order such as Figure 2 above, and asked to assign to each diagram one of the following 7 statements:

1. X is very far to the right of 0.
2. X is far to the right of 0.
3. X is not so far to the right of 0.
4. X is to the right of 0.
5. X is not so close to the right of 0.
6. X is close to the right of 0.
7. X is very close to the right of 0.

The lines in the diagrams were again of 2 lengths and 2 different positions of the 0 in the shorter line were used. The diagrams have fixed X's at 7 distances from 0. Hence there are 7 x 3 = 21 diagrams. Each was presented twice.

5. RESULTS

Consider first the distribution of the responses from the 24 subjects when told to place an X far to the right of 0 on the long line. The distances were measured by the number of quarter inches. The results were:

Distance (in $\frac{1}{4}$'' units):	7	8	9	10	11	12	13	Total
Number of subjects who moved that distance:	1	3	13	6	10	8	7	48

This is shown in Figure 4. The dip at 10 is too sharp to be due to chance. A bimodal distribution seems to be present. The maximum distance was 13, and the right-hand part of the distribution resembles a uniform distribution with a threshold at 11. It might be the case that we drew our sample from a population in which 1 + 3 + 13 + 6 or 23 out of 48 or about half, were estimators, and 10 + 8 + 7 or 25 out of 48, or the other half, were thresholders. It does not support the belief that most people behave in the way conceptualized by fuzzy set theory. Just about as many people seem to behave in the way conceptualized by decision theory.

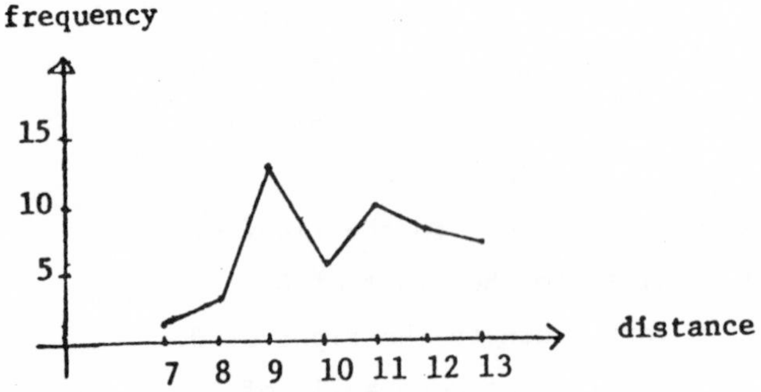

Figure 4

Recall that each person responded to the same diagram twice at different times, with distractions in between. It is interesting to note how many people responded the same way or differently both times. This is shown next.

Difference in distance moved on both trials:	0	1	2	3	4	5	Total
Number of subjects who obtained that difference:	4	7	5	2	4	2	24

Again we have a bimodal distribution. It seems as if some people in the same category come from a population which estimates consistently, while others are very inconsistent. It is noteworthy that the distances chosen by the 4 perfectly consistent subjects were 9, 9, 9 and 11; that is, three of them came from the "estimator" population, and the fourth may have.

From the remaining data in part 1, the following conclusions can be drawn. Recall that the responses were the distance from 0 at which an X was placed.

1. The responses or distances decreased according to the following order of the stimuli in the verbal instruction: very far, far, not so close, Ø, not so far, close, very close. The responses are consistent (transitive). Here Ø refers to the absence of any adjective, as in "put an X above the point on the line which is _ to the right of the 0." (The blank was left blank.) The reversal of "not so close" and "not so far" is perhaps a little surprising, but understandable, because "not so close" is semantically similar to "far." (It is actually ambiguous.)
2. The response to "very far" is the maximum length of the line, independently of line length or the 0's location. Similarly, the response to "very close" is an X right next to 0, independently of line length and the 0's location. There is very little variance over the subjects.
3. The variance of the responses to diagrams in which the 0 is at the center of the line is less than the variance when the 0 is to the left of the center. The latter diagram offers a longer maximum distance over which to distribute X's. To test whether this is due to the eccentric location of 0 or the length, we can compare the responses on the long line with the

responses on the short line with the 0 entered. The difference is not very significant. Hence, it seems to be the eccentricity of the 0 that accounts for the increased variance.

4. The variance is greatest when the blank is not filled in, i.e., for the Ø stimulus. Indeed the variance increases as we move from either extreme (very far, very close) towards Ø.

5. Line length (context) does not affect the responses for extreme stimuli, such as very far and very close, but the mean response to other stimuli is scaled down. The ratio of the line lengths was 9/13. The mean responses were:

Stimulus	far	not so close	Ø	not so far	close
Long line (0 at center):	10.0	6.5	4.5	4.3	2.4
Short line (0 at center):	6.6	4.9	3.3	3.4	2.1
Ratio:	.66	.75	.73	.81	.89

At least for "far," the responses shrank by .66 from the long to the short line, which were in a ratio of about .69. It seems as if the subjects scaled down in direct proportion to the line lengths.

From the data in part 2, the following conclusions may be drawn. Here the response is the selection of a phrase, such as "X is far to the right of 0," from 7 such choices to marked lines presented as stimuli.

6. Most subjects select "far" and "close," while fewer subjects select "very far" and "very close."

7. Very few subjects select "not so far" and "not so close" and the blank, Ø, is used least often of all.

The data shows remarkable uniformity for a psychological

experiment, even though 2 subjects place X's to the left of 0 in some cases.

6. CONCLUSIONS

One out of two people seem to behave, when asked to place an X far to the right of a mark on a line, as if they interpret "far distances" as a fuzzy set with a grade-of-membership assignable to "d ε F" that increases continuously with d. Measuring that grade-of-membership by observing how frequently they placed the X at distance d from the mark appears to be useful for connecting fuzzy set theory with psychology. Context, in the form of a line of limited length, affects the response in nearly direct proportion to the line lengths. Of course, this conclusion is not likely to hold as the line length increases indefinitely. The linearity of the relation between the response and the context (line length) is probably local.

Most of these conclusions are hypotheses supported by evidence. But more experimentation is required to establish them more firmly and to delimit the range of variables over which they hold. On the whole, fuzzy set theory does seem appropriate for conceptualizing certain aspects of the behavior of perhaps half the population.

REFERENCES

Badre, A. N., "On Hypotheses and Representational Shifting in Ill-Defined Problem-Situations," Ph.D. Thesis, University of Michigan, Ann Arbor, Mich., 1973.

Dreyfuss-Raimi, G., Robinson, J., Kochen, M. and Badre, A. N., "On the Psycholinguistic Reality of Fuzzy Sets: Effect of Context and Set," (MHRI manuscript available from M. Kochen; to be submitted to the Proc. of Chicago Linguistic Circle, April 1975).

Kochen, M. and Badre, A. N., "On the Precision of Adjectives

Which Denote Fuzzy Sets," Journal of Cybernetics, forthcoming (1974).

Kochen, M. and Badre, A. N., "Question-Asking and Shifts of Representation in Problem-Solving," American Journal of Psychology, September 1974.

Kochen, M., Badre, A. N. and Badre, B., "The Nature of Recognizing and Formulating Mathematical Problems," Instructional Science, 3, #3, 1974.

Lakoff, G., "Linguistics and Natural Logic," In D. Davidson and G. Harman (Eds.), Semantics of Natural Language, Dordrecht, Holland: D. Reidel Publishing Co., 1971.

Rosch, E. H., "On the Internal Structure of Perceptual and Semantic Categories," In T. E. Moore (Ed.), Cognitive Development and the Acquisition of Language, New York: Academic Press, 1973, pp. 111-144.

Rosch, E. H., "Universals and Cultural Specifics in Human Categorization," In R. Brislin, S. Bochnor and W. Bonner (Eds.), Cross-Cultural Perspectives on Learning, Sage Press, in press 1974.

Zadeh. L. A., "A Fuzzy Theoretic Interpretation of Linguistic Hedges," Journal of Cybernetics, 2, 1972, pp. 4-34.

Zadeh, L. A., "The Concept of a Linguistic Variable and Its Application to Approximate Reasoning," Memorandum No. ERL-M411, October 15, 1973, Electronics Research Lab., University of California, Berkeley.

EXPERIMENTAL APPROACH TO FUZZY SIMULATION OF MEMORIZING, FORGETTING AND INFERENCE PROCESS

M. Kokawa, K. Nakamura and M. Oda
Automatic Control Laboratory
Nagoya University
Nagoya, Japan

1. INTRODUCTION

The human subjective function of decision-making is generally attended with ambiguity. And then the "fuzzy algebra" seems suitable to make a good expression of human subjectivity instead of probability-theoretical expression which has been used in the conventional theory of stochastic decision-making process [1][2]. From this point of view, the authors have studied the fuzzy-theoretical formulation of human decision-making process through a psycho-engineering experiment using playing cards in a card game.

This paper is an approach to making fuzzy models of the memorizing-, forgetting- and inference processes which are essentially important in the human decision-making process [3]. In Section 2 are presented the block diagram expression of a whole model of human decision making system and mapping expressions of experience process, memory process, and inference process based on memory [4]. Section 3 is a fine description of fuzzy formulation of human decision-making. Two types are considered there, one is the decision-making based on memory only, and the other is the one depending on both memory and inference based on memory. These processes are expressed by fuzzy relations. Section 4 is the experimental study to make a fuzzy model of memory process. It is found that the certainty degree of memory is a value of the membership function with its ambiguity, and varies with time either in the collapsing manner (forgetting) or in the emphasizing manner (sharpening). Thus, the certainty degree of memory and its ambiguity is a time function dependent on several subjective or objective factors governing the difficulty of memorizing. In Section 5 is analyzed the inference process through the two kinds of inference experiments, and is clarified the relation among the quantity of memory effect, the quantity of available information and the degree of inference, in the inference process [5].

2. PROPOSITION OF A WHOLE MODEL OF HUMAN DECISION-MAKING PROCESS

The whole structure of human decision-making is modelled by the block diagram of Figure 1 [3][6].

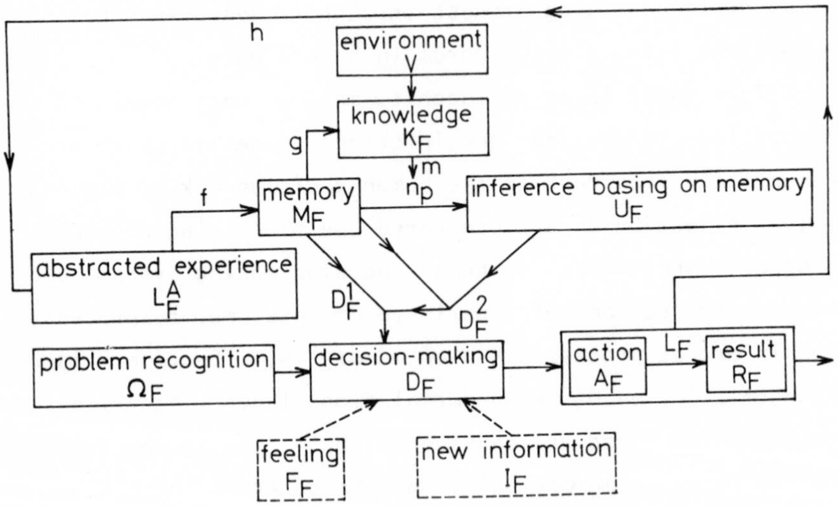

Figure 1. A whole model of decision-making

i) Experience

Experience is something or knowledge acquired by one's

own senses or introspection or its acquisition process.

Facing a new problem human thinks that what kind of problem it is (recognition and classification of the problem), what kinds of actions have been taken in the past for the same or similar problems to the present one and what kinds of results have been obtained (remembrance and re-evaluation of the past experience), what kinds of environmental changes have happended (addition of new environmental changes), and so on [7].

Now, make the following notations:

ω: problem

$a(\omega)$: action for problem ω

$r(a)$: result of action $a(\omega)$

$\Omega = \{\omega\}$: set of problems

$A = \{a(\omega)\}$: set of actions

$R = \{r(a)\}$: set of results

Then the experience can be defined as an association of A and R, where a relation L from A into R can also be defined as $L:A \to R$, and is changeable with the environment.

ii) Memory

The human memory is to maintain and recall only some of the experiences which are selected to do a specified role. Memory has the following four levels, i.e. (1) Memorizing, (2) Retention, (3) Recall and (4) Recognition. And the difficulty or the degree of memory is usually governed by the following few factors: a) intensity of association with experiences, b) depth of interest, c) depth of impression, d) conspicuousness, e) peculiarity and f) originality. It is most probable that a human does not memorize all his experiences, but only the strongly impressed ones which should be termed as abstracted experiences.

Noting that

L: experience

L^A: abstracted experience

M: memory

Then abstracted experience L^A and memory M are given by mappings h, from L to L^A and f, from L^A to M respectively, i.e. $h: L \rightarrow L^A$, $f: L^A \rightarrow M$, therefore $f \circ h: L \rightarrow M$.

iii) <u>Inference Based on Memory</u>

Inference is an important function for decision-making, which is developed on the base of the memorized experiences. The inference U is generated from M by a mapping n_p, i.e. $n_p: M \rightarrow U$. Where n_p is a mapping developed by knowledge under the specified environmental stimulus.

3. FUZZY FORMULATION OF DECISION-MAKING PROCESS

Here, we will consider the formulation of the decision-making process by referring some definitions on fuzzy algebra [8] ∿ [11] necessary for the argument in this section.

Since the transitive law is not generally satisfied in human decision-making, the decision-making sequence can not be expressed by the equivalence relation. Then, it is proposed to describe the process in terms of n-step fuzzy relation defined by the notion of fuzzy composition and combination degree defined as a limit of the n-step fuzzy relation when $n \rightarrow \infty$. Relation having a combination degree whose value   specified threshold is an equivalence relation [12]. If a deicision-making sequence can be expressed by the fuzzy partial ordering, then the decision-making can be accomplished by using an extended form of Szpilrajn's theorem, which is derived for the partially ordered set [13].

Thus, all the functions and states in the

decision-making process should be considered to concern with fuzziness, which are represented by suffix F.

Then abstracted experience L_F^A which is mapped through h from L_F can be shown by the fuzzy matrix.

$$L_F^A = h(L_F) = [\mu_{L_F^A}(a_i, r_j)] \tag{3.1}$$

where

$\{a_i,\ i = 1,2,\ldots,n\} \in A_F$
$\{r_j,\ j = 1,2,\ldots,\ell\} \in R_F$

$\mu_{L_F^A}$: membership function of fuzzy set of abstracted experiences L_F^A

After the operation of mapping f, the information on L_F^A is all transferred to the memory M_F, which is shown as

$$M_F = f(L_F^A) = [\mu_{M_F}(a_i, r_j)] \tag{3.2}$$

μ_{M_F}: membership function of memory set M_F

Inference based on memory U_F which is given from M_F by the mapping n_p can be also expressed as

$$U_F = n_p(M_F) = [\mu_{U_F}(a_i, r_j)] \tag{3.3}$$

μ_{U_F}: membership function of inference set U_F

Now, denote D_F^1 and D_F^2 by the relation from Ω_F (set of problems) to A_F and the relation defined by (3.6) respectively.

$$D_F^1 = [\mu_{D_F^1}(\omega_k, a_i)] \tag{3.4}$$

$$D_F^2 = [\mu_{D_F^2}(a_i, r_j)] \tag{3.5}$$

where

$\{\omega_k,\ k = 1,2,\ldots,m\} \in \Omega_F$

$$\mu_{D_F^2}(a_i, r_j) = \max\{\mu_{M_F}(a_i, r_j),\ \mu_{U_F}(a_i, r_j)\} \tag{3.6}$$

Notice that a bigger one* between the corresponding two elements in M_F and U_F is to be selected as the element of D_F^2.

It is probably true to consider that the decision matrix D_F is a composition of D_F^1 and D_F^2, i.e.

$$D_F = D_F^1 \circ D_F^2 \tag{3.7}$$

so elements of D_F can be decided by (3.8) with the definition of composition of fuzzy relation.

$$\mu_{D_F}(\omega_k, r_j) = \max_{a_i \in A_F} \min [\mu_{D_F^1}(\omega_k, a_i), \mu_{D_F^2}(a_i, r_j)],$$

$$k = 1,2,\ldots,m, i = 1,2,\ldots,n, \; j = 1,2,\ldots,\ell. \tag{3.8}$$

If we have a problem which is the same one experienced in the past and for which it is memorized that a good result is obtained by taking a corresponding action, then we have the same action which was taken in the past. If the decision is made by memory only, then the decision matrix D_F is equal to D_F^1.

4. FUZZY SIMULATION OF MEMORIZING- AND FORGETTING PROCESSES

4.1. Experimental Methods for Fuzziness of Memory

In order to make a fuzzy model of a decision making process, an interesting experiment using playing cards is schemed. In this section is first explained the experiments for the study of memorizing processes and the time behavior of memory. Memory experiments are done by the method of reproduction, with the following procedure.

*This decision is in the optimistic case. But in the pessimistic case a smaller one will be taken as the element of D_F^2, then "max" in the equation (3.6) is to be substituted by "min".

1) Certain number of cards (for instance 10 or 15 cards) are sequentially shown to subjects (five ∿ seven junior high school students including two girls) for a certain period of time (3 sec./card, 6 sec./card, etc.).

2) After a certain time (0 min., 15 min., 30 min., 60 min., 7 days, etc.), each subject is asked to make an entry of 0.2, 0.4, 0.6, 0.8, or 1.0 by his own certainty degree of memory (depending on his subjectivity) in the cells on a test form, where the number of cells is the same as the number of displayed cards.

(Notes) The subject is instructed to make an entry of 1.0 only for the cards which are certainly confirmed of its display.

3) After then, the experimenter makes an entry in the corresponding cell of the form ○ for a correct answer, × for incorrect, or □ for forgotten, and also the order of the displayed cards. For example, ♠ 3 in Table 1 indicates that the degree of memory is 0.6, the order of the displayed card is the 8th, and the answer is correct. (A full example is shown in Table 1.)

(Memory Experiments) No. 7-6-1

Date : 10 month 13 date 1972 year

Subject : sex(male, female), age 14 , name H. A.

	A	2	3	4	5	6	7	8	9	10	J	Q	K	Jo
♠			0.6 8○					0.2 ×			1.0 ×		1□	
♣	1.0 7○			0.2 ×				2□		0.2 5○				
♦					1.0 4○					1.0 10○	6□			
♥						1.0 3○		0.8 ×	9□					

Experimental Conditions and Results

Displayed number of cards : 10 sheets

Displayed time : 6 sec./sh.

Passed time after experiment : 13 date 20 : 30′ −) 13 date 20 : 30′ = 0 D 0 H 0 M

Rate of correct answer : 6 / 10

Table 1.
Experimental Form

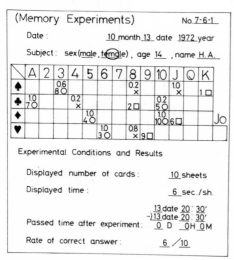

4) The procedures 1) ∿ 3) are repeated by changing the displayed time, number of displayed cards, and the class of cards (experiments 1 ∿ 11).

4.2. Results of Experiments

The results of experiments are illustrated in Figure 2 ∿ Figure 5. Figure 2 explains the memorizing process, which is composed of the functions h and f defined by (3.1) and (3.2) respectively and their combined function f∘h. The figure in x-μ plane in Figure 2(c) shows the certainty degree of memory and the ambiguity of fuzziness of memory. That is, the difference between the maximum and minimum values of the degree of the initial memory of each card can be considered as the ambiguity of the mean value for an averaged subject (cf. the detailed explanation of Section 5.2).

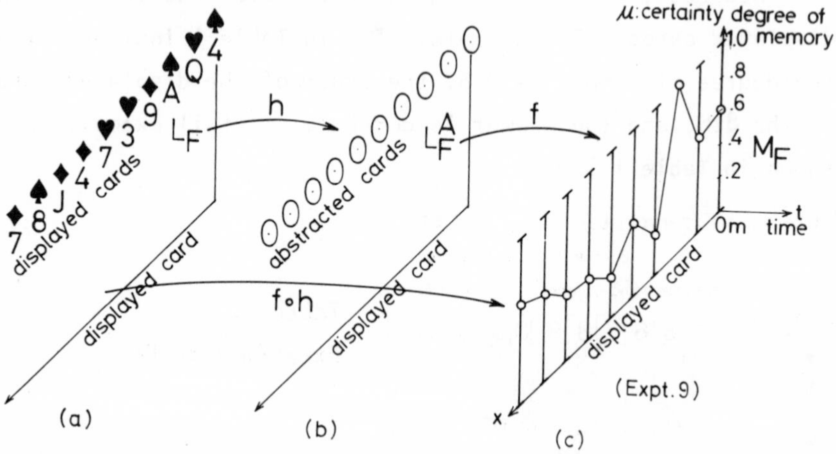

Figure 2. Expression of mappings h, f, and f ∘ h.

The certainty degree of memory decreases, in other words, the fuzziness of memory increases as time passes. Figure 3 is a schematic figure showing the time variation of the certainty degree of memory. P_1 indicates an ordinary forgetting

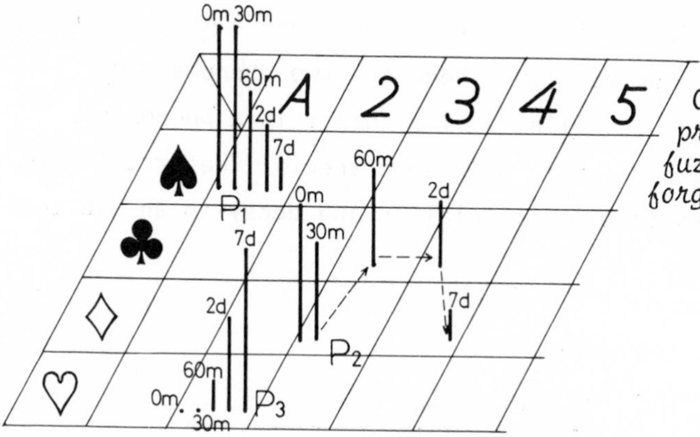

Figure 3. Conceptual process of fuzziness of forgetting

process. P_2 is a special model of forgetting process in which the degree of memory decreases as time goes as well as the information of memorized element (number, sort, color, etc.) is partially discharged. P_3 shows the process where the card is not recalled for a while after the displaying experience, and thereafter begins to appear in memory and the certainty degree of memory is gradually emphasized. The process of P_3 is concerned with the sharpening phenomena and the reminiscnese phenomena.

As described in Section 2, the memory is usually dependent upon six factors of (a) ∿ (f) relating to the difficulty of memory, and varies with time. Therefore, the memory behavior, that is, memorizing-forgetting process can be shown, in general, as the membership time function $\mu(a,b,c,d,e,f,t)$. In Figure 4 ∿ Figure 6 which are the experimental results for time behavior of the certainty degree of memory, the membership function is simply shown by $\mu(x,t)$, where x is the displayed card, and $0 \leq \mu(x,t) \leq 1.0$. As is clear from those figures, the certainty degree of memory of joker and honor are high, while that of plain cards are low. The ambiguity of the mean value of memory degree as shown in Figure 2(c) is naturally variable with time, but the time behavior of ambiguity is not drawn in Figure 4 ∿ Figure 6 for simplicity. We can find there the two phenomena of "sharpening" and "levelling". The former is a phenomenon emphasizing and developing only some special features, while the latter is to lose its speciality and decay to an averaged form [4].

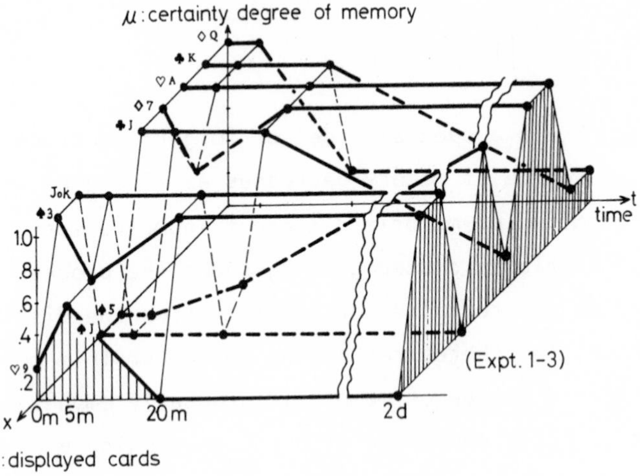

Figure 4. Fuzzy process of forgetting

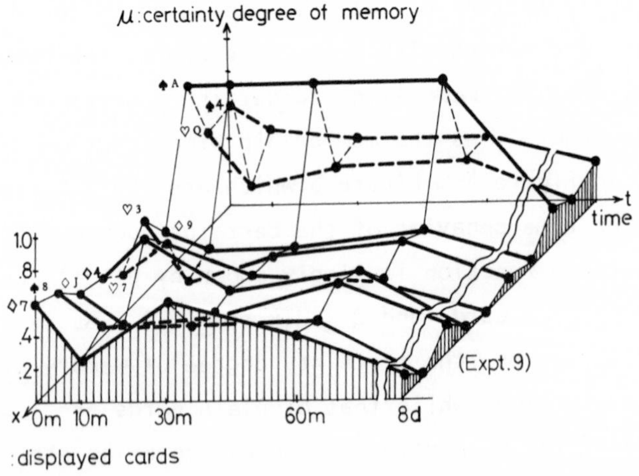

Figure 5. Averaged fuzzy process of forgetting

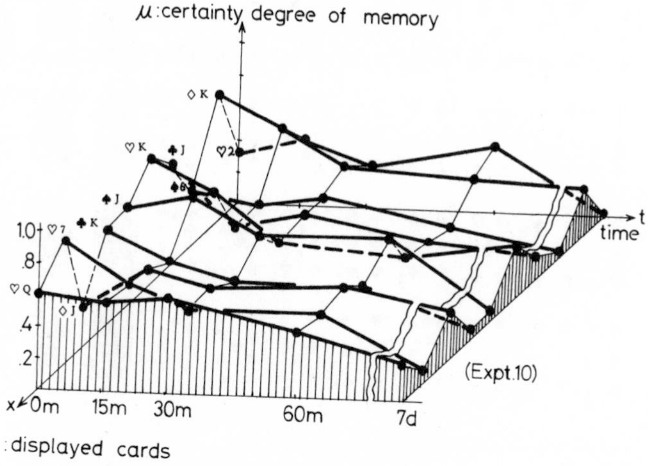

Figure 6. Averaged fuzzy process of forgetting

5. FUZZY SIMULATION OF INFERENCE PROCESS BASED ON MEMORY

5.1. Fuzzy-theoretical Experiments of Inference [17]

Here is explained the experiments for making a fuzzy-theoretic and information-theoretic model of human inference process. In this experiment, the playing card is also used as a material. Number of subjects participated is ten in total (five males and five females whose ages cover 19 through 29 years). Test arrangements of the cards are of seven kinds, an example of which is shown in Figure 7.

Experiments are executed by the following procedure.

1. An experimenter selects one of the seven test arrangements of the cards (4x13=52 cards).

2. At first, all cards are arranged backside up.

3. The experimenter instructs a subject to find the card of a specific name.*

* The name of the card instructed is determined in a way of preventing the experimenter's subjectivity.

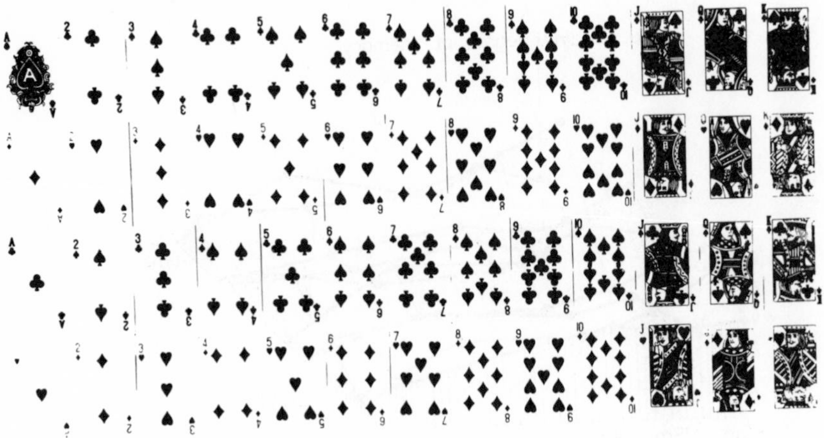

Figure 7. An example of the arrangment of the card (Arrangement 4).

4. At each time, the subject selects just one card which he believes to be the instructed one, and tells the experimenter its certainty degree in a numerical value [0,1], and then turns up the card.

5. The experimenter notes in the cells on the test form (cf. Table 2) the following items, i.e. the name of the instructed card, the name and the certainty degree of the card selected by the subject, and also the correctness of the selected name.

Test: Arrangement (5) - Case (II)
Date: 7 month 12 date 1973 year
Subject: sex (male, female), age (19 yrs.), name (K.H.)

Trial Order	1	2	3	4	5	6	···	17	18	···
Instructed Card by Experimenter	♠8	♥7	♠6	♥7	♠8	♥5	···	♥J		
Certainty Degree of Selected Card	0.1	0.1	0.1	0.3	0.5	0.7	···	1.0		
Turned-up Card by Subject	♣6	♦7	♠6	♥7	♠8	♥9	···	♥J		
Correctness of Turned-up Card	×	×	○	○	○	×	···	○		

Table 2. Form of fuzzy inference experiment

6. Procedures 3, 4, and 5 are repeated until the subject can guess correctly the instructed card with its

certainty degree of 1.0, in other words, until the subject can find completely the arrangement rule of the cards.

The experiments consist of two cases for every subject: in Case I the turned-up card is kept faceside out, and in Case II the turned-up card is immediately returned over as it was. (There is an interval of more than two weeks between Case I and Case II.) In Case I, the arrangement rule of the cards can be inferred from seeing the turned-up cards, which means it is not necessary for a subject to keep the cards in mind since all the turned-up cards can be considered to be completely memorized, while in Case II the turned-up cards must be memorized in his mind for the subject to infer the arrangement rule of the cards. From the results of the two cases, we can make a comparative investigation of the inferences with memorizing effort (Case II) and without memorizing effort (Case I).

5.2. Results of Experiments

The experimental results are shown in Figure 8 ∿ Figure 11. Figure 8 is the result of Case I, and Figure 9 is that of Case II.

Figure 8 and Figure 9 show the certainty degree answered by the subject in each trial. The value shown by sign ○ in the figure is the average value of ten subjects for the same number of trial order. As shown in the figures, the value of certainty degree expands over a range around the average value. The range of distribution for ten subjects may be regarded as the ambiguity of the mean value for an averaged subject.

5.3. Relation among the Quantities of Memory Effect, Information, and Inference

The difference between the results of Case I (cf. Fig. 8)

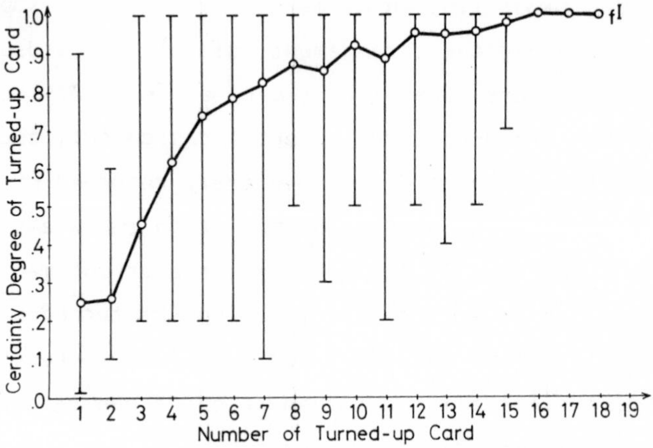

Figure 8. Mean value of ten subjects in Case I-4.

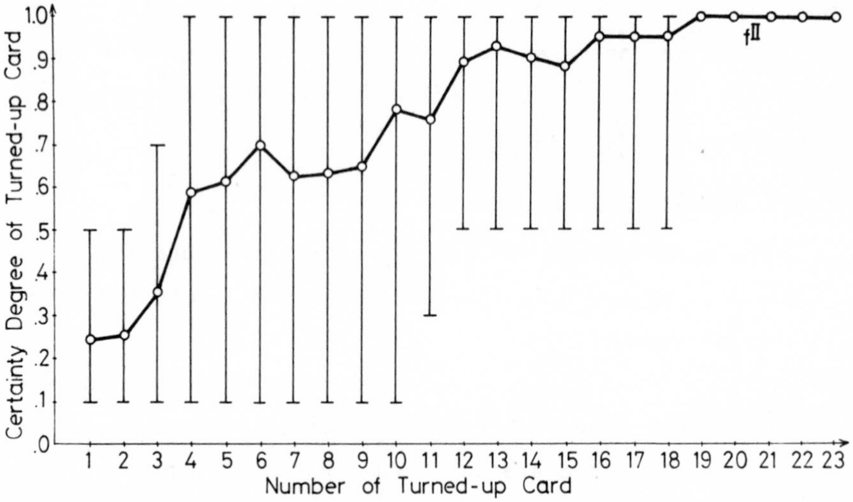

Figure 9. Mean value of ten subjects in Case II-4.

and Case II (cf. Fig. 9) can be considered to correspond the quantity of memory effect to the inference, q. Because, in Case II against Case I, it is necessary for a subject to memorize the turned-up cards in order to infer the arrangement rule, so the certainty degree of the turned-up card in Case II is less worth by the quantity of fuzziness caused by memorizing than that in Case I.

The degree of memory effect q_i at the ith trial is defined as

$$q_i = \begin{cases} f_i^{I}-f_i^{II}, & f_i^{I}-f_i^{II} \geq 0 \\ 0 \quad , & f_i^{I}-f_i^{II} < 0 \end{cases}, \quad i=1,2,\ldots,N^{II} \tag{5.1}$$

where f_i^{I} and f_i^{II} are the certainty degree of the ith trial card in Case I and that of Case II respectively, and N^{II} is the total number of turned-up cards before the subject finds the arrangement rule completely in Case II.

Then, the quantity of memory effect to the inference is given by

$$q = \sum_{i=1}^{N^{II}} q_i \Delta i, \quad i=1,2,\ldots,N^{II} \tag{5.2}$$

where Δi is the interval between the ith and i+1th trials (cf. Fig. 10).

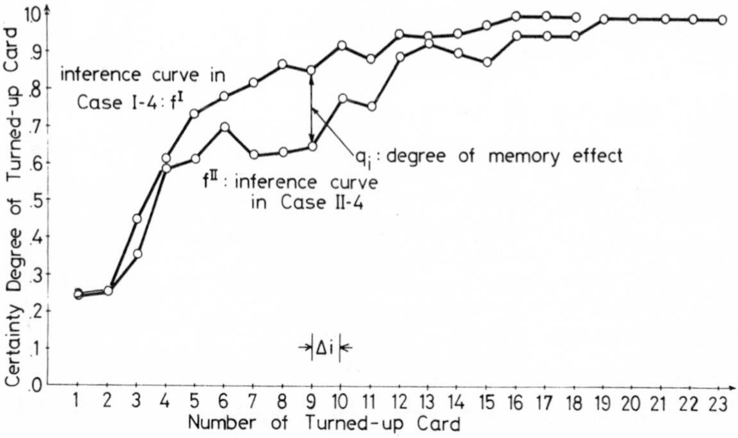

Figure 10. Degree and quantity of memory effect (Mean value in Cases I-4 and II-4).

Next, we consider the quantity of information of the turned-up card. The subject will select the card by the inference based on the information, which is to be measured depending upon the certainty degree of the cards turned up in the preceding trials. In Case I, the difference between the certainty degree of (i+1)th trial and the quantity of information obtained in the preceding trials of 0 $\sim$ ith is considered to be the degree of inference.

The quantity of information of the card is calculated as follows. In this paper, the nonprobabilistic entropy P, [14] and [15], basing on the membership function $f(x_j)$ of the event x_j is used, i.e.

$$p = -k \sum_{j=1}^{N'} \{f(x_j)\ln f(x_j) + \overline{f}(x_j)\ln \overline{f}(x_j)\} \tag{5.3}$$

where, in our case, K is a positive constant, N' is the number of turned-up cards before the subject finds the arrangement rule of the cards completely in Case I, x_j is the turned-up card at the jth trial, and $\overline{f}=1-f$.

The reason to use such an entropy P defined above is as follows. Equation (5.3) is defined so that the quantity of information $\{f\ln f + \overline{f}\ln\overline{f}\}$ has the maximum value for the event x_j when the state is the most ambiguous, i.e. $f(x_j)=0.5$, [14]. In our experiments, Fig. 8, Fig. 9, and other experimental results with different arrangements show that the state is the most ambiguous when $f(x_j)\doteqdot 0.5$. For example, in Fig. 9, the difference between the maximum and minimum values of the certainty degree at each trial is biggest when $f(x_j)=0.6\sim 0.8$. Since the difference between the maximum and minimum values at each trial can be considered as the ambiguity of the mean value for an averaged subject, and the most ambiguous point is the one where the difference is the biggest.

By the above reasons, the quantity of information can be calculated by f^I of Fig. 8 and Equation (5.3), which is normalized by $K=1/_N I$; then P is as shown the curve in Fig. 11.

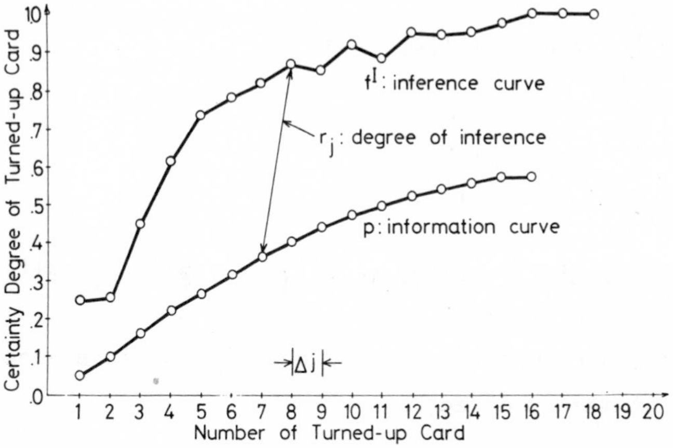

Figure 11. Quantity of information, degree of inference, and quantity of inference.

Since the results of Case I, f^I can be considered to be developed by the inference based on the quantity of available information, then the quantity of inference is given by subtracting the entropy of available information calculated by Equation (5.3) from the results f^I. Therefore, the degree of inference r_j is defined as

$$r_j = f_j^I - P_{j-1}^I, \quad j=1,2,\ldots,N^I \tag{5.4}$$

where P_{j-1}^I is the quantity of information of cards turned-up until the j-1th trial.

$$P_{j-1}^I = -\frac{1}{N^I}\sum_{\ell=1}^{j-1}\{f^I(x_\ell)\ln f^I(x_\ell) + \overline{f}^I(x_\ell)\ln\overline{f}^I(x_\ell)\}, \quad K=1/N^I \tag{5.5}$$

Then the quantity of inference r is given by

$$r = \sum_{j=1}^{N^I} r_j\Delta_j \; , \quad j=1,2,\ldots,N^I \tag{5.6}$$

where Δ_j is the interval between the jth and j+1th trials.

In the above analysis, if f^{II} is used in place of f^{I}, then the quantity

$$v = \sum_{j=1}^{N^{II}} v_j \Delta_j \ , \ j=1,2,\ldots,N^{II}$$

$$= \sum_{j=1}^{N^{II}} (f_j^{II} - P_{j-1}^{II})\Delta_j \tag{5.7}$$

where $P_{j-1}^{II} = -\dfrac{1}{N^{II}} \sum_{\ell=1}^{j-1} \{f^{II}(x_\ell) \ln f^{II}(x_\ell) + \bar{f}^{II}(x_\ell) \ln \bar{f}^{II}(x_\ell)\}$
$$\tag{5.8}$$

will mean the composition of the quantity of inference and the quantity of memory effect.

6. CONCLUSIONS

The remarkable results of a series of fuzzy experiments developed in this paper are summarized as follows:

1. The certainty degree of memory is a membership function and has its ambiguity which corresponds to the distribution range of mean values answered by subjects.

2. The certainty degree of memory and its ambiguity are both time functions dependent on few factors governing the difficulty of memorizing.

3. Analyzing the results of the two kinds of ingenious experiments for analysis of inference process, and introducing the nonprobabilistic entropy of information, can be clarified the relation among the quantity of memory effect, the quantity of information and the quantity of pure inference.

The general formation of h, f, g, n_p, f^{I}, and f^{II} remains as a difficult problem in future. The study of the relation among knowledge, quantity of information, and quantity of inference will also be needed, where the quantity of information of the event used in a general decision-making is greatly effected by an individual's knowledge [16]. The research of decision-making with additional effects by the introduction of feeling and new information will be reported in future.

REFERENCES

[1] D. B. Yudin, "Decision-Making in Complex Situations", IEEE Trans. Engineering Cybernetics, No. 2, pp. 209-222, 1970.

[2] E. A. Patrik and Y. L. Shen, "Interactive Use of Problem Knowledge for Clustering and Decision-Making", IEEE Trans. Computers, Vol. C-20, pp. 216-221, 1971.

[3] M. Kokawa, K. Nakamura, and M. Oda, "A Formulation of Human Decision-Making Process", Res. Rept. of Auto. Cont. Lab., Nagoya Univ., Vol. 19, pp. 3-10, 1972.

[4] M. Kokawa, K. Nakamura, and M. Oda, "Fuzzy Expression of Human Experience-to-Memory Process", Res. Rept. of Auto. Cont. Lab., Nagoya Univ., Vol. 20, pp. 27-33, 1973.

[5] M. Kokawa, K. Nakamura, and M. Oda, "Fuzzy-Theoretical Approaches to Forgetting Process and Inference", Res. Rept. of Auto. Cont. Lab., Nagoya Univ., Vol. 21, pp. 1-10, 1974.

[6] L. P. Schrenk, "Aiding the Decision Maker-a Decision Process Model", IEEE Trans. Man-Machine Systems, Vol. MMS-10, No. 4, pp. 204-218, 1969.

[7] H. Mieno, "Man-Machine Systems on Creative Thinking", The Japanese Journal of Ergonomics, Vol. 6, No. 4, pp. 175-180, 1971.

[8] L. A. Zadeh, "Fuzzy Sets", Information and Control, Vol. 8, pp. 338-353, 1965.

[9] M. Mizumoto, "Fuzzy Algebra and Its Application 1 (Mathematical Basis of Fuzzy Algebra)", Mathematical Sciences, Japan, pp. 50-57, April 1970.

[10] M. Mizumoto, "Fuzzy Algebra and Its Application 7 (Fuzzy Matrix)", Mathematical Sciences, Japan, pp. 78-83, January 1971.

[11] M. Mizumoto, "Fuzzy Algebra and Its Application 8 (Fuzzy Matrix)", Mathematical Sciences, Japan, pp. 68-72, February 1971.

[12] S. Tamura, S. Higuchi, and K. Tanaka, "Pattern Classification Based on Fuzzy Relations", IEEE Trans. Systems, Man, and Cybernetics, Vol. SMC-1, No.1, pp.61-66, 1971.

[13] L. A. Zadeh, "Similarity Relations and Fuzzy Orderings", Information Sciences, Vol. 3, No. 2, pp. 177-200, 1971.

[14] A. DeLuca and S. Termini, "A Definition of a Nonprobabilistic Entropy in the Setting of Fuzzy Sets Theory", Inf. and Cont., Vol. 20, pp. 301-312, 1972.

[15] R. M. Capocelli and A. DeLuca, "Fuzzy Sets and Decision Theory", Inf. and Cont., Vol. 23, No. 5, pp. 446-473, 1973.

[16] V. A. Trapeznikov, "Man in the Control System", Automation and Remote Control, Vol. 33, No. 2, pp. 171-179, 1972.

[17] M. Kokawa, K. Nakamura and M. Oda, "Fuzzy Theoretical and Concept Formational Approaches to Memory and Inference Experiments", Trans. of the Institute of Electronics and Communication Engineers of Japan, Vol. 57-D, No. 8, pp. 487-493, 1974.

ON FUZZY ROBOT PLANNING*

J. A. Goguen
Computer Science Department
University of California
Los Angeles, California 90024 U.S.A.

INTRODUCTION

This is not another theoretical paper, giving some aspects of the theory of "fuzzy X's" (where for this author, X has included automata, topological spaces, and convex sets). Rather, it is an attempt to lay foundations for applying fuzziness in highly practical problems of control and communication. There is little mathematics here, and only a few things from computer science; the goals and ideas are treated as more important than their technical expressions. There is a lot of "philosophy." After considerable experimentation with far drier approaches, I am convinced that an exposition like that in this paper helps more than it hinders, though it does do both.

Unfortunately, there are no experimental results to report. I delayed writing this in the hope that there would be, but things just have not worked out that way. Perhaps experimental data will exist by the time you read this. In any case, they are forthcoming; all our major decisions on how to proceed with implementation have been made, and it is "just" a matter of getting it done.

Getting what done? This paper discusses a system for using fuzzy hints to get through a maze. Our basic contention, which the running system will hopefully prove, is that

*Research supported by NSF Contract GK-42112 while the author was at UCLA, and also at Naropa Institute and the University of Colorado, each in Boulder, Colorado.

fuzziness, rather than being a problem, can be very useful in practical situations.

Section 1 defines the notion of robustness, and argues for its importance. Section 2 discusses some problems with and approaches to, natural language understanding. Section 3 describes the system itself in a general way, including the "intermediate representation language" idea, and also discusses how to evaluate system performance. Section 4 gives more details of how it works, including the "hedge algebra." Section 5 contains some conclusions plus some ideas for further work. One particular point is that the system in a sense embodies [Zadeh 1973]'s idea of (fuzzy) "linguistic variables." I apologize if I have been too enthusiastic or repetitious about certain points.

We do assume previous familiarity with fuzzy sets (see [Zadeh 1965]), though we certainly do not use anything as fancy as [Goguen 1967]. It might help to have some familiarity with the "robot planning" literature in artificial intelligence, primarily (perhaps) for the contrast of its highly syntax and logic-based approach with the present paper. Some computer science background will help at certain points in Section 4.

This work grew out of the author's earlier (and grander) ideas for a "hierarchically organized metaphor using robot" HOMEUR [Goguen 1972], and it owes much of its present form to the questions, comments and contributions of the UCLA "Fuzzy Robot Users Group," especially D. DeAngelis, A. Gershman, K. Kim, R. Pottinger, E. Shaket, and J. Tardo, each of whom the author very much wishes to thank. He also wants to thank the University of Colorado and Naropa Institute, both in Boulder, Colorado for assistance and inspiration during the preparation of this report.

1. ROBUSTNESS

In this paper robustness means the ability to respond without program modification to slightly perturbed, or to somewhat inexactly specified situations. This ability would seem to be very useful in a variety of applications and is moreover characteristic of the way people cope with their environment.

We first illustrate the significance of "slightly perturbed ... situations" in the above definition. Say a given person knows how to get through a complex intersection in a very efficient way. He will not be distrubed if road construction alters, even greatly alters, certain features in essential ways. But if an essential feature is altered in an essential way, he will have to rethink the situation. In assembly line automation, it would be highly desirable for each step to perform its function in spite of the inevitable inaccuracies of the positions of objects coming down the line. Standard programming ("rotate 6.14 degrees clockwise, raise 2.02 centimeters") of robot arms etc. certainly does not have this property. See [Ambler-Barrow-Brown-Burstall-Popplestone 1973] for an interesting approach which does permit flexibility and seems to embody fuzziness.

Hints are an important illustration of what we mean by "inexactly specified situations" in the definition of robustness. We may give directions for getting around in a city, "Go about ten blocks north until you see a drugstore at a stop light, then turn rightish, ..." which only very vaguely describe the situation, but which are entirely adequate in the context where they are to be executed.

A key point here is that the inexactness of the description is not a liability; on the contrary, it is a blessing in the sufficient information can be conveyed with less effort. That sufficient information can be conveyed with less effort. inexactness makes for greater efficiency!

The work reported in this paper is aimed at producing computer programs which exhibit robustness in situations similar to those described above. Our methods include: fuzzy sets; careful hierarchical (recursive) organizations; use of procedures to represent knowledge; actions, etc.; and a semantic (i.e., meaning) orientation, including a general goal orientation in the system itself (to run the maze).

We believe that such an approach has wide applicability in areas of problem solving and natural language understanding. The pilot project we discuss in detail is to design a "robot" to accept and use vague hints about how to run a maze (or, one can think of "hunting for buried treasure" with a "treasure map").

2. NATURAL LANGUAGE UNDERSTANDING

We reject the view that a rigid syntactic foundation is necessary for natural language understanding. A person can often understand sentences (for example from a non-native speaker) which are quite ill-structured in a formal grammatical sense; e.g.,

(1) "The view that will to be rejected, is that it is necessary a foundation syntactic rigidly, language for to understand to have to have."

In fact, natural language understanding seems to be a rather *robust* affair. Perhaps it is like the navigation through a familiar but complex intersection mentioned in the previous section: small changes in non-crucial features are easily tolerated. (It is interested to speculate about just what "small" means in this situation; i.e., about the "topology of English.")

We take for granted the idea that most words have inherent fuzziness, that no rigid boundaries can be drawn for their use. This is very clear with colors: there is a linear continuum of hues between red and yellow. At no point is there a clear separation; and declaring intermediate values to be "orange" doesn't help either, since the boundaries of orange are equally unclear. A more amusing example is to imagine a continuum of physical objects between some given chair and table, constructed by letting the chair back shrink while its seat expands and flattens, and its legs become higher. There will be some strange objects in this continuum which cannot clearly be assigned to either class.

The extent to which such phenomena can be modelled by fuzzy sets is another question. Quite likely, fuzzy algorithms [Zadeh 1968] and "linguistic variables" ([Zadeh 1973]) come closer than pure fuzzy sets, tho those are still very useful. Probably we have not yet seen the ultimate model, and anyway, the choice of model should depend on the precise use to which it is put.

It is evident that context has a tremendous effect on meaning, and that it will not do to treat words, or even sentences, in isolation. Given the proper context, any word can have any meaning. It can always be done crudely, by saying "In the following, the word X shall mean ...". More limited effects can be achieved more subtly, by complex interactions between words of the type often called "metaphorical." For example[1.] in

1. Footnote on bottom of next page.

(2) "The landscape of frozen forms is a jagged one, here and there rising to great heights of morphemes piled upon morphemes, in between sinking to levels only one or two morphemes deep."

"frozen forms" refers to what are often called "idioms" (perhaps "now standardized figures of speech" is clearer).

I suggest that in cases such as (2), we chose the meaning which maximizes the overall meaningfulness in the given context. For another, simpler example, "green ideas" may not make much sense until we think of green as "not yet fully developed; untried; not yet ripe" etc. On the other hand it might refer to the work of a certain painter, who expressed certain ideas in the color green. Context can easily render one or the other the most meaningful; or something else entirely different. This kind of optimization over a set of given fixed meanings is not enough to explain everything; we must also consider "non-literal" meanings. Thus "frozen" in the Bolinger quotation does not refer to some physical substance below its melting point, and its sense of "fixed, no longer producing a fresh response" does not appear in my dictionary. One may not wish to have all senses stored in a dictionary; and anyway we have to handle "fresh" uses by some sort of "understanding" process rather than by look-up. Metaphorical language seems to dig "inside" the meanings of words, and to pick out some characteristics while ignoring others; see [Reddy 1972].

Under such a maximizing mechanism, there may well not be any "meaningless" or "nonsensical" sentences. For example, by providing enough context to pick appropriate secondary or metaphorical meanings, we can make fine sense of the following

1. D. Bolinger, "The Atomization of Meaning," Language, XLI, 1965, p. 571, as quoted by [Reddy 1972].

sentence declared "nonsensical" by [Chomsky 1957],

(3) "Colorless green ideas sleep furiously."

It might mean

(4) "Boring untried ideas are doing nothing in X's mind but there is much activity (such as speaking, writing, etc.) about them."

in the context of a few sentences setting a suitable scene. Incidentally, [Chomsky 1957] (and [Minsky 1974] which quotes it) declare that

(5) "Furiously sleep ideas green colorless."

is treated quite differently from (4). But I think that in a suitable context, where (4) is meaningful, (5) will be treated the same[2.] as (4).

This suggests that the robustness of human language understanding rests on a <u>semantic</u> foundation which is more important than the syntactic structure. Moreover, a computer program which insists on producing an exact parse tree will probably do a lot of unnecessary extra work. For example, this seems to be true of [Winograd 1972]. We believe that fuzziness, robustness, etc. should make things <u>easier</u>, rather than harder. But, we do not claim to know how this works for natural language in an algorithmic sense; it seems to be a difficult problem, and quite possibly its solution will radically change the capability of natural language understanding systems. We do claim some insight into what is needed: fuzziness; the maximization principle; and semantic components of words. See also [Goguen 1969, 1972, 1974].

It might be noted that much of artificial intelligence seems to have suffered from an excess concern with syntax at

2. In fact, from the point of view of poetry (that is of rhythm, sound, balance &c.), quite possibly (5) is better than (4)!

the expense of semantics. Many AI programs merely conduct exhaustive searches through spaces of formal descriptions, without any idea of the "meaning" of what they are doing. It is a very important problem to get beyond this, and perhaps a very difficult one.

We do not suggest, however, that syntax is useless. It seems to be a kind of scaffolding which aids in the assembly of the total semantic structure from the various words as building blocks. There may even be situations where a rigid syntax is appropriate, such as present day computer languages. Yet the vision of a robust programming language which would compensate for small errors, has considerable attractiveness, especially for non-professional computer users.

3. SYSTEM OVERVIEW AND OBJECTIVES

The particular "fuzzy robot planning" system we are working on has the following general structure

$$\text{English subset} \xrightarrow{\quad I \quad} \text{IRL} \xrightarrow{\quad C \quad} \text{QLISP}$$

where: IRL stands for an "intermediate representation language"; I is a (semantic) interpreter into IRL; C is a "compiler"; and QLISP (see [Reboth-Sacerdoti 1973]) is a particularly attractive programming language in which we can express "meanings" as algorithms.

The subset of English will be suitable for expressing appropriate kinds of hints. As suggested in the previous

3. One alleged advantage of rigidity is that it helps to catch errors in program conception. (This is, for example, visible in LISP, where beginner's conceptual errors often show up as incorrect parentheses.) One would hope for robust languages which still retain this feature.

section, it should not be too rigidly defined. We are still unsure of the best way to proceed with this subset and its function I, although there are some crude but effective strategies available. Our main focus has been on various IRL's and C's, of various complexities, to get a concrete feel for the general problems involved. The IRL's are rigidly syntactically defined. Of course, they are intended to express (to embody and to convey) definitely fuzzy information, such as "fairly far almost South."

The way in which we intend the system to "run" a maze is as follows: it should start by knowing nothing except the starting point and that there is a goal. When it reaches a node, it will be told the choices of movement which it has at the node (e.g. straight, left, or right), and if there are landmarks, appropriate information about them. Of course, the entire maze will be stored somewhere in the system, but it will not be available to the searching algorithm.

The idea behind having the final output appear in an algorithmic language is that, if H is a hint for maze M, then C(H) will be run on M after compilation, and C(H) can also be run on M', M", etc., perturbations of the original M, to test for robustness. The same C(H) should work almost as well for mazes for which H is only an "almost appropriate hint." In future versions, we might want to make use of the possibilities of C(H) being an algorithm to permit it to stop and request more information, to manipulate a data base at run time, and so on (any dynamic action).

It is remarkable how a very simple hint can provide accurate guidance through a very complex maze. Assuming IRL and C rich enough, consider the following, in which the correct path can be found with no mis-steps whatsoever!

(H) *Go fairly far East, until you pass a wriggle in the path, and then go Southeast.*

(M)

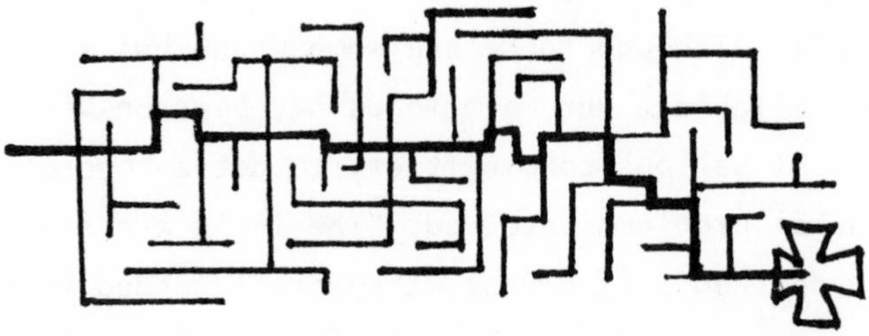

Of course, we can also have landmarks (such as "past the large tree"), making IRL a bit more complex, but simplifying the task of searching (or "running") in actual practice. Note that it is possible to give hints H which are very "misleading" for M, in that C(H) has to do a lot of backtracking. For example, C(H) on

(M_1)

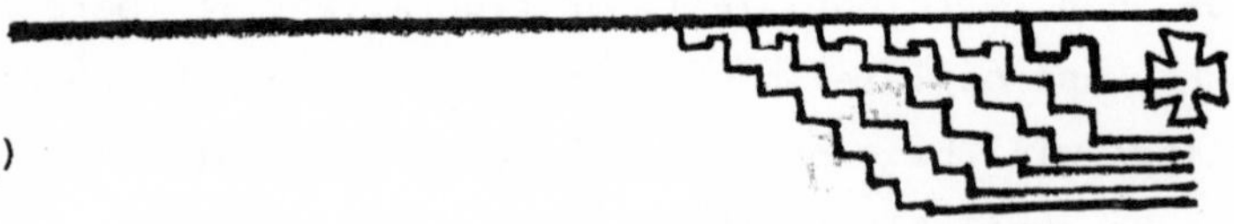

might end up exploring all of each of the non-straight paths before finding the goal.

The reader may see now that there are many subtle issues here, such as: how far to go before giving up on a path which becomes increasingly unpromising; how to measure "promise"; how to decide when one "segment" of a hint (such as "... next go fairly far East then...") has been completed; how the measure "fairly far"; and so on. In fact, even for very simple IRL's, the problem of finding the most efficient C(H) for each given H seems to be very difficult. We do not even attempt this.

There are also the problems of constructing a good H for a given M, and of measuring when M' is "close to" M. We do not discuss these in this paper. But it will be important that H is a good hint for M when we run C(H) on M, and that M' be close to M when we run C(H) on M' as a perturbation of M. In general, intuition is a reliable and sufficient guide for our present purposes.

Despite the mathematical difficulties of finding an "optional" C(H), we have available a good criterion for "adequacy": compare C(H) on M with the performance B(H) of a human being B given H, on M. Of course, B will have to be given access to M only step-by-step, just as C is (rather than all-at-once as the drawing above). We needn't expect C(H) to do better than B(H); if it doesn't do much worse, we will be happy.

The "meta-objective" is to give a concrete example in which fuzziness is used in making a problem easier; e.g., to show that simple fuzzy hints do better than rigid exact directions for maze solving, because (1) they are shorter, and (2) they still apply if the maze is slightly different.

It is clear that we are not discussing a "robot" in the literal sinse, but in the metaphorical sense of a way of coping with distances, directions, landmarks, and so on. However, both "fuzzy" and "planning" in this paper's title are fairly literal. Fuzziness appears in the hints, in the mazes, in the construction of the system, and in our theorizing about it. "Planning" refers to the construction of C(H) from H, and is similar to other "robot planning" projects in artificial intelligence.

4. SOME DETAILS

While our system is not yet programmed, many details of its construction are clear. First, there is not just one IRL and C, but several, embodying more and more complex principles of description of worlds. Moreover, there are many possible worlds to which our idea of compiling hints into search algorithms might be applied.

We are starting with a very simple IRL called IRL_0, which we now describe. An element of IRL_0 is just a sequence of fuzzy vectors, where a *fuzzy vector* is a pair consisting of a fuzzy length and a fuzzy direction. By a "*fuzzy length*" we mean a fuzzy set of lengths: more precisely, a function $L:[0,300] \longrightarrow [0,1]$, where [0,300] is the set of path lengths we are considering, between zero and 300 (actually only integral values will come up) and [0,1] is the unit interval, all real numbers between 0 and 1. L(n) will tell the appropriateness of L referring to the length n. By a *fuzzy direction* we mean a fuzzy set of directions, that is, a function $D:S^1 \longrightarrow [0,1]$, where S^1 is the unit circle, represented for us with degrees, [0,360] with 0=360. D(d) tells the appropriateness of D referring to d. Combining these, <L,D> gives a fuzzy value for each vector <n,d> by multiplication, $L(n)\cdot D(d)$. A more general definition of fuzzy vector would be a fuzzy set of vectors <n,d>, or easier, a function $R^2 \longrightarrow [0,1]$ representing a fuzzy set of vectors <x,y> in the Cartesian plane (R is the real numbers). Our pairs <L,D> describe a sub-class of these in a particularly simple and convenient way.

(In algebraic terms, IRL_0 is the free semigroup generated by the fuzzy vectors, with concatenation as the operation.)

We intend to generate these fuzzy vectors from an algebra of hedges. A few predicates will be given fixed meanings as fuzzy sets, such as "long", "short", and the basic directions (North, Southeast, etc.); these are constants. A number of hedges will be implemented as unary operators. For example, "very" in "very long" is a squaring operator in [Zadeh 1972];

more precisely, (very long)(x) = very (long (x)) = $(\text{long}(x))^2$; "not" work similarly. Sometimes "long" might be a comparative concept, "x is long compared to y" (note that y might actually be an implicit part of the present context). Then "long" is a binary operator in the hedge algebra. We save further details for later publications. [Lakoff 1973] described some limitations of [Zadeh 1972]'s scheme, and it will be interesting to see how this works out in practice.

IRL_1 is a modest extension of IRL_0 permitting free use of "OR" as a binary corrective. (More precisely, IRL_1 is the free algebra generated by fuzzy vectors as constants, and and OR as binary operations; note that A.B, representing "first A then B", already appears in IRL_0.) The kind of English sentence this represents is

> "First go Southeast a short way, or else Northeast, and then go pretty far South; or you could just go far straight South."

Further refinements would permit iterations, while loops, until's, landmarks, etc. See example (H) in Section 3.

It is possible and interesting to include hedge algebra directly into IRL's. This leads to a more complex algebra of descriptions of fuzzy paths; in effect it is a concrete embodiment of [Zadeh 1973]'s notion of linguistic variables! I think that this opens up some quite interesting and potentially significant areas of research. A typical element of such an algebra might be

<VERY(FAR),ALMOST(SOUTH)> • <RATHER(NEAR), SOUTHEAST>.

The purpose of IRL's is to provide a clear target language for the English input, and a source for the compilation of search algorithms. Underlying all this must be some definite choice of a <u>world</u> in which the search problems are to be set and solved. There is considerable scope for choice here;

but clearly the IRL chosen must be adequate for the world. We chose as a basic world a plane square-ruled grid, 100 by 100, and we have developed some convenient ways of representing mazes within this grid. Thus, (M) of section 3 is just the kind of thing we consider.

Now suppose we have H in IRL_0 and M of the sort described above. How will C(H) deal with M? Again, there are many possibilities. We describe in some detail one way of producing algorithms from hints; let this compiler be denoted C_1.

The basic idea is that $C_1(H)$ is block structured, with one block for each vector in H. In computer science, block structuring refers to a nesting of the parts of an algorithm. In this case, we cannot be sure that we have gotten the correct segment in M for the ith vector of H until we have checked that the $i+1^{st}$ is correct. Thus, the part of $C_1(H)$ which matches the ith vector in H to a segment will have to contain the part for the $i+1^{st}$; this leads to the nesting structure. In fact, not until the last segment can we be sure: then either we find the goal, or we don't. Interestingly enough, each block of code has exactly the same internal structure (except the first and the last - and they can be made the same by a trick): assuming some given starting point, they look for a reasonably straight segment of about the right length in about the right direction; success activates a search for the next segment (or the goal), to verify this segment; on the other hand, failure is reported to the next block up, so it can try to find a better starting point.

Of course, there are lots of details to keep track of. We must be sure not to repeat mistakes. "Failure" is not absolute, but at a certain level, since the match between the vector in H and the segment in M is anyway only fuzzy; thus, these levels will have to be kept back of, and sometimes readjusted. "Straightness" will be one of the most important criteria for remaining in a block, and has to be measured in some suitable fuzzy way. We are entitled to assume that H is reasonable, i.e., not misleading or contrived. But still, it would be nice if $C_1(H)$ would always find the solution, even if H is way off; this is easily achieved by providing resources to remember failures and to backtrack.

We can now see more clearly a good standard of performance for $C_1(H)$; it should do <u>significantly</u> better than a breadth first (or depth first) search (see [Nilsson 1971]) when H fits M. Of course, $C_1(H)$ will do more computation than these methods, since it is trying to use H, and they are not. But on the average, the blind search methods will make many mistakes, and do a lot more searching (when H matches M).

The advantage of each block of $C_1(H)$ having the same structure, is that C_1 can be given a simple recursive structure. It should be easy to write and fast to run on H.

There is a pretty similar compiler for IRL_1; the nesting of blocks is no longer just one-within-one, since the OR feature provides multiple options for success.

We have chosen QLISP as object code for compilation because of its convenient and powerful features. It has built in backtracking, pattern matching, and theorem proving. Of course, it runs slower than ordinary LISP (or PL/I, or whatever), though we might use just LISP for the implementation of C_1, which is comparatively simple.

Unfortunately, I cannot report precise experimental results at this writing but it will probably not have escaped the reader's attention that I am pretty confident of success.

5. CONCLUSIONS AND EXTENSIONS

The main point we have been trying to make is that fuzziness, far from being a difficulty, is often a convenience, or even an essential, in communication and control processes.

It might be noted that in ordinary human communications, the ability to stretch and modify word meanings is essential. There are many more situations occurring in life than we have ready-made tags for. Even so simple a word as "chair" has all kinds of readily visible complexities to its use. It has ambiguity, in that it has more than one distinct area of application (in addition to the usual, we have "Would the chair recognize my motion now?" and "Would you like to chair this meeting?"). Vagueness (or fuzziness) we have already discussed in Section 2; it is closely related to generality, the possibility of referring to more than one object. In fact without generality, language would be almost impossible. Imagine if we had to give each chair a new proper name before we could talk about it! As far as "stretchiness" is concerned, note that some people make a living designing objects they call "chairs," but in which other people might sit with only the greatest reluctance. The concept of "chair" is constantly evolving, in fact. While our system does not (yet) exhibit evolution, it should clearly exhibit the usefulness of vagueness even in purely mechanical situations.

We believe this is only a beginning, that there are very many uses for vagueness in special languages for controlling processes. For example, consider navigation systems and machine tool controllers. It is not just that the fuzziness is easier for humans (the way they usually describe processes), but it is actually more efficient for the machines too! We must give up the idea that "more exact" is always better, at least in communication situations.

We have already mentioned some extensions we are considering for our work. Obviously we can experiment with more complex IRL's, and other kinds of compilers into search algorithms. We can design fancy evaluation experiments, and we can apply them also to human performance in these tasks. Perhaps we could figure out the kinds of algorithms people actually use. We could also develop system capabilities for interaction, dialogue, display, and learning, each of which would add a new dimension and raise interesting questions.

An area where more research is needed, is the translation from English into IRL's, and the design of IRL's maximally appropriate for this process. Our initial crude approach will be closer to cataloguing some translations of IRL's into English.

We also have mentioned adding "landmarks" to the problem; that is, providing further information than just what the alternatives are at each node, and whether or not the goal has been reached. There could be a property list at each node, giving (for example) a COLOR. This could of course be fuzzy. Or it might be exact (a wavelength) while the description of what we are looking for is vague (e.g., RED). Some nodes might be labelled DRUGSTORE or GASSTATION, and we might consider giving fuzzy information about these landmarks at nearby nodes (e.g., "possible drugstore three squares straight ahead"). A "tower" might be more or less visible from various distances. And so on and so on. The question of how to "best" use such information in searching needs further investigation. (Note this information is available only at run time, not at compile time.)

The approach could be applied to entirely different types of problem from maze running, ranging from machine tool programming to game playing. Landmarks would be much more important here than for the simple world of IRL_0.

I regard the fact that there are so many questions and possibilities as highly encouraging. Obviously, this is only an initial approach to a large territory.

REFERENCES

[Ambler-Barrow-Brown-Burstall-Popplestone 1973]. "A Versatile Computer - Controlled Assembly System," Proc. Third International Joint Confr. on Artificial Intelligence, Stanford, Ca., pp. 298-307.

[Chomsky 1957] Syntactic Structures, Mouton, the Hague.

[Goguen 1967] "L-Fuzzy Sets", Jnl. Math. Analysis and Applications, 18, pp. 145-174.

[Goguen 1969] "The Logic of Inexact Concepts," Synthese, 19, pp. 325-373.

[Goguen 1972] "Hierarchical Inexact Data Structures in Artificial Intelligence Problems," Proc. Fifth Hawaii International Conference on Systems Sciences, Honolulu, pp. 345-347.

[Goguen 1974] "Concept Representation in Natural and Artificial Languages: Axioms, Extensions and Applications for Fuzzy Sets," International Jnl of Man-Machine Studies, 6.

[Lakoff 1973] "Hedges: A Study in Meaning Criteria and the Logic of Fuzzy Concepts" Jnl. of Philosophical Logic, 2, pp. 458-508.

[Minsky 1974] "A Framework for Representing Knowledge", MIT Artificial Intelligence Memo No. 306, Cambridge, Ma.

[Nilsson 1971] Problem-Solving Methods in Artificial Intelligence, McGraw-Hill.

[Reddy 1972] "Reference and Metaphor in Human Language", Ph.D. Dissertation, Dept. of English, University of Chicago.

[Reboth-Sacerdoti 1973] "A Preliminary QLISP Manual," Technical Note 81, Artificial Intelligence Center, Stanford Research Institute, Stanford, Ca.

[Winograd 1972] Understanding Natural Language, Academic Press.

[Zadeh 1965] "Fuzzy Sets", Information and Control, 8, pp. 338-353.

[Zadeh 1968] "Fuzzy Algorithms", Information and Control, 12, pp. 94-102.

[Zadeh 1972] "A Fuzzy Set-Theoretic Interpretation of Hedges", Memorandum M-335, Electronics Research Laboratory, Univ.

of Calif. at Berkeley.
[Zadeh 1973] "Outline of a New Approach to the Analysis of Complex Systems and Decision Processes" IEEE Transactions on Systems, Man and Cybernetics, Vol. SMC-3.

AN APPROACH TO PATTERN RECOGNITION AND ASSOCIATIVE MEMORIES USING FUZZY LOGIC

Masamichi Shimura
Faculty of Engineering Science
Osaka University, Toyonaka, Osaka
JAPAN

ABSTRACT

The present paper discusses a classification algorithm of multicategory classifiers which is based on fuzzy logic and can be used even if statistical independency of pattern vectors is violated. The algorithm is comparatively simple and requires an unusually short time for learning. We also consider the memory systems that recall entities stored in an associative manner. Entities are stored in the form of fuzzy matrix and are recalled by fuzzy logic according to the conditional probability of each category. A computer simulation is made in English character reading and its results are presented.

1. INTRODUCTION

Events occurring in the real world can be classified into three categories; deterministic, probabilistic and fuzzy events. In a pattern recognition system, patterns are usually probabilistic rather than deterministic because patterns themselves occur probabilistically or patterns are generally affected by random noise from poor printing, dirt on the paper, etc. If statistical information regarding input patterns is available, statistical or probabilistic decision making is useful and powerful in recognition of the patterns. Therefore, statistical approaches to pattern recognition have been discussed and widely adopted by many authors. However, the situation is often encountered where no a priori information is available or where the theory of a large number cannot be used. Furthermore, in a learning system, its probability model must be inferred from the training set. That is, the training patterns as well as the test patterns must satisfy some statistical assumptions such as probability distributions. In a practical situation, however, one could hardly expect the distribution of the patterns to correspond precisely to the distribution assumed. In addition, it is often unreasonable to assume the statistical independency of the components of pattern vectors, whereas the independency has been assumed in most pattern recognition systems. Chow [2],[3] has presented the non-linear categorizer under the assumption that each element depends only on its nearest neighbors.

From various points of view as mentioned above, pattern recognition is considered essentially fuzzy, because there exists no precise boundary between categories. Particularly in the case where the number of sample patterns is small and where the statistical independency cannot be assumed, a fuzzy decision process is useful in classification of unknown patterns and yields rather simple methods of classification.

Fuzzy set concepts and fuzzy algorithms proposed by Zadeh have been developed since 1965, and they have been applied to various fields; pattern recognition, automata theory, control systems, language, etc. Zadeh [8], [9], [10] has discussed the advantage of using the fuzzy sets concept in engineering systems and studied its algorithms. Wee and Fu [7] have formulated a class of fuzzy automata and discussed its application to pattern recognition with non-supervised learning. In their pattern recognition system, the decision is made according to the grade of membership functions of each category and the learning is performed by the fuzzy automata.

The present paper deals with a class of pattern recognition using fuzzy logic. Input patterns are usually given as deterministic data although they may contain some fuzziness, and the output decision is also deterministic. Generally, patterns are presented on mosaic cells corresponding to retina, and each cell gives a logic "1" output if the pattern lies mainly in the cell and a logic "0" output otherwise. It is sometimes difficult to decide whether the pattern lies mainly in the cell or not. From this point of view, it is preferable to employ multi-valued logic for feature extraction instead of two-valued logic, although the former is comparatively difficult to be handled, particularly in hardware. Thus we assume that input data are deterministic. The classification process should be fuzzy instead. We consider, therefore, the fuzzy decision process under the assumption where the input and output of the systems are deterministic.

We note that probability often reflects the degree of belongingness of patterns to each category in pattern recognition systems, because a measure of uncertainty is often associated with it. In our system, therefore, the classification criteria are based on the maximization of the possible certainty of correct classification according to the probability.

Some of the advantages of our system are as follows:

(i) The system is flexible and easily extendable.

(ii) The learning algorithm is simple and requires a comparatively short time for learning.

(iii) Classification can be made even in the case where no statistical property of input patterns is available. (The learning is essentially nonparametric.)

(iv) The system is realizable with hardware.

Association is a well known function of the human brain to accumulate information and read-out the accumulated entities. Nakano [6] has proposed a model of associative memory called the Associatron, which is considered a digitized Perceptron. Kohonen [4] studied the analog memory similar to the Associatron using the correlation matrix. The associative memory considered here consists of fuzzy matrices based on the probability of occurrence of each component. In the memory, information is stored in a distributed fashion so that stored information can be recalled even when a part of the memory is destroyed. A computer simulation is made with English characters and two types of associative memories are presented.

2. MULTICATEGORY PATTERN CLASSIFICATION

As is described in the introduction, the decision of categories is deterministic although the classification process is fuzzy. Thus, the decision is made based on the maximum grade of membership. That is, if ${}^{\alpha}f(X) = \max_{\mu} {}^{\mu}f(X)$, then pattern X is decided to be from category C_{α}, where ${}^{\mu}f(X)$ is a grade of membership of category C_{μ} $(\mu=1,\ldots,R)$ for pattern X.

Let $X = (x_1,\ldots,x_n)$ be an n-dimensional (preprocessed) pattern vector taking value +1 or -1 and ${}^{\mu}Y = ({}^{\mu}y_1,\ldots,{}^{\mu}y_n)$ be an n-dimensional direct fuzzy output for category C_{μ}. Also, let ${}^{\mu}F = \{{}^{\mu}f_{ij}\}$ be a fuzzy matrix for category C_{μ}.

Now we will consider the simple fuzzy classifier with an external teacher as shown in Figure 1. Preprocessed pattern X is applied to fuzzy processor G consisting of fuzzy matrices ${}^{\mu}G = \{{}^{\mu}g_{ij}\}$. Since the output of this processor ${}^{\mu}Y = ({}^{\mu}y_1, \ldots, {}^{\mu}y_n)$ should be fuzzy, the non-fuzzy output for category C_μ, η_μ is assumed to be the maximum of ${}^{\mu}y_j$, i.e.

$$\eta_\mu = \max_j {}^{\eta}y_j. \tag{1}$$

As described before, pattern X is decided to be from category C_α if

$$\eta_\mu = \max_\mu \eta_\mu. \tag{2}$$

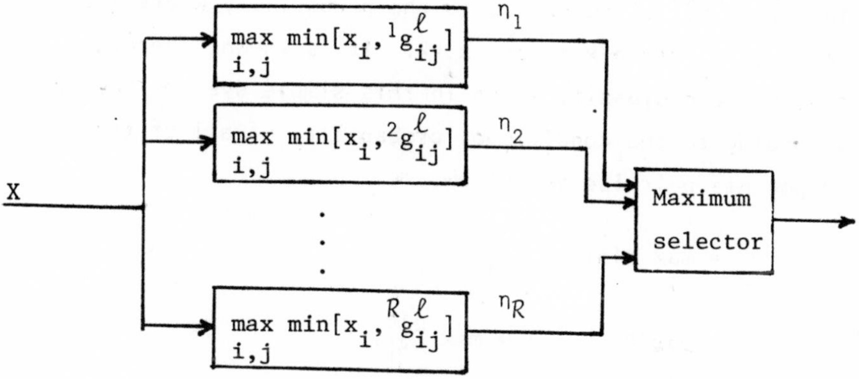

Figure 1. The Structure of the Simple Classification System.

Define the direct fuzzy output ${}^{\mu}y_j$ by

$${}^{\mu}y_j = \max_i \min(x_i, {}^{\mu}g^{\ell}_{ij}). \tag{3}$$

The learning behavior is reflected by having fuzzy matrices ${}^{\mu}G^{\ell}$ and their elements ${}^{\mu}g^{\ell}_{ij}$ are obtained by the iterative process as follows:

$$^{\mu}g^{\ell}_{ij}(t+1) = \alpha^{\ell}_{ij}(t)\,^{\mu}g^{\ell}_{ij}(t) + (1-\alpha^{\ell}_{ij}(t))\phi(X \in C_{\mu}). \quad (4)$$

where $^{\mu}g^{\ell}_{ij}(t)$ denotes $^{\mu}g^{\ell}_{ij}$ at the t^{-th} iteration in a learning phase, and $\phi(\theta) = 1$ if θ is true and 0 otherwise. Set

$$\alpha^{\ell}_{ij}(t) = 1 - \frac{1}{N(t,\ x_i=1,\ x_j=\ell)}, \quad (5)$$

where $N(t,\ x_i=1,\ x_j=\ell)$ is the number of patterns of which the elements $x_i=1$ and $x_j=\ell$ up to the t^{-th} iteration. Then we have

$$^{\mu}g^{\ell}_{ij} \equiv \lim_{t \to \infty} {^{\mu}g^{\ell}_{ij}}(t) = \mathrm{prob}(X \in C_{\mu} | x_i=1,\ x_j=\ell). \quad (6)$$

That is, $^{\mu}g^{\ell}_{ij}(t)$ converges to the probability where X belongs to C_{μ} given $x_i=1$ and $x_j=\ell$. It is known, therefore, that pattern classification in this simple system is made according to the conditional probability of all of the components pairwise as follows:

$$\begin{aligned} ^{\mu}y_j &= \max_{i} \min(x_i,\ ^{\mu}g^{\ell}_{ij}) \\ &= \max_{\substack{i \\ (x_i=1)}}(0,\ ^{\mu}g^{\ell}_{ij}) = \max_{i}\ ^{\mu}g^{\ell}_{ij} \\ &= \max_{\substack{i \\ (x_i=1)}} \mathrm{prob}(X \in C_{\mu} | x_i, x_j) \equiv y_j^{(1)}. \end{aligned} \quad (7)$$

Next, we consider matrices $^{\mu}H^{\ell} = \{^{\mu}h^{\ell}_{ij}\}$ which are obtained by the iterative manner similar to that in obtaining $^{\mu}G^{\ell}$ such as

$$^{\mu}h^{\ell}_{ij} = \lim_{t \to \infty}{=}\ ^{\mu}h^{\ell}_{ij}(t)$$

$$\mu_{h_{ij}^{\ell}}(t+1) = \beta_{ij}^{\ell}(t)\mu_{h_{ij}^{\ell}}(t) + (1-\beta_{ij}^{\ell}(t))\phi(X \in C_{\mu}), \quad (8)$$

where

$$\beta_{ij}^{\ell}(t) = 1 - \frac{1}{N(t,\ x_i=0,\ x_j=\ell)}\ , \quad (9)$$

Using $\mu_{H^{\ell}}$, we define $\mu_{y_j'}$ as

$$\mu_{y_j'} = \min_i \max(x_i,\ \mu_{h_{ij}^{\ell}}). \quad (10)$$

Since

$$\mu_{h_{ij}^{\ell}} = \mathrm{prob}(X \in C_{\mu} | x_i=0,\ x_j=\ell), \quad (11)$$

we have

$$\begin{aligned}\mu_{y_j'} &= \min_{\substack{i \\ (x_i=0)}} \max(1,\ \mu_{h_{ij}^{\ell}}) \\ &= \min_{\substack{i \\ (x_i=0)}} \mu_{h_{ij}^{\ell}} \\ &= \min_{\substack{i \\ (x_i=0)}} \mathrm{prob}(X \in C_{\mu}|x_i,x_j). \quad (12)\end{aligned}$$

It is reasonable to introduce the compound system defined as

$$\mu_{y_j} = \max[\max_{\substack{i \\ (x_i=1)}} \mathrm{prob}(X \in C_{\mu}|x_i,x_j),\ \max_{\substack{i \\ (x_i=0)}} \mathrm{prob}(X \in C_{\mu}|x_i,x_j)]. \quad (13)$$

Considering that

$$\min_{\substack{i \\ (x_i=0)}} \mathrm{prob}(X \in C_{\mu}|x_i,x_j) = 1 - \max_{\substack{i \\ (x_i=0)}} \mathrm{prob}(X \in C_{\mu}|x_i,x_j),$$

we have from (13)

$$\mu_{y_j} = \max_i \text{prob}(X \in C_\mu \; x_i, x_j) \quad (14)$$

$$= \max(y_j^{(1)}, y_j^{(0)}), \quad (15)$$

where

$$y_j^{(0)} = 1 - \mu_{y_j'} \; . \quad (16)$$

Note that

$$\mu_{g_{ij}^1} = \mu_{g_{ji}^1}$$
$$\mu_{h_{ij}^0} = \mu_{h_{ji}^0} \quad (17)$$

and that

$$\mu_{g_{ij}^0} = 1 - \mu_{h_{ij}^1} \; . \quad (18)$$

From (7) and (10), we can write

$$y_j^{(1)} = \max[\max_{\substack{i \\ (x_i=0)}} \min(x_i, \mu_{g_{ij}^1}), \max_{\substack{i \\ (x_i=1)}} \min(x_i, \mu_{g_{ij}^0})]$$

$$y_j^{(0)} = 1 - \min[\min_{\substack{i \\ (x_i=0)}} \max(x_i, \mu_{h_{ij}^1}), \min_{\substack{i \\ (x_i=0)}} \max(x_i, \mu_{h_{ij}^0})]$$

$$= \max[\max \min(1-x_i, 1-\mu_{h_{ij}^1}), \max \min(1-x_i, 1-\mu_{h_{ij}^0})].$$

Here we introduce trigonometric matrices $\overline{G}$ and $\overline{H}$ in order to omit redundant elements such as

$$\begin{cases} \mu_{\bar{g}^1_{ij}} = \begin{cases} \mu_{g^1_{ij}} & \text{for } i \geqq j \\ 0 & \text{for } i < j \end{cases} \\ \mu_{\bar{h}^0_{ij}} = \begin{cases} \mu_{h^0_{ij}} & \text{for } i \geqq j \\ 1 & \text{for } i < j \end{cases} \\ \mu_{\bar{g}^0_{ij}} = \begin{cases} \mu_{g^0_{ij}} & \text{for } i < j \\ & \text{for } i \geqq j \end{cases} \\ \mu_{\bar{h}^1_{ij}} = \begin{cases} \mu_{h^1_{ij}} & \text{for } i < j \\ 1 & \text{for } i \geqq j. \end{cases} \end{cases} \tag{20}$$

Using these matrices, the following fuzzy output μ_{y_j} is obtained from (15) and (19).

$$\begin{aligned} y_j &= \max[\mu_{y_j}(1), \mu_{y_j}(0)] \\ &= \max[\mu_{\xi_j}(1), \mu_{\xi_j}(0)] , \end{aligned} \tag{21}$$

where

$$\begin{aligned} \mu_{\xi_j}(1) &= \max[\max_{\substack{i \\ (x_i=1)}} \min(x_i, \mu_{\bar{g}^1_{ij}}),\ \max_{\substack{i \\ (x_i=0)}} \min(1-x_i, 1-\mu_{\bar{h}^1_{ij}})] \\ \mu_{\xi_j}(0) &= \max[\max_{\substack{i \\ (x_i=1)}} \min(x_i, \mu_{\bar{g}^0_{ij}}),\ \max_{\substack{i \\ (x_i=0)}} \min(1-x_i, 1-\mu_{\bar{h}^0_{ij}})]. \end{aligned} \tag{22}$$

Equation (22) can be rewritten as follows:

$$\begin{aligned} \mu_{\xi_j}(\ell) &= \max[\max \min(x_i, \mu_{\bar{g}^\ell_{ij}}),\ \max \min(1-x_i, 1-\mu_{\bar{h}^\ell_{ij}})] \\ &= \frac{1}{2} + \frac{1}{2} \max[\max_i \min(2x_i-1, 2\mu_{\bar{g}^\ell_{ij}}-1), \\ &\qquad -\min_i \max(2x_i-1, 2\mu_{\bar{h}^\ell_{ij}}-1)]. \end{aligned} \tag{23}$$

Let $X' = (x'_1,\dots,x'_n)$ be the new pattern vector of which the components $x'_i = 2x_i - 1$ take the value 1 or -1, and $^{\mu}F^{\ell}$ be fuzzy matrices such as $^{\mu}F^{\ell} = {}^{\mu}\overline{G}^{\ell'} + {}^{\mu}\overline{H}^{\ell'} - 1$ $(\ell = 2\ell' - 1)$. Then considering (20), we have

$$^{\mu}y_j = \max[\max \min(x'_i,\ {}^{\mu}f^{\ell}_{ij}),\ -\min \max(x'_i,\ {}^{\mu}f^{\ell}_{ij})]. \tag{24}$$

Also, we have the learning algorithm of obtaining matrices $^{\mu}F^{\ell}$ as follows:

$$^{\mu}f^{\ell}_{ij}(t+1) = \begin{cases} \alpha^{1}_{ij}(t)\,^{\mu}f_{ij}(t) + (1-\alpha^{1}_{ij}(t))\phi(X' \in C_{\mu}) & \text{for } i \geqq j \\ \alpha^{-1}_{ij}(t)\,^{\mu}f_{ij}(t) - (1-\alpha^{-1}_{ij}(t))\phi(X' \in C_{\mu}) & \text{for } i < j \end{cases} \tag{25}$$

where

$$\alpha^{m}_{ij}(t) = 1 - \frac{1}{N(t, x'_i x'_m = m)} . \tag{26}$$

Note that $^{\mu}f^{\ell}_{ij}$ is positive if $i \geqq j$, for $\ell = 1$ or if $i < j$ for $\ell = -1$, and negative otherwise, and $|^{\mu}f^{\ell}_{ij}|$ takes value between 0 and 1.

The structure of the compound system is shown in Figure 2.

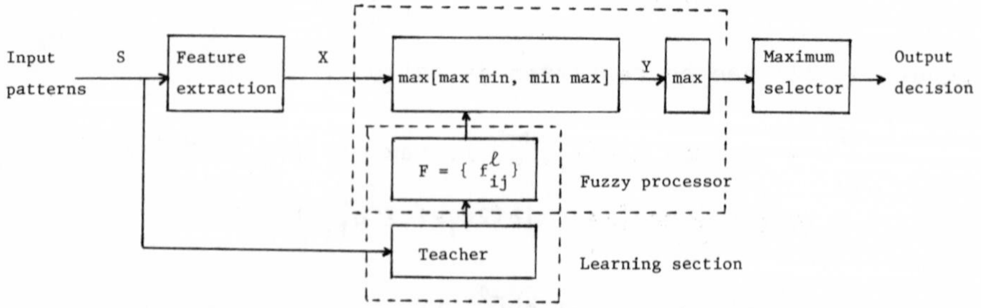

Figure 2. The Structure of the Compound Classification System.

3. ASSOCIATIVE MEMORIES I

Associative memories are considered distributed memories as discussed in the introduction. Let $X^{\mu} = (x_1^{\mu},\ldots,x_n^{\mu})$ be n-dimensional input patterns, where x_j takes value 1 or -1. The input patterns are accumulated on memory elements in the form of fuzzy matrix F, and recalled by the key input $Q^{\mu} = (q_1^{\mu},\ldots,q_n^{\mu})$ corresponding to pattern X^{μ}. In the recalling process, q_i is given to the associative memory, taking value 1, -1 or 0. Known elements take the value 1 or -1 and unknown (or forgotten) elements take the value 0. Figure 3 illustrates pattern A given on the 5 x 6 mosaic cells, where shaded, white and zero cells indicate 1, -1 and 0, respectively. Even if the incomplete pattern as shown in Figure 3(b) is applied to the memory as a key input, the associative memory should recall the accumulated pattern shown in Figure 3(a). In this case key input Q^{μ} can be considered as a variation of accumulated pattern X^{μ}. In the associative memory, therefore, forgotten elements of a pattern are recalled by using n'(<n) known elements. The structure of the associative memories is shown in Figure 4.

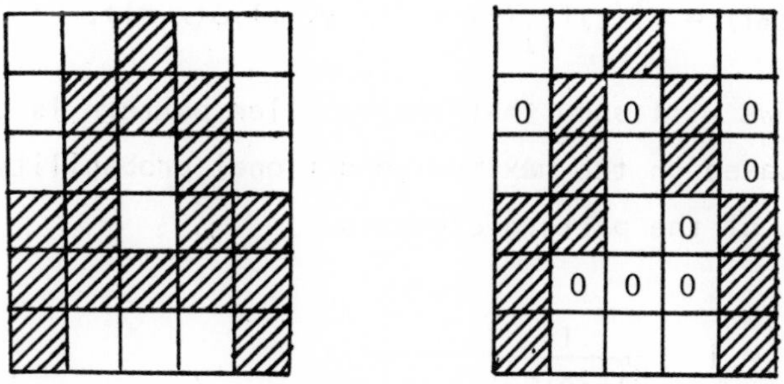

(a) (b)

Figure 3. (a) Original Pattern and (b) Key Input.

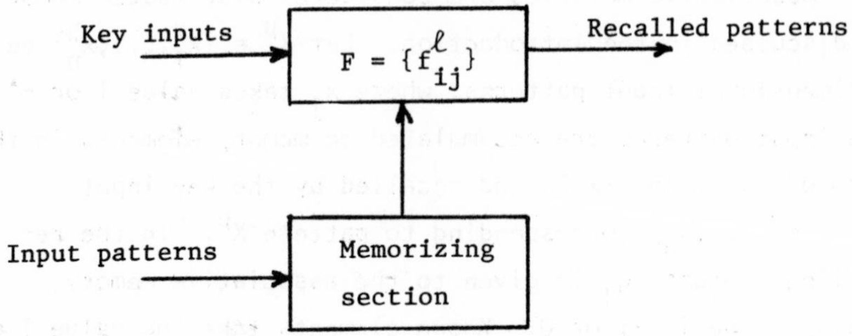

Figure 4. The Structure of the Associative Memory.

Let $Z^{\mu} = (z_1^{\mu},\ldots,z_n^{\mu})$ be the recalled pattern when key input Q^{μ} is applied. The recall of a particular z_j^{μ} is made by

$$z_j = \psi[\max \min(|q_i^{\mu}|, f_{ij}^{\ell}) + \min \max(1-|q_i^{\mu}|, f_{ij}^{\ell})] \qquad (27)$$

where

$$\psi(\theta) = \begin{cases} 1 & \text{if } \theta > 1 + \delta \\ 0 & \text{if } 1 - \delta \leqq \theta \leqq 1 + \delta \\ -1 & \text{if } 1 - \delta > \theta . \end{cases}$$

Fuzzy matrices F are obtained iteratively as

$$f_{ij}^{\ell}(t+1) = \gamma_i^{\ell}(t) f_{ij}^{\ell}(t) + (1 - \gamma_i^{\ell}(t))\phi(x_j = 1). \qquad (28)$$

Now, we will show that the recalled pattern is constructed based on the maximum conditional probability of occurrence of the pattern elements.

Set

$$\gamma_i^{\ell}(t) = 1 - \frac{1}{N(t, x_i = \ell)} \quad , \qquad (29)$$

where $N(t, x_i=\ell)$ is the number of patterns of which the elements $x_i = \ell$ up the t^{-th} iteration in the learning phase. Then, we have

$$f_{ij}^{\ell} = \lim_{k \to \infty} f_{ij}^{\ell}(t) = \text{prob}(q_j=1 | q_i=\ell) \ . \tag{30}$$

From (28) and (30),

$$\max_i \min(|q_i|, f_{ij}^{\ell}) = \max_i(0, f_{ij}^{\ell}) = \max_i \text{prob}(q_j=1|q_i)$$

$$\min_i \max(1-|q_i|, f_{ij}^{\ell}) = \min_i(1, f_{ij}^{\ell}) = \min_i \text{prob}(q_i=1|q_i)$$

$$= 1 - \max_i \text{prob}(q_j=-1|q_i) \ .$$

Note that $|q_i|$ becomes 1 if the i^{-th} element is known and 0 otherwise. Therefore,

$$z_j^{\mu} = \psi[\max_i \text{prob}(q_j=1|q_i) - \max_i \text{prob}(q_j=-1|q_i) + 1]$$

or

$$z_j^{\mu} = \begin{cases} 1 & \text{if} \quad \max_i \text{prob}(q_j=1|q_i) > \max_i \text{prob}(q_j=-1|q_i) + \delta \\ 0 & \text{if} \quad |\max_i \text{prob}(q_j=1|q_i) - \max_i \text{prob}(q_j=-1|q_i)| \leqq \delta \\ -1 & \text{if} \quad \max_i \text{prob}(q_j=-1|q_i) > \max_i \text{prob}(q_j=1|q_i) + \delta \ . \end{cases} \tag{31}$$

That is, forgotten elements are recalled by comparison of the maximum probabilities of being 1 and -1.

Consider the special case where complete pattern X is applied as a key input. In this case,

$$f_{jj}^{\ell} = \lim_{t \to \infty} f_{jj}^{\ell}(t) = \begin{cases} 1 & \text{for } \ell = 1 \\ 0 & \text{for } \ell = -1 \ . \end{cases} \tag{32}$$

Since

$$\max_i \min(|q_i^{\mu}|, f_{ij}^{\ell}) = 1$$

$$\min_i \max(1-|q_i^{\mu}|, f_{ij}^{\ell}) = 0 \ ,$$

z_j^μ is always 1 even if the corresponding element q_j^μ is -1. To circumvent this incorrect recall, we use new fuzzy matrices $F^{\ell'}$ of which the element $f_{ij}^{\ell'}$ is

$$f_{ij}^{\ell'} = (1-2\varepsilon)f_{ij}^{\ell} + \varepsilon \qquad \text{for } i \neq j, \tag{33}$$

where $0 \leqq \varepsilon \leqq 1$. Using these matrices, therefore, we have

$$z_j^\mu = \psi[\max\min(|q_i^\mu|, f_{ij}^{\ell'}) + \min\max(1-|q_i^\mu|, f_{ij}^{\ell'})]$$

$$= \begin{cases} \psi[1 + c_1] & \text{for } \ell = 1 \\ \psi[c_2 + 0] & \text{for } \ell = -1, \end{cases}$$

where $c_1 \leqq \varepsilon$ and $c_2 \leqq 1 - \varepsilon$. If $\varepsilon > \delta$, then

$$z_j^\mu = \begin{cases} 1 & \text{if } q_j^\mu = 1 \\ -1 & \text{if } q_j^\mu = -1. \end{cases} \tag{34}$$

This means that elements of the recalled pattern are those of the corresponding key input or that known elements give the same output as themselves.

If we set $\varepsilon = 0$ and $\delta > 0$, however, equation (34) does not hold. In this case, the recalled pattern may be the corresponding original pattern even if the key input is partially wrong. That is, the memory may have the ability to correct input patterns according to a set of conditional probabilities.

4. ASSOCIATIVE MEMORIES II

Here we consider another type of associative memory that has the ability to memorize sequences of patterns. Such a memory is called a dynamic associative memory. The structure of such a memory is similar to that discussed before but has feedback loops as shown in Figure 5. In a learning

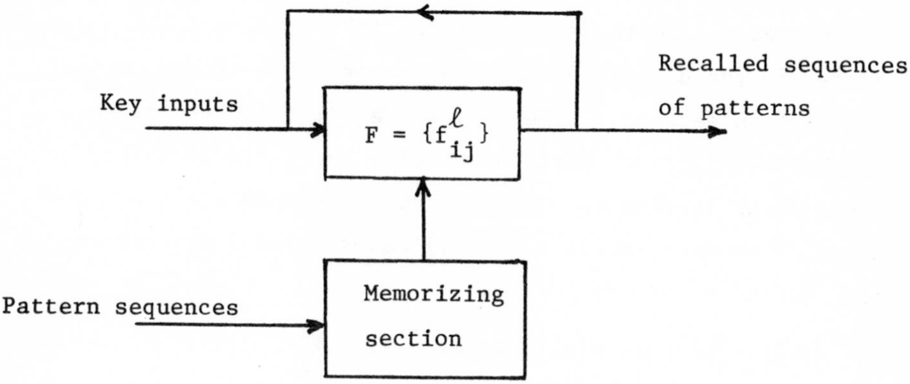

Figure 5. The Structure of the Dynamic Associative Memory.

phase, patterns are presented to the memory system in a given sequence. Each pattern in the sequence is recalled by a key pattern which is usually the top in the sequence. If a sequence of characters A, B, C, D, E is stored, then the output of the system is B, C, D, E when key pattern A' corresponding to A is given.

It can be conjectured that the probability of correct recall generally becomes large with increase of memory capacity used. However, one of the most important things in constructing such an associative memory is that the memory capacity should be small, so long as the probability of correct recall is comparatively high.

Now, we define the grade of membership f_j and g_j, where $z_j(t+1)$ takes values 1 and -1, respectively, by

$$f_j(t+1) = \max_i \min[|q_i(t)|, f^{\ell}_{ij}] \tag{35}$$

$$g_j(t+1) = \max_i \min[|q_i(t)|, 1-f^{\ell}_{ij}]$$

$$= 1 - \min_i \max[1-|q_i(t)|, f^{\ell}_{ij}],$$

where q_i is the i^{-th} element of input pattern Q(t) and f^{ℓ}_{ij} is the grade of membership where $z_j(t+1) = 1$ for $q_i(t) = \ell$. Since, in the dynamic associative memory, the stored contents are pattern sequences, the (t+1)st recalled pattern depends on not only the t^{-th} pattern but also the patterns which appeared up to time t. Then, let us introduce the n^{-th} order grade of membership where $z_j(t+1)$ takes value 1 defined in the iterative form as follows:

$$f_j^{(n)}(t+1) = \max[\max_i \min[|q_i(t)|, {}^{(1)}f^{\ell}_{ij}], f_j^{(n-1)}(t)]$$

$$f_j^{(n-1)}(t) = \max[\max_i \min[|q_i(t-1)|, {}^{(2)}f^{\ell}_{ij}], f_j^{(n-2)}(t-1)]$$

$$\vdots \qquad\qquad \vdots$$

$$f_j^{(1)}(t-n+2) = \max[\max_i \min[\ q_i(t-n+1)\ , {}^{(n)}f_{ij}], f_j^{(0)}(t-n+1)]$$

$$f_j^{(0)}(t-n+2) = 0.$$

In the same manner, the n^{-th} order grade of membership, $g_j^{(n)}(t+1)$ where $z_j(t+1)$ takes value -1 is defined as

$$g_j^{(n)}(t+1) = 1 - h_j^{(n)}(t+1)$$

$$h_j^{(n)}(t+1) = \min[\min_i \max[1-|q_i(t)|, {}^{(1)}f^{\ell}_{ij}], h_j^{(n-1)}(t)]$$

$$\vdots \qquad\qquad \vdots \qquad\qquad (37)$$

$$h_j^{(k)}(t-n+k) = \min[\min_i \max[1-|q_i(t-n+k-1)|, {}^{(n-k+1)}f^{\ell}_{ij}], h_j^{(k-1)}(t-n+k-1)]$$

$$\vdots \qquad\qquad \vdots$$

$$h_j^{(0)}(t-n+1) = 1.$$

Since the outputs are non-fuzzy, the recalled pattern Z(t+1) is decided by

$$z_j(t+1) = \begin{cases} 1 & \text{if } f_j^{(n)}(t+1) > g_j^{(n)}(t+1) + \delta \\ 0 & \text{if } |f_j^{(n)}(t+1) - g_j^{(n)}(t+1)| < \delta \\ -1 & \text{if } f_j^{(n)}(t+1) < g_j^{(n)}(t+1) - \delta. \end{cases} \tag{39}$$

The learning algorithm of obtaining ${}^{(n)}f_{ij}^{\ell}$ is

$${}^{(n)}f_{ij}(t+1) = \alpha_i^{\ell}\, {}^{(n)}f_{ij}^{\ell}(t) + (1-\alpha_i^{\ell})\phi(x_j(t+1)=1), \tag{40}$$

where

$$\alpha_i^{\ell}(t) = 1 - \frac{1}{N(t,x_i=\ell)} \qquad (\ell=1 \text{ or } -1)$$

$N(t,x_i=\ell)$ = the number of patterns of which the element $x_i = \ell$ up to t.

Therefore, the values of ${}^{(k)}f_{ij}^{\ell}$ after learning becomes

$${}^{(k)}f_{ij}^{\ell}(M) = \text{prob}(x_j(t+1)=1\,|\,x_i(t-k+1)=\ell), \tag{41}$$

where M is the number of patterns appeared in the sequences.

The schematic diagram of the dynamic associative memory is shown in Figure 6. The notations used in the figure are as follows:

$Y = \{|x_i|\}$, $\overline{Y} = \{1-|x_i|\}$, ${}^{(n)}F = \{{}^{(n)}f_{ij}^{\ell}\}$, ${}^{(n)}H = \{{}^{(n)}h_{ij}^{\ell}\}$,

$\bar{\circ}$ = max min and $\underline{\circ}$ = min max.

5. COMPUTER SIMULATION

5.1. Pattern Classification

The patterns used in the computer experiments are

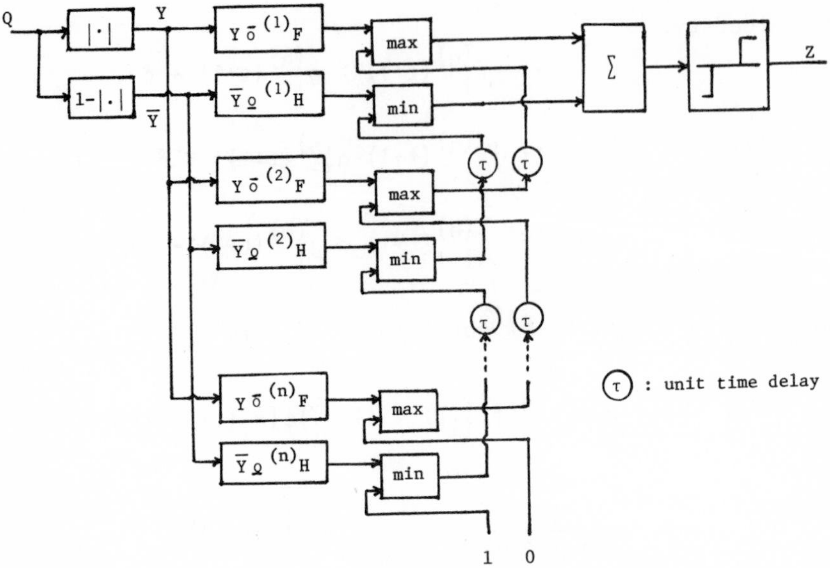

Figure 6. The Structure of the Dynamic Associative Memories.

handwritten English characters A, B,...,Z. They are given on 9 x 11 mosaic cells as shown in Figure 7(a). Each mosaic cell gives a logic +1 output if the letter lies mainly in the cell and a logic 0 output otherwise. From each pattern on the mosaic cells $S = (s_{11}, \ldots, s_{119})$, twenty-five features $x_1, x_2, \ldots, x_{25}$ as shown in Figure 7(b) are extracted to reduce the dimensions.

The original 9 x 11 features should be used for general pattern recognition. Since the probability of correct recognition depends upon the variation of handwritten characters, we made several experiments to study the recognition performance of the classifier. Examples of handwritten characters used are shown in Figure 8.

Firstly, we made a computer simulation of classifying five characters A, B, C, D, E using the original (no preprocessed) patterns on 5 x 6 mosaic cells. Twenty-five (five

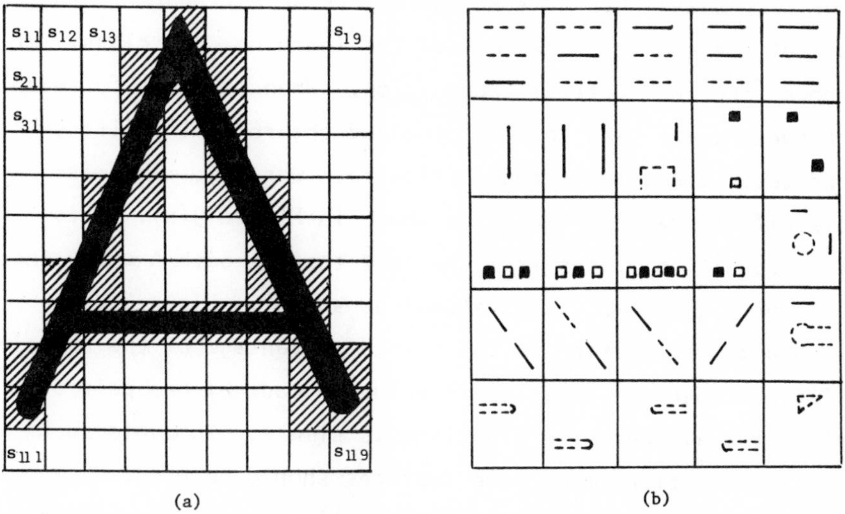

Figure 7. *(a) Sensory Units -- 9 x 11 Mosaic Cells. The Output of the Shaded Cells is +1 and the Output of White Cells is -1.*
(b) Selected Features. —— Indicates +1 and --- Indicates -1.

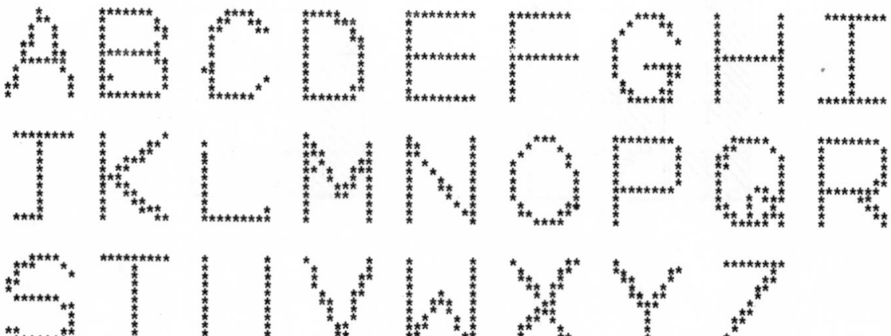

Figure 8.
Examples of English Characters Used in Our Experiments.

for each class) handwritten characters were prepared. When all of the characters were used for learning, the classifier showed the 100% correct recognition. When ten (two for each) characters are used for learning and the remaining fifteen for test, the correct recognition rate was 93%. The misclassified character was D as shown in Figure 9(a). This misrecognition arises from the fact that in both A and D the pair of s_{13} and s_{53} take value 1 as shown in Figure 3(a). Such a misrecognition is the main disadvantage of the classifier, because the decision is actually made by one pair of elements, although the algorithm is simple. To avoid such a misrecognition, much more features should be used. However, this type of error is not peculiar to the classifier discussed here, and it can be considered that the classifier still has comparatively good performance in classifying handwritten characters.

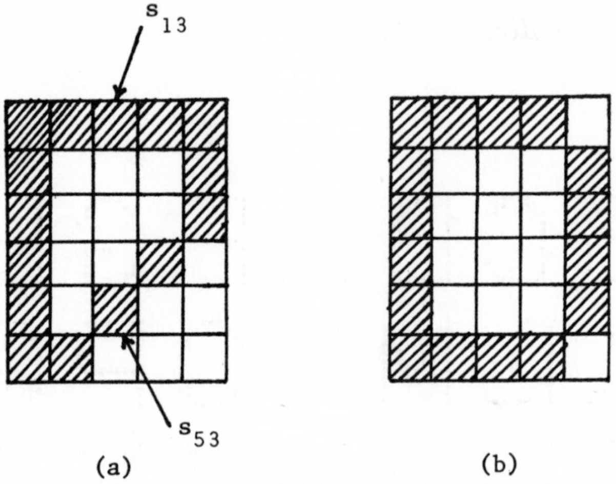

Figure 9. (a) Misrecognized Character D and (b) Original Character D.

Secondly, 780 English characters A, B,...,Y, Z (30 for each character) were tested using the filter detecting lines and points of the patterns. The thirty extracted features are shown in Figure 7(b). The result is that the probability of correct recognition was 93%, although random classification of them is 1/26 x 100 = 3.9%. The mean values and variances are given in Table 1. Figure 10 plots the probability of correct recognition vs. the number of categories.

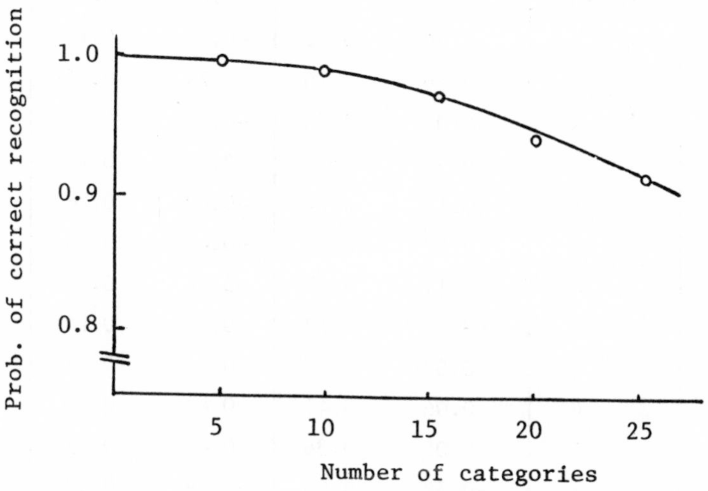

Figure 10. The Probability of Correct Recognition vs the Number of Categories.

Thirdly, we used 26 x 3 = 78 characters written by one person. In handwritten character reading, it is comparatively easy for the classifier to recognize characters of a particular person, because their variation is much smaller so long as he does not intentionally write in a different hand.

Since the learning time of the classifier proposed is quite short, the following procedure is useful for the practical application of the classifier.

patterns	25 features		99 features	
	M	D	M	D
A	-0.16	0.35	-0.71	0.16
B	0.33	0.15	-0.77	0.06
C	-0.34	0.29	-0.76	0.06
D	0.06	0.21	-0.79	0.09
E	0.02	0.20	-0.67	0.09
F	-0.22	0.27	-0.68	0.15
G	-0.21	0.49	-0.74	0.13
H	0.23	0.18	-0.65	0.12
I	-0.60	0.12	-0.79	0.18
J	-0.56	0.16	-0.59	0.26
K	-0.04	0.26	-0.74	0.23
L	-0.34	0.32	-0.71	0.17
M	0.11	0.32	-0.69	0.28
N	0.02	0.31	-0.65	0.25
O	-0.11	0.43	-0.85	0.17
P	-0.04	0.17	-0.66	0.14
Q	0.02	0.40	-0.76	0.24
R	0.06	0.41	-0.65	0.31
S	-0.05	0.36	-0.63	0.32
T	-0.58	0.08	-0.69	0.23
U	-0.05	0.37	-0.80	0.19
V	-0.31	0.26	-0.74	0.23
W	-0.12	0.33	-0.68	0.30
X	-0.10	0.28	-0.66	0.32
Y	-0.47	0.18	-0.73	0.21
Z	-0.22	0.14	-0.68	0.19

$M = \frac{1}{n}\sum_{i}^{n} x_i$: mean value $\quad D^2 = \frac{1}{n}\sum_{i}^{n} (x_i - M)^2$: variance

Table 1 The Mean Values and Variance of the Patterns.

Step 1 Before one uses the classifier, all his handwritten characters should be presented to it.

Step 2 The classifier is organized by the learning of the presented characters.

Step 3 The classifier is ready to be used.

Step 4 When another person uses the classifier, the classifier must relearn his handwriting.

In our experiment, a graphic display with a light pen was used as an input device and all English characters A, B, ...,Z were presented three times for learning. Our experimental results are shown in Figure 11.

5.2. Associative Memories

In the experiments, we used five numerals 1, 2, 3, 4, 5 and five English characters A, B, C, D, E. These characters are given on 5 x 6 mosaic cells and the output of each cell is directly applied to the memory system as a feature. Examples of recalled patterns and their key inputs are illustrated in Figure 12. Figure 13 shows the accuracy of recalled patterns ρ as a function of the correlation coefficient of memorized patterns and the corresponding key patterns γ

$$\rho = \frac{1}{R} \sum_{\mu=1}^{R} \frac{1}{4n} \sum_{i=1}^{n} (x_i^{\mu}+1)(z_i^{\mu}+1)$$

$$\gamma = \frac{1}{R} \sum_{=1}^{R} \frac{1}{n} \sum_{i=1}^{n} x_i^{\mu} q_i^{\mu} .$$

From Figure 13, it is known that the associative memory can recall memorized patterns with probability 0.9 even if one third of all elements are missed.

Characters	780 chars	26 chars particular person	Characters	780 chars	26 chars particular person
A	1.0		N	0.93	
B	1.0		O	0.73	
C	1.0		P	1.0	
D	0.9		Q	0.7	*
E	0.87		R	1.0	
F	1.0		S	1.0	*
G	0.8	*	T	0.93	
H	1.0		U	1.0	
I	0.87		V	1.0	
J	1.0		W	0.93	*
K	0.57	*	X	0.93	
L	0.97		Y	0.87	
M	0.9		Z	1.0	
			mean value	0.93	

(a)

Correct characters	A	D	E	I	Q	K	N
Handwritten characters	A	D	E	I	Q	K	N
Output decision	R	O	B	J	O	X	I

(b)

Figure 11. (a) Probabilities of Correct Recognition and (b) Examples of Misrecognized Characters.

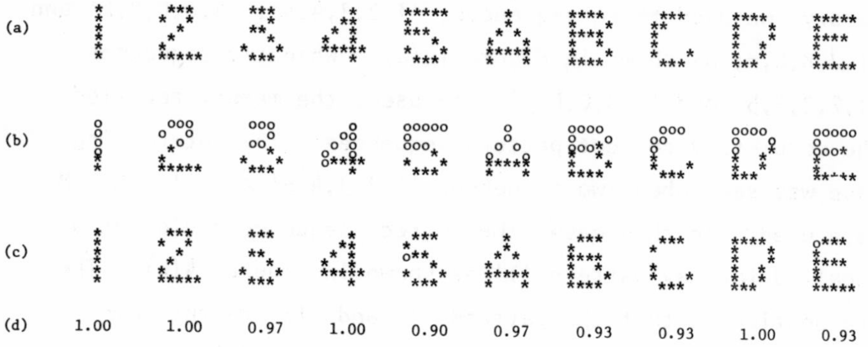

Figure 12. (a) Original Characters, (b) Key Inputs, (c) Recalled Patterns and (d) The Probability of Correct Recall.

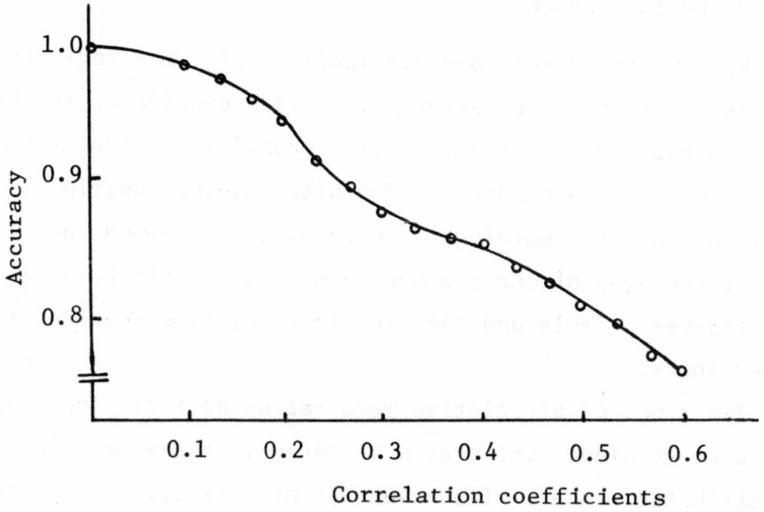

Figure 13. Accuracy of Recalled Patterns v.s. Correlation Coefficients.

In the computer simulation of the dynamic associative memory, we used three sequences, '1,2,3,4,5', 'A,B,C,D,E' and 'I,J,K,L,M; as shown in Figure 14(a). When two sequences '1,2,3,4,5' and 'A,B,C,D,E' were used, the memory recalled the completely correct pattern sequences. The interesting case was seen when two sequences '1,2,3,4,5' and 'I,J,K,L,M' were used. In this case, the correct sequence could not be recalled for key pattern '1' as shown in Figure 14(b). The explanation is that the pattern '1' and 'I' are the same. Thus the memory cannot determine whether the following pattern is '2' or 'B' and gives the pattern similar to both '2' and 'B'. However, if the other patterns, e.g. '5' or 'M' were given as a key input, the memory could recall the correct sequences as shown in Figure 14(c).

6. CONCLUDING REMARKS

We have discussed some applications of fuzzy logic to learning systems. The pattern classifier considered in this paper is based on the conditional probability on the assumption of pairwise dependence. The discriminant function is nonlinear and the learning is nonparametric. Among the major advantages of our approach are that the algorithm is comparatively simple and that the learning time required is rather short.

Two types of associative memories using fuzzy logic have been also studied. Entities are stored in these memories in a distributed manner. The advantage of this type of memory is the ability of memorizing a large number of patterns although the accuracy of recalled patterns decreases with increasing the number.

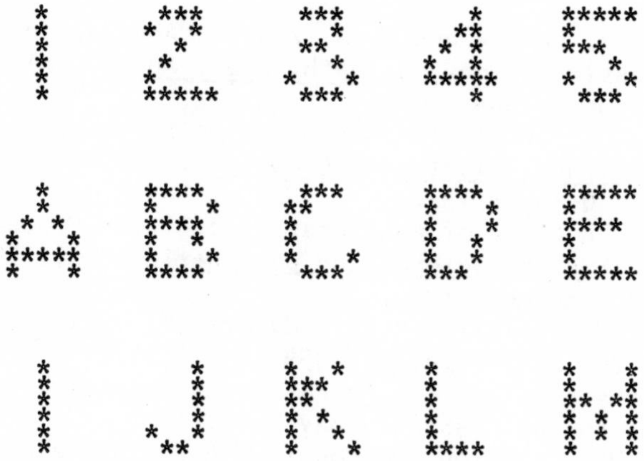

(a)

(b)

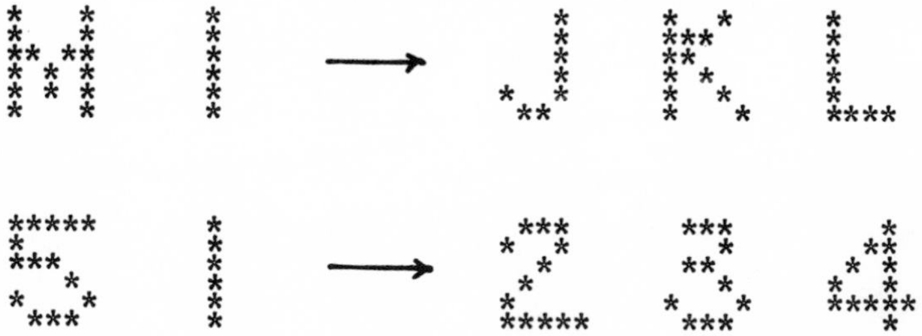

(c)

Figure 14. (a) *Stored Sequences,*
(b) *and* (c) *Recalled Sequences.*

REFERENCES

1. R. Bellman, R. Kalaba and L. A. Zadeh, "Abstraction and Pattern Classification", J. Math. Anal. Appl., Vol. 13, pp. 1-7, January 1966.
2. C. K. Chow, "A Recognition Method Using Neighbor Dependence", IRE Tráns. Electronic Computers, Vol. EC-11, pp. 683-690, October 1972.
3. C. K. Chow, "A Class of Nonlinear Recognition Procedures", IEEE Trans. Systems Science and Cybernetics, Vol. SSC-2, pp. 101-109, December 1966.
4. T. Kohonen, "Correlation Matrix Memories", IEEE Trans. Computers, Vol. C-21, pp. 353-359, April 1972.
5. G. Nagy, "State of the Art in Pattern Recognition", IEEE Procs., Vol 56, pp. 836-862, May 1968.
6. K. Nakano, "Associatron - A Model of Associative Memory", IEEE Trans. Systems, Man and Cybernetics, Vol. SMC-2, pp. 380-388, July 1972.
7. W. G. Wee and K. S. Fu, "A Formulation of Fuzzy Automata and Its Application as a Model of Learning Systems", IEEE Trans. Systems Science and Cybernetics, Vol. SSC-5, pp. 215-223, July 1969.
8. L. A. Zadeh, "Fuzzy Sets", Information and Control, Vol. 8, pp. 338-353, June 1965.
9. L. A. Zadeh, "Fuzzy Sets and Systems", Proceedings of the 1965 Symposium on System Theory, Brooklyn, New York.
10. L. A. Zadeh, "Fuzzy Algorithms", Information and Control, Vol. 12, pp. 99-102, February 1968.

BIBLIOGRAPHY ON

FUZZY SETS AND THEIR APPLICATIONS

1. Zadeh, L.A. (1965), Fuzzy sets, Inf. and Control 8,338-353.

2. Zadeh, L.A. (1965), Fuzzy sets and systems, Proc. Symp. on System Theory, Poly. Inst. of Brooklyn, 29-39.

3. Bellman, R.E., Kalaba,R., and Zadeh, L.A. (1966), Abstraction and pattern classification, Jour. Math. Anal. and Appl. 13, 1-7.

4. Klaua, D. (1966), Grundbegriffe einer mehrwertigen Mengenlehre, Mber. Dt. Akad. Wiss.8, 782-802.

5. Zadeh, L.A. (1966), Shadows of fuzzy sets, Prob. in Trans. of Inf. 2, 37-44.

6. Chang, C.L. (1967), Fuzzy sets and pattern recognition, Thesis, Univ. of Calif., Berkeley.

7. Goguen, J.A. (1967), L-fuzzy sets, Jour. Math. Anal. and Appl. 18, 145-174.

8. Paz, A. (1967), Fuzzy star functions, probabilistic automata and their approximation by non-probabilistic automata, Jour. of Comp. and System Sci. 1, 371-390.

9. Wee, W.G. (1967), On a generalization of adaptive algorithms and applications of the fuzzy set concept to pattern classification, Tech. Rep. 67-7, Dept. of Elec. Eng., Purdue Univ., Lafayette, Indiana.

10. Chang, C.L. (1968), Fuzzy topological spaces, Jour. Math. Anal. and Appl 24, 182-190.

11. Gentilhomme, Y. (1968), Les ensembles flous en linguistique, Notes on Theoretical and Applied Linguistics 5, Bucarest, Roumania.

12. Goguen, J.A. (1968), Categories of fuzzy sets: Applications of non-Cantorian set theory, Thesis, Univ. of Calif., Berkeley.

13. Hirai, H., Asai, K., and Katajima, S. (1968), Fuzzy automaton and its application to learning control systems, Memo. Fac. of Eng., Osaka City University, 10, 67-73.

14. Nasu, M., and Honda, N. (1968), Fuzzy events realized by finite probabilistic automata, Inf. and Control 12, 284-303.

15. Netto, A.B. (1968), Fuzzy classes, Notices of the Amer. Math. Soc. 68T-H28, 945.

16. Santos, E.S. (1968), Maximin automata, Inf. and Control 13, 363-377.

17. Santos, E.S., and Wee, W.G. (1968), General formulation of sequential machines, Inf. and Control 12, 5-10.

18. Tsichritzis, D. (1968), Fuzzy properties and almost solvable problems, Tech. Rep. 70, Dept. of Elec. Eng., Princeton Univ., Princeton, N.J.

19. Zadeh, L.A. (1968), Fuzzy algorithms, Inf. and Control 12, 94-102.

20. Zadeh, L.A. (1968), Probability measures of fuzzy events, Jour. Math. Anal. and Appl. 23, 421-427.

21. Chang, S.S.L. (1969), Fuzzy dynamic programming and the decision making process, Proc. 3d Princeton Conf. Inf. Sci. and Systems, 200-203.

22. Goguen, J.A. (1969), The logic of inexact concepts, Synthese 19, 325-373.

23. Fu, K.S., and Li, T.J. (1969), Formulation of learning automata and games, Inf. Sci. 1, 237-256.

24. Marinos, P.N. (1969), Fuzzy logic and its application to switching systems, IEEE Trans. on Comp. C18, 343-348.

25. Mizumoto, J., Toyoda, J., and Tanaka, K. (1969), Some considerations of fuzzy automata, Jour. of Comp. and System Sci. 3, 409-422.

26. Ruspini, E.R. (1969), A new approach to clustering, Inf. and Control 15, 22-32.

27. Santos,E.S.(1969), Maximin sequential-like machines and chains, Math. Systems Theory 3, 300-309.

28. Tsichritzis,D.(1969), Measures on countable sets, Tech. Rep. 8, Univ. of Toronto, Dept. of Comp. Sci., Canada.

29. Wee,W.G., and Fu,K.S.(1969), A formulation of fuzzy automata and its application as a model of learning systems, IEEE Trans. System Sci. and Cyber. SSC-5, 215-223.

30. Watanabe,S.(1969), Modified concepts of logic, probability and information based on generalized continuous characteristic function, Inf. and Control 15, 1-21.

31. Zadeh,L.A.(1969), Biological application of the theory of fuzzy sets and systems, in "Biocybernetics of the Central Nervous System," L.D.Proctor, ed., Little Brown and Co., Boston, 199-212.

32. Lee,E.T., and Zadeh,L.A.(1969), Note on fuzzy languages, Inf. Sci. 1, 421-434.

33. Bellman,R.E., and Zadeh,L.A.(1970), Decision-making in a fuzzy environment, Management Science 17,B-141-B-164.

34. Borisov,A.N., and Kokle,E.A.(1970), Recognition of fuzzy patterns by feature analysis, in Cybernetics and Diagnostics, No.4, Riga, U.S.S.R.

35. Esogbue,A.M.O.(1970), Dynamic programming and fuzzy allocation processes, Tech. Memo 202, Oper. Res. Dept., Case Western Reserve Univ., Cleveland, Ohio.

36. Gitman,I.(1970), Organization of data; A model and computational algorithm that uses the notion of fuzzy sets, Thesis, McGill Univ., Montreal, Canada.

37. Gitman,I., and Levine,M.D.(1970), An algorithm for detecting unimodal fuzzy sets and its applications as a clustering technique, IEEE Trans. on Computers C 19, 583-593.

38. Kitajima,S., Asai,K.(1970), Learning controls by fuzzy automata, Jour. of JAACE 14, 551-559.

39. Klaua,D.(1970), Stetige Gleichmächtigkeiten continuierlich-wertiger Mengen, Mber. Dt. Akad. Wiss. 12, 749-758.

40. Lee,E.T., and Zadeh,L.A.(1970), Fuzzy languages and their acceptance by automata, Proc. 4th Princeton Conf. Inf. Sci. and Systems, 399.

41. Lee,S.C., and Lee,E.T.(1970), Fuzzy neurons and automata, Proc. Fourth Princeton Conf. on Inf. Sci. and Sys.381-385.

42. Mizumoto,M., Toyoda,J., Tanaka,K.(1970), Fuzzy languages, Trans. IECE 53-c, 333-340.

43. Otsuki,S.(1970), A model for learning and recognizing machine, Inf. Processing 11, 664-671.

44. Ruspini,E.R.(1970), Numerical methods for fuzzy clustering, Inf. Sci. 2, 319-350.

45. Santos,E.S.(1970), Fuzzy algorithms, Inf. and Control 17, 326-339.

46. Smith,R.E.(1970), Measure theory on fuzzy sets, Thesis, Dept. of Math., Univ. of Saskatchewan, Saskatoon, Canada.

47. Tanaka,K., Toyoda,J., Mizumoto,M., Tsuji,H.(1970), Fuzzy automata theory and its application to automatic controls, Jour. of JAACE 14, 541-550.

48. Asai,K., and Kitajima,S.(1971), Learning control of multimodal systems by fuzzy automata, Pattern Recognition and Model Learning, Plenum Press, New York.

49. Asai,K., and Kitajima,S.(1971), A method for optimizing control of multimodal systems using fuzzy automata, Inf. Sci. 3, 343-353.

50. Bremermann,H.J.(1971), Cybernetic functionals and fuzzy sets, IEEE Symp. Record, Systems, Man and Cyber.,248-253.

51. Brown,J.G.(1971), A note on fuzzy sets, Inf. and Control 18, 32-39.

52. Chang,C.L.(1971), Fuzzy algebra, fuzzy functions and their applications to function approximation, Div. of Computer Res. and Tech., National Institutes of Health, Bethesda, Maryland.

53. Chang,S.K.(1971), Automated interpretation and editing of fuzzy line drawings, Proc. Spring Joint Comp. Conf., 393-399.

54. Chapin,E.W.(1971), An axiomization of the set theory of Zadeh, Notices of the American Math. Soc., Notice 687-02-4, 753.

55. De Luca,A., and Termini,S.(1971), Algorithmic aspects in complex systems analysis, Scientia 106, 659-671.

56. Fujisake,H.(1971), Fuzziness in medical sciences and its processing, Proc. of Symp. on Fuzziness in Systems and its Processing, Prof. Group of System Engineering of SICE.

57. Gottwald,S.(1971), Zahlbereighskonstruktionen in einer mehrwertigen Mengenlehre, Ztschr. Math. Logik Grdl. Math. 17, 145-188.

58. Honda,N.(1971), Fuzzy sets, Jour. of IECE,54, 1359-1363.

59. Hormann,A.M.(1971), Machine-aided value judgments using fuzzy-set techniques, SP-3590, System Dev. Corp., Santa Monica, Calif.

60. Lee,E.T., and Chang,C.L.(1971), Some properties of fuzzy logic, Inf. and Control 19, 417-431.

61. Mizumoto,M.(1971), Fuzzy automata and fuzzy grammars, Thesis, Fac. of Eng. Sci., Osaka Univ., Osaka, Japan.

62. Mizumoto,M.(1971), Fuzzy sets theory, 11th Prof. Group Meeting on Control Theory of SICE.

63. Mizumoto,M.,Toyoda,J.,Tanaka,K.(1972), General formulation of formal grammars, Trans. IECE 54-c, 600-605,1971.

64. Moisil,G.C.(1971), Role of computers in the evolution of science, Proc. Int. Conf. on Science and Society,Belgrade, Yugoslavia, 134-136.

65. Poston,T.(1971), Fuzzy geometry, Thesis, Univ. of Warwick, England .

66. Rosenfeld,A.(1971), Fuzzy groups, Jour. Math. Anal. and Appl. 35, 512-517.

67. Sugeno,M.(1971), On fuzzy nondeterministic problems, Annual Conference Record of SICE.

68. Tahani,V.(1971), Fuzzy sets in information retrieval, Thesis, Dept. of Elec. Eng. and Comp. Sci., Univ. of Calif., Berkeley.

69. Tamura,S., Niguchi,S., and Tanaka,K.(1971), Pattern classification based on fuzzy relations, IEEE Trans. on Systems, Man and Cyber. SMC-1, 61-66; also in Trans. IECE 12, 937-944.

70. Tamura,S.(1971), Fuzzy pattern classification, Proc. of Symp. on Fuzziness in Systems and its Processing, Prof. Group of System Engineering of SICE.

71. Terano,T.(1971), Fuzziness and its concept, Proc. of Symp. on Fuzziness in Systems and its Processing, Prof. Group of System Engineering of SICE.

72. Zadeh,L.A.(1971), Toward a theory of fuzzy systems, in Aspects of Network and System Theory, R.E.Kalman and N.De Claris, eds., 469-490.

73. Zadeh,L.A.(1971), Quantitative fuzzy semantics, Inf. Sci. 3, 159-176.

74. Zadeh,L.A.(1971), Similarity relations and fuzzy orderings, Inf. Sci 3, 177-200.

75. Zadeh,L.A.(1971), Human intelligence vs. machine intelligence, in Proc. Int. Conf. on Science and Society, 127-133, Belgrade, Yugoslavia.

76. Zadeh,L.A.(1971), Toward fuzziness in computer systems - Fuzzy algorithms and languages, in Architecture and Design of Digital Computers, 9-18, G. Boulaye, ed., Dunod, Paris.

77. Asai,K., and Kitajima,S.(1972), Optimizing control using fuzzy automata, Automatica 8, 101-104.

78. Borghi,O.(1972), On a theory of functional probability, Revista Un. Mat. Argentina 26, 90-106.

79. Borisov,A.N., Woolf,G.N., and Osis,Ja.Ja.(1972), Application of the theory of fuzzy sets to state identification of complex systems, in Cybernetics and Diagnostics, No.5, Zinantne Press, Riga, U.S.S.R.

80. Capocelli,R.M., and De Luca,A.(1972), Measures of uncertainty in the context of fuzzy sets theory, Atti del IIe Congresso Nationale di Cibernetica di Casciana Terme, Pisa, Italy.

81. Chang,S.K.(1972), On the execution of fuzzy programs using finite-state machines, IEEE Trans. Elec. Comp. C-21, 241-253.

82. Chang,S.S.L., and Zadeh,L.A.(1972), Fuzzy mapping and control, IEEE Trans. Systems, Man and Cyber. SMC-2.,30-34.

83. Chang,S.S.L.(1972), Fuzzy mathematics, man, and his environment. IEEE Trans. on Systems, Man, and Cyber. SMC-2, 92,-93.

84. De Luca, A., and Termini,S.(1972), A definition of nonprobabilistic entropy in the setting of fuzzy set theory, Inf. and Control 20, 301-312.

85. De Luca,A., and Termini,S.(1972), Algebraic properties of fuzzy sets, Jour. Math. Anal. and Appl. 40, 373-386.

86. Furukawa,M., Nakamura,K., Oda,M.(1972), Fuzzy model of human decision-making process, Annual Conf. Records of JAACE.

87. Hendry,W.L.(1972), Fuzzy sets and Russell's paradox, Los Alamos Sci. Lab., Univ. of Calif., Los Alamos, New Mex.

88. Kitajima,S., Asai,K.(1972), Learning model of fuzzy automaton with state-dependent output (3), Annual Joint Conf. Records of JAACE.

89. Lee,E.T.(1972), Fuzzy languages and their relation to automata, Thesis, Dept. of Elec. Eng. and Comp. Sci., Univ. of Calif., Berkeley.

90. Lee,R.C.T.(1972), Fuzzy logic and the resolution principle, Jour. of A.C.M. 10, 109-119.

91. Lee,E.T.(1972), Proximity measures for the classification of geometric figures, Jour. of Cybernetics 2,43-59.

92. Lientz,B.P.(1972), On time-dependent fuzzy sets, Inf. Sci.4, 367-376.

93. Mizumoto,M., Toyoda,J., and Tanaka,K.(1972), General formulation of formal grammars, Inf. Sci. 4, 87-100; also in Trans. IECE 54-c, 600-605,1971.

94. Mizumoto,M., Toyoda,J., Tanaka,K.(1972), L-fuzzy logic, Res. on Many-Valued Logic and Its Applications, Kyoto Univ., Japan.

95. Mizumoto,M., Toyoda,J., Tanaka,K.(1972), Formal grammars with weights, Trans. IECE 55-d, 292-293.

96. Moisil,G.C.(1972), Essais sur les logiques non chrysippiennes, Pub. of Roumanian Acad. of Sci., Bucarest, Roumania.

97. Mukaidono,M.(1972), On some properties of fuzzy logic, Tech. Rep. on Automaton of IECE.

98. Mukaidono,M.(1972), On the B-ternary logical function--A ternary logic with consideration of ambiguity, Trans. IECE 55-d, 355-362.

99. Nakata,H., Mizumoto,M., Toyoda,J., Tanaka,K.(1972), Some characteristics of N-fold fuzzy CF grammars, Trans.IECE 55-d, 287-288.

100. Preparata,F.D., and Yeh,R.T.(1972), Continuously valued logic, Jour. of Comp. and System Sci. 6, 397-418.

101. Santos,E.S.(1972), Max-product machines, Jour. Math. Anal. and Appl. 37.

102. Santos,E.S.(1972), On reductions of maximin machines, Jour. Math. Anal. and Appl. 40, 60-78.

103. Shimura,M.(1972), Application of fuzzy functions to pattern classification, Trans. IECE 55-d, 218-225.

104. Siy,P., and Chen,C.S.(1972), Minimization of fuzzy functions, IEEE Trans. on Computers C-21,100-102.

105. Sugeno, M.(1972), Fuzzy measures and fuzzy integrals, Trans. SICE, 218-226.

106. Sugeno,M.(1972), Evaluation of similarity of patterns by fuzzy integrals, Annual Conf. Records of SICE.

107. Tanaka,K.(1972) Analogy and fuzzy logic, Mathematical Sciences.

108. Tanaka,H., Okuda,T., Asai,K.(1972), On the fuzzy mathematical programming, Annual Conf. Records of SICE.

109. Terano,T.(1972), Fuzziness of systems, Nikka-Giren Engineers, 21-25.

110. Tsuji,H., Mizumoto,M., Toyoda,J., Tanaka, K.(1972), Interaction between random environments and fuzzy automata with variable structures, Trans. IECE 55-d, 143-144.

111. Wong,G.A., Shen,D.C.(1972), On the learning behavior of fuzzy automata, in Proc. Int. Congress on Cyber. and Systems, Oxford Univ., Oxford.

112. Woodhead,R.G.(1972), On the theory of fuzzy sets to resolve ill-structured marine decision problems, Dept. of Naval Arch. and Shipbuilding, Univ. of Newcastle upon Tyne, G.B.

113. Zadeh,L.A.(1972), A rationale for fuzzy control, Jour. of Dynamic Systems, Measurement and Control 94, Series G, 3-4.

114. Zadeh,L.A.(1972), On fuzzy algorithms, ERL Memo M-325.

115. Zadeh,L.A.(1972), A fuzzy-set-theoretic interpretation of linguistic hedges, Jour. of Cybernetics 2, 4-34.

116. Zadeh,L.A.(1972), Fuzzy languages and their relation to human and machine intelligence, Proc. of Int. Conf. on Man and Computer, Bordeaux, France, 130-165, S. Karger, Basel.

117. Zadeh,L.A.(1972), Linguistic Cybernetics, Proc. of the Int. Symposium on System Science and Cybernetics, Oxford Univ., England.

118. Chang,S.S.L., Zadeh,L.A.,(1972), Fuzzy mapping and control, IEEE Trans. on Systems, Man and Cyber. SMC-2, 30-34.

119. Becker,J.M.(1973), A structural design process, Thesis, Dept. of Civil Eng., Univ. of Calif., Berkeley, Calif.

120. Bellman,R.E., Giertz,M(1973), On the analytic formalism of the theory of fuzzy sets, Inf. Sci. 5, 149-156.

121. Bellman,R.E., Marchi,E.(1973), Games of protocol: The city as a dynamic competitive process, Tech. Rep. RB73-36, Univ. of Southern Calif., Los Angeles, Calif.

122. Bezdek,J.C.(1973), Fuzzy mathematics in pattern classification, Thesis, Center for Appl. Math., Cornell Univ.,Ithaca, N.Y.

123. Bossel,H.H., Hughes,B.B.(1973), Simulation of value-controlled decision-making, Rep. SRC 73-11, Systems, Res. Center, Case Western Reserve Univ., Cleveland, Ohio.

124. Capocelli,R.M., De Luca,A.(1973), Fuzzy sets and decision theory, Inf. and Control 23, 446-473.

125. Conche,B., Jouault,J.P., Luan,P.M.(1973), Application des concepts flous à la programmation en languages quasi-naturels, Séminarie Bernard Roy, Univ. Paris-Dauphine.

126. Conche,B.(1973), Eléments d'une méthode de classification par utilisation d'un automate flou, J.E.E.F.L.N., Univ. Paris-Dauphine.

127. Cools,M., Peteau,M.(1973), STIM 5: Un programme de stimulation inventive utilisant la théorie des sous-ensembles flous, IMAGO Disc. Paper, Univ. Catholique de Louvain, Belgium.

128. Dal Cin,M.(1973), Fuzzy-state automata, their stability and fault-tolerance, Inst. of Sci., Univ. of Tubingen, Germany.

129. Endo,Y., Tsukamoto,Y.(1973), Apportion models of tourists by fuzzy integrals, Annual Conf. Records of SICE.

130. Engel,A.B., Buonomano,V.(1973), Towards a general theory of fuzzy sets I, Inst. de Mat. Univ. Estadual de Campinas, Brasil.

131. Fung,L.W., Fu,K.S.(1973), An axiomatic approach to rational decision-making based on fuzzy sets, Elec. Eng. Rep., Purdue Univ. Lafayette, Indiana.

132. Furukawa,M., Nakamura,K., Oda,M.(1973), Fuzzy variant process of memories, Annual Conf. Records of SICE.

133. Gluss,B.(1973) Fuzzy multistage decision-making, Int. Jour. Control, U.K. 17, 177-192.

134. Goguen,J.A.(1973), The fuzzy Tychonoff theorem, Jour. of Math. Anal. and Appl., 734-742.

135. Gottinger,H.W.(1973), Toward a fuzzy reasoning in the behavioral science, Cybernetica, 113-135.

136. Gusev,L.A., Smirnova,I.M.(1973), Fuzzy sets: theory and applications (a survey), (in Russian), Avtomatika i Telemekhanika, 66-85.

137. Jakubowski,R., Kasprzak,A.(1973), Application of fuzzy programs to the design of machining technology, Bull. Polish Acad, Sci. 21, 17-22.

138. Kalmanson,D, Stegall,F.(1973), Recherche cardio-vasculaire et theorie des ensembles flous, La Nouvelle Presse Medicale 41, 2757-2760. Also, Amer. Jour. of Cardiology 35.

139. Kandel,A.(1973), On minimization of fuzzy functions, IEEE Trans. on Computers C-22, 826-832.

140. Kandel,A.(1973), Comment on an algorithm that generates fuzzy prime implicants by Lee and Chang, Inf. and Control 22, 279-282.

141. Kaufmann,A, Cools,M., Dubois,T.(1973), Stimulation inventive dans un dialogue homme-machine utilisant la méthode des morphologies et la théorie des sous-ensembles flous, IMAGO Disc. Paper 6, Univ. Catholique de Louvain, Belgium.

142. Kaufmann,A.(1973),"Introduction à la Théorie des Sous-Ensembles Flous," (Fuzzy Sets Theory), Masson, Paris.

143. Kitagawa,T.(1973), Three coordinate systems for information cience Approaches, Inf. Sci. 15, 159-169.

144. Kitagawa,T.,(1973), Biorobots for simulation studies of learning and intelligent controls, U.S. - Japan Seminar on Learning Control and Intelligent Control, Gainesville, Fla.

145. Kling,R.(1973), Fuzzy planner: reasoning with inexact concepts in a procedural problem-solving language, Jour. of Cybernetics 3, 1-16.

146. Kotoh,K., Hiramatsu,K.(1973), A representation of pattern classes using the fuzzy sets, Trans. IECE 56-d, 275-282.

147. Lakoff,G.(1973), Hedges: a study in meaning criteria and the logic of fuzzy concepts, Jour. of Phil. Logic 2, 458-508.

148. Mizumoto,M., Toyoda,J., Tanaka,K.(1973), N-fold fuzzy grammars, Inf. Sci. 5, 25-43, also in Trans. IECE 54-c, 856-857, 1971.

149. Mizumoto,M., Toyoda,J., Tanaka,K.(1973), Examples of formal grammars with weights, Information Processing Letters 2, 74-78.

150. Nazaroff,G.J.(1973), Fuzzy topological polysystems, Jour. Math. Anal. and Appl. 41, 478-485.

151. Negoita,C.C.(1973), On the application of the fuzzy sets separation theorem for automatic classification in information retrieval systems, Inf. Sci. 5, 279-286.

152. Prugovecki,E.(1973), A posulational framework for theories of simultaneous measurement of several observables, Found. of Phys. 3, 3-18.

153. Ragade,R.K.(1973), On some aspects of fuzziness in communication, I. Fuzzy entropies, Systems Res. and Planning, Systems Eng., Bell-Northern Research,Canada.

154. Ragade,R.K.(1973) A multiattribute perception and classification of visual similarities, Systems Res. and Planning Papers, S-001-73, Bell-Northern Research, Canada.

155. Ruspini,E.H.(1973), New experimental results in fuzzy clustering, Inf. Sci. 6, 273-284.

156. Serizawa,M.(1973), A search technique of control rod pattern of smoothing core power distributions by fuzzy automaton, Jour. of Nuclear Sci. and Technology 10.

157. Shimura,M.(1973), Fuzzy sets concept in rank-ordering objects, Jour. of Math. Anal. and Appl. 43, 717-733.

158. Siy,P.(1973), Fuzzy logic and handwritten character recognition, Thesis, Dept. of Elec. Eng.,Univ. of Akron, Ohio.

159. Sugeno,M., Terano,T.(1973), An approach to the identification of human characteristics by applying fuzzy integrals, Proc. 3rd I.F.A.C. Symp. on Identification and System Parameter Estimation.

160. Sugeno,M(1973), Constructing fuzzy measure and grading similarity of patterns by fuzzy integrals, Trans. SICE 9, 359-367.

161. Tamura,S., Tanaka,K.(1973), Learning of fuzzy formal language, IEEE Trans. on Systems, Man, and Cyber. SMC-3, 98-102.

162. Tanaka,H., Okuda,T., Asai,K.(1973), Fuzzy mathematical programming, Trans. SICE 9, 109-115.

163. Tsuji,H., Mizumoto,M., Toyoda,J., Tanaka,K.(1973), Linear fuzzy automation, Trans. IECE 56-a, 256-257.

164. Tsukamoto,Y., Iida,H.(1973), Evaluation models of fuzzy systems, Annual Conf. Records of SICE.

165. Vincke,P.(1973), Une application de la théorie des graphes flous, Univ. Libre de Bruxelles.

166. Vincke,P.(1973), La théorie des ensembles flous, Memorie, Faculté de Sci.,Univ. Libre de Bruxelles.

167. Wong,CK.(1973), Covering properties of fuzzy topological spaces, Jour. Math. Anal. and Appl. 43, 697-704.

168. Yeh,R.T.(1973), Toward an algebraic theory of fuzzy relational systems, Tech. Rep. TR-25., Dept. Comp. Sci., Univ. of Texas at Austin.

169. Zadeh,L.A.(1973), A system-theoretic view of behavior modification, in "Beyond the Punitive Society," 160-169. H. Wheeler, ed., W. Freeman Co., San Francisco.

170. Zadeh,L.A., (1973), Outline of a new approach to the analysis of complex systems and decision processes, IEEE Trans. on systems, Man and Cyber. SMC-3, 28-44.

171. Zadeh,L.A.(1973), The concept of a linguistic variable and its application to approximate reasoning, ERL Memo M-411, to appear in Information Sciences.

172. Arbib,M.A., Manes,E.G.(1974), Fuzzy morphisms in automata theory, Proc. First Int. Symp. on Category Theory Applied to Computation and Control, 98-105.

173. Aubin,J.P.(1974), Fuzzy games, MRC Tech. Summary Rep. 1480, Math. Research Center, Univ. of Wisconsin-Madison, Madison.

174. Barnev,P., Dimitrov,V., Stanchev,V.(1974), Fuzzy system approach to decision-making based on public opinion investigation through questionnaries, Inst. of Math. and Mech., Bulgarian Acad. of Sci., Sofia, Bulgaria.

175. Bezdek,J.C.(1974), Numerical taxonomy with fuzzy sets, Jour. Math. Biology. 1, 57-71.

176. Chen,C.(1974), Realizability of communication nets: An application of the Zadeh criterion, IEEE Trans. on Circuits and Systems CAS-21, 150-151.

177. Dimitrov,V., Wechler,W., Barnev,P.(1974), Optimal fuzzy control of humanistic systems, Inst. of Math. and Mech., Bulgarian Acad. of Sci., Sofia, Bulgaria, and Dept. of Math., Tech. Univ., Dresden, GDR.

178. Dubois,T.(1974), Une methode d'evaluation par les sous-ensembles flous appliquee a la simulation, IMAGO Disc. Paper 13, Univ. Catholique de Louvain, Belgium.

179. Dunn,J.C.(1974), Some recent investigations of a new fuzzy partitioning algorithm and its application to pattern classification problems, Center for Appl. Math., Cornell Univ.,Ithaca, N.Y.

180. Fellinger,W.L.(1974), Specifications for a fuzzy system modeling language, Thesis, Oregon State Univ., Corvallis.

181. Fung,L.W., Fu,K.S.(1974), The kth optimal policy algorithm for decision making in fuzzy environments, in "Identification and System Parameter Estimation," ed. by P. Eykhoff, North Holland Publ., 1052-1059.

182. Fung,L.W., Fu,K.S.(1974), Characterization of a class of fuzzy optimal control problems, Proc. 8th Princeton Conf. on Inf. Sci. and Systems.

183. Giles,R.(1974), Lukasiewicz's logic and fuzzy set theory, Queen's Univ. Math. Preprint 1974-29, Kingston, Canada.

184. Goguen,J.A.(1974), Concept representation in natural and artificial languages: axioms, extensions and applications for fuzzy sets, Int. Jour. Man-Machine Studies 6, 513-561.

185. Goodman,J.S.(1974), From multiple balayage to fuzzy sets, Inst. Mat., Univ. of Florence, Italy.

186. Gottwald,S.(1974), Fuzzy topology: product and quotient theorems, Jour. Math. Anal. Appl. 45, 512-521.

187. Hutton,B.(1974), Normality in fuzzy topological spaces, Dept. of Math., Univ. of Auckland, New Zealand.

188. Kandel,A.(1974), On the minimization of incompletely specified fuzzy functions, Inf. and Control 26, 141-153.

189. Kandel,A.(1974), Codes over languages, IEEE Trans. on Systems, Man, and Cyber. SMC-4, 135-138.

190. Kandel,A., Obenauf,T.A.(1974), On fuzzy lattices, CSR 128., Comp. Sci. Dept., New Mexico Tech., Socorro, New Mexico.

191. Kandel,A., Yelowitz,L.(1974), Fuzzy chains, IEEE Trans. on Systems, Man, and Cyber. SMC-4, 472-475.

192. Kim,H.H., Mizumoto,M., Toyoda,J., Tanaka,K(1974), Lattice Grammars, Trans. IECE 57-d.

193. LeFaivre,R.(1974), Fuzzy problem solving, Tech. Rep. 37, Madison Acad. Comp. Center, University of Wisconsin.

194. Lowen,R.(1974), Topologie flous, C.R. Acad. des Sciences 278, 925-928, Paris.

195. Lowen,R.(1974), A theory of fuzzy topologies, Thesis, Dept. of Math.,Free Univ. of Brussels, Belgium.

196. Malvache, N., and Vidal,P.(1974),Application des systèmes flous à la modélisation des phénomenes de prise de décision et d'appréhension des informations visuelles chex l'homme, A.T.P.-C.N.R.S., #1K05, Paris.

197. Malvache,N., Willayes,D.(1974), Représentation et minimisation de fonctions flous, Doc. Centre Univ. de Valenciennes, France.

198. MacVicar-Whelan,P.J.(1974), Fuzzy sets, the concept of height, and the hedge very, Tech. Memo. 1, Phys. Dept., Grand Valley State Colleges, Allendale, Mich.

199. Meseguer,J., Sols,I.(1974), Automata in semimodule categories, Proc. First Int. Symp. on Category Theory Applied to Computation and Control, 196-202.

200. Negoita,C.V., Ralescu,D.A.(1974),"Fuzzy Sets and Their Applications,"Technical Press, Bucarest, Roumania.

201. Okada,N., Tamachi,T(1974), Automated editing of fuzzy line drawings for picture description, Trans. IECE 57-a, 216-223.

202. Prugovecki,E.(1974), Measurement in quantum mechanics as a stochastic process on spaces of fuzzy events, Dept. of Math., Univ. of Toronto, Canada.

203. Prugovecki,E.(1974), Fuzzy sets in the theory of measurement of incompatible observables, Found. of Phys. 4, 9-18.

204. Reisinger,L(1974), On fuzzy thesauri, Proc. Comp. Stat., Vienna.

205. Sanchez,E.(1974), Equations de relations flous, These de doctorat en biologie humaine, Faculte de Medecine de Marseille, Marseille, France.

206. Serfati,M.(1974), Algebres de Boole avec une introduction a la theorie des graphes orientes et aux sous-ensembles flous, Ed. C.D.U., Paris.

207. Sugeno,M.(1974), Theory of fuzzy integrals and its applications, Thesis, Tokyo Institute of Tech., Tokyo.

208. Van Velthoven,G.D.(1974), Application of fuzzy sets theory to criminal investigation, Thesis, Univ. of Louvain, Belgium.

209. Thomason,M.G.(1974), Finite fuzzy automata, regular fuzzy languages, and pattern recognition, Dept. Elec. Eng., Duke Univ. Durham, North Carolina.

210. Warren,R.H.(1974), Optimality in fuzzy topological polysystems, Appl. Math. Res. Lab., Wright-Patterson AFB, Ohio.

211. Wechler,W., Dimitrov,V.(1974), R-fuzzy automata, Proc. IFIP Congress 3, 657-660.

212. Wong,C.K.(1974), Fuzzy points and local properties of fuzzy topology, Jour. Math. Anal. and Appl. 46, 316-328.

213. Wong,C.K.(1974), Fuzzy topology: product and quotient theorems, Jour. Math. Anal. Appl. 45, 512-521.

214. Yeh,R.T., Bang,S.Y.(1974), Fuzzy relations, fuzzy graphs, and their applications to clustering analysis, Inst. of Comp. Sci. and Comp. Appl., Rep. SESLTC-3, Univ. of Texas at Austin.

215. Zadeh,L.A.(1974), A fuzzy-algorithmic approach to the definition of complex or imprecise concepts, ERL Memo M474, Univ. of Calif., Berkeley, Calif.

216. Zadeh,L.A.(1974), Fuzzy logic and approximate reasoning, ERL Memo M-479, to appear in Synthese.

217. Zimmermann,H.J.(1974), Optimization in fuzzy environments, Tech. Rep., Institute for Operations Res., Tech. Hochschule, Aachen, Germany.

218. Mamdani,E.H.(1974), Application of fuzzy algorithms for control of simple dynamic plant, Proc. IEE 121, 1585-1588.

219. Albin,M.(1975), Fuzzy sets and their application to medical diagnosis, Ph.D.Thesis, Dept. of Math., Univ. of Calif., Berkeley, Calif.

220. Blin,J.(1975), Fuzzy relations in group decision, US-Japan Seminar on Fuzzy Sets and Their Applications, Berkeley, Calif.,1974; to appear in Jour. of Cybernetics.

221. Dunn,J.C.(1975), Some recent investigations of a new fuzzy partitioning algorithm and its applications to pattern classification problems, US-Japan Seminar on Fuzzy Sets and Their Applications, Berkeley, Calif., 1974; to appear in Jour. of Cybernetics.

222. Hung,N.T.(1975), Information fonctionnelle et ensembles flous, Seminar on Questionnaires, Univ. of Paris 6, Paris.

223. Hutton,B.(1975), Normality in Fuzzy Topological Spaces, Jour. of Math. Anal. and Appl. 50, 74-79.

224. Jouault,J.P., Luan,P.M.(1975), Application des concepts flous a la programmation en languages quasi-naturels, Inst. Informatique d'Entreprise, C.N.A.M., Paris.

225. Kandel,A.(1975), On the properties of fuzzy switching functions, US-Japan Seminar on Fuzzy Sets and Their Applications, Berkeley, Calif.,1974; to appear in Jour. of Cybernetics.

226. Kaufman,F.(1975), A survey of fuzzy sets theory and applications to languages, automata and algorithms, US-Japan Seminar on Fuzzy Sets and Their Applications, Berkeley, Calif.,1974; to appear in Jour. of Cybernetics.

227. Knopfmacher,K.(1975), On measures of fuzziness, Jour. Math. Anal. and Appl. 49, 529-534.

228. LeFaivre,R.(1975), The representation of fuzzy knowledge, US-Japan Seminar on Fuzzy Sets and Their Applications, Berkeley, Calif.,1974; to appear in Jour. of Cybernetics.

229. Lee,E.T.(1975), An application of fuzzy sets to the classification of geometric figures and chromosome images, US-Japan Seminar on Fuzzy Sets and Their Applications, Berkeley, Calif.,1974; to appear in Inf. Sci.

230. Lee,S.C.(1975), Fuzzy sets and neural networks, US-Japan Seminar on Fuzzy Sets and Their Applications, Berkeley, Calif.,1974; to appear in Jour. of Cybernetics.

231. Mamdani,E.H., and Assilian,S.(1975), An experiment in linguistic synthesis with a fuzzy logic controller, Int. Jour. Man-Machine Studies 7, 1-13.

232. Santos,E.S.(1975), Realization of fuzzy languages by probabilistic, max-product, and maximin automata, Inf. Sci. 8, 39-53.

233. Santos, E.S.(1975), Fuzzy automata and languages, US-Japan Seminar on Fuzzy Sets and Their Applications, Berkeley, Calif.,1974; to appear in Inf. Sci.

234. Saridis,G.(1975), Fuzzy notions in nonlinear system classification, US-JapanSeminar on Fuzzy Sets and Their Applications, Berkeley, Calif.,1974; to appear in Jour. of Cybernetics.

235. Weiss,M.D.(1975), Fixed points, separation and induced topologies for fuzzy sets, Jour. of Math. Anal. and Appl. 50, 142-150.

236. Zimmermann,H.J., Gehring,H.(1975), Fuzzy information profile for information selection, Congress Book 11,

Fourth Inst. Congress., ed. AFCET, Paris.

237. Rieger,B.B.(1975), On a tolerance topology model of natural language meaning, Tech. Rep. Germanic Inst. Tech. Hochschule, Aachen, Germany.

238. Sols,I.(1975), Topology in complete lattices and continuous fuzzy relations, Zaragoza Univ., Zaragoza, Spain.